Ecological Studies, Vol. 213

Analysis and Synthesis

Edited by

M.M. Caldwell, Washington, USA
G. Heldmaier, Marburg, Germany
R.B. Jackson, Durham, USA
O.L. Lange, Würzburg, Germany
H.A. Mooney, Stanford, USA
E.-D. Schulze, Jena, Germany
U. Sommer, Kiel, Germany

Ecological Studies

Further volumes can be found at springer.com

Donald McKenzie • Carol Miller
Donald A. Falk

Editors

The Landscape Ecology of Fire

 Springer

Editors
Donald McKenzie
Pacific Wildlife Fire Sciences Laboratory
Pacific Northwest Research Station
U.S. Forest Service
Seattle, WA 98103-8600
USA
dmck@u.washington.edu

Carol Miller
Aldo Leopold Wilderness
Research Institute
Missoula, MT 59801
USA
cmiller04@fs.fed.us

Donald A. Falk
School of Natural Resources
and the Environment
University of Arizona
Tucson, AZ 85721-0001
USA
dafalk@u.arizona.edu

ISBN 978-94-007-0300-1   e-ISBN 978-94-007-0301-8
DOI 10.1007/978-94-007-0301-8
Springer Dordrecht London Heidelberg New York

Cover illustration: Post-fire landscape pattern revealing the varying effects of forest management
(clearcut logging) on wildfire severity in the Tripod Complex Fire (2006) in north-central Washington
state, USA. Photo taken by and © Christina Lyons-Tinsley (University of Washington) and used with her
permission.

Printed on acid-free paper

Springer is part of Springer Science+Business Media (www.springer.com)

Dedication

In early 2007, Lara Kellogg and I (McKenzie) drafted an outline for what would become this book. Theretofore, she had completed a graduate degree with me and worked as a geospatial analyst. She had never done anything remotely akin to editing a technical book, but took the task with a balance of humility and confidence to which many of us aspire.

Lara was most at home in a vertical landscape of sky, rock, and ice whose remoteness and intensity most of us visit only in our dreams. Unlike many others of her persuasion, however, she was equally agile in the virtual landscape of points, pixels, and polygons. Having barely begun what surely would have been a creative and productive career as a landscape ecologist, her work on the spatial correlation structure of fire-history records set a standard for much future work in the field.

In April 2007 we lost Lara to the mountains she loved most, in the Alaska wilderness. She was orders of magnitude larger than life, and we thank her for the inspiration she provides us, in both our work and our daily lives, as we see this book to completion.

Foreword

In the mid 1980s I was asked to create a fire regime map of the Selway-Bitterroot Wilderness Area for the Bitterroot National Forest fire management staff. The well known fire historian Steve Barrett had already completed most of the work by synthesizing all available fire history results by forest habitat type, so I figured it would be easy to create a map of habitat types and then assign fire regimes to each habitat type. However, when the mapped fire regimes were compared to actual fire history field data, I found that the map's accuracy was disturbingly low, ranging from 40% to 60%. At first I thought that low accuracies were a result of inaccurate habitat type mapping, but subsequent revisions of the habitat type map that increased accuracies to over 80% did nothing to improve the accuracy of the fire regime map. I searched and searched for answers to this dilemma but in the end, I gave up and sent the map to the Bitterroot National Forest with a warning about its low accuracy. It wasn't until years later after reading Forman and Godron's Landscape Ecology book that I fully understood the profound influence of spatial and temporal context on fire regimes. It was clear that fire regimes are the manifestation of spatial factors, such as topography, wind, and patch characteristics, as they interact with antecedent climate, fuels, vegetation and humans across the landscape, and fire regimes would be difficult, if not impossible, to understand, let alone predict, without a spatiotemporal foundation.

Landscape ecology is the "glue" that holds ecosystem theory together and nowhere is that more evident than in the study of wildland fire ecology. Fire is one of those unique and complex processes that operates across multiple scales of space and time because its ignition and spread are dictated by diverse factors of climate, weather, fuels, and topography, which also operate at different scales. It wasn't until the field of landscape ecology burst onto the ecological scene in the early 1980s that the missing pieces of wildland fire dynamics fell easily into place. The concepts of scale, resolution, and extent fit perfectly into fire science and they helped explain new and exciting phenomena that would have never been discovered without a context of space. In my experience, it is only in the framework of landscape ecology that the many varied aspects of fire regimes can be explored and explained using the extensive body of fire history data collected by the many dedicated scientists. Moreover, as I learned in the Bitterroot project, it is difficult to map fire regimes across a landscape without a basic knowledge of landscape ecology fundamentals, and the identification of the appropriate scale, landscape extent, time frame, and spatial variability allows a more accurate depiction and prediction of fire regimes across large areas.

It would be difficult to overemphasize the impact that landscape ecology has had on wildland fire science, yet there have been few comprehensive summaries or syntheses of the integration of landscape ecology and wildland fire in the literature. It is the concepts of landscape ecology that make fire science much easier to understand, interpret, and apply. Particularly valuable is a physical or mechanistic approach to landscape fire ecology, where biophysical drivers such as climate, energy flux, and plant ecophysiology are used to build a more "unified theory of the ecology of fire." Fire processes and their interactions are dynamic and we should never assume that there is such a thing as an "equilibrium condition"; wildland fire ecology exhibits non-linear behavior that in turn produces non-equilibrium responses, which is important to consider when attempting to apply fire science to management issues.

I believe that the next major advances in the field of wildland fire science will be in two areas: (1) the study of the variability of fire across spatiotemporal scales, and (2) the linkage of fire regimes with the biophysical processes that control them. Scaling laws, self-organized criticality, and power laws, along with semi-variance and geostatistical analyses, represent exciting new advances in understanding fire's spatial and temporal variability. But we must first understand the multi-scaled basic physical processes that influence fire dynamics if we are to understand wildland fire and manage its effects. This is more important than ever as we are faced with rapid and uncertain changes in climate, the coarsest and arguably most powerful driver of fire regimes.

In the end, the complexity of landscape fire dynamics must eventually be synthesized to a level where it can be understood and applied by natural resource management. Fire history and spatially explicit historical fire regimes are now being used by many managers to quantify the historical range and variability of landscape characteristics, and this envelope of variability is then used to prioritize, design, and implement management actions at multiple scales. This book presents essential information and some useful applications of landscape fire ecology for natural resource management. I only wish I had this book when I was spending long days and nights trying to improve that Selway-Bitterroot fire regime map.

March 19, 2010   Robert E. Keane

Preface

This is a book about fire on landscapes. We explore fire as a contagious spatial process from a number of perspectives, including fundamental theory, fire-climate interactions, interactions with other ecological processes, and ecosystem management. Along the way we visit traditional domains of landscape ecology such as scaling, pattern-process interactions, and the complex interplay of top-down and bottom-up controls on ecosystem dynamics. We devote considerable space to theoretical considerations, particularly cross-scale modeling and landscape energetics, which we believe are under-represented in the current literature on landscape ecology of fire and other disturbances. In the remainder of the book, we look at fire climatology in an explicitly spatial context, examine four case studies of fire dynamics, two topical and two geographic in focus, and discuss issues facing fire management under rapid global change.

Our geographic focus is western North America (Fig. 1). This not only reflects the expertise of the editors and authors, but also allows us to look at a single large and diverse bioregion from multiple perspectives. Moreover, fire regimes in western North America are relatively less modified by humans than many other fire-prone landscapes around the world. Western North America is endowed with expanses of uninhabited areas over which we have ample opportunity to observe fire at a variety of scales. This facilitates our examining the interactions of climate, vegetation, and fire; fire extent, severity, and spatial pattern; and fire's interactions with other disturbances such as insect outbreaks and with other ecological processes such as invasions of landscapes by non-native plants.

Fire regimes in western North America, and the western United States in particular, have evolved in a mostly temperate climate, ranging from maritime to continental, and from wet to arid. Topography is very diverse, ranging from flat to extremely rugged, with elevations from below sea level to greater than 4,000 m. Human-induced changes in the fire regime range from essentially none (subalpine and other systems with stand-replacing fire regimes) to significant (Native American burning, twentieth-century fire exclusion, human-facilitated spread of invasive non-native species). Major vegetation types include semi-arid grasslands, chaparral, semi-arid woodlands, and a wide range of conifer and mixed forests. Western North America therefore encompasses many (though not all) of the major

Fig. 1 Locations in the western USA of study sites analyzed or referred to in individual chapters of the book. Chapter numbers are in parentheses. Map color schemes here and elsewhere in the book draw substantially upon ideas at http://colorbrewer2.org/, developed by C.A. Brewer, Dept. of Geography, Pennsylvania State University

fire-regime types of Earth's fire-prone ecosystems, and we believe that the more general inferences from this book will have wide applicability around the world.

Section I focuses on the concepts of ecosystem energetics, scaling, and resilience. In Chap. 1, we outline a potential theoretical framework for landscape fire based on ecosystem energetics. This chapter provides a lens through which succeeding chapters may be viewed. We explore how the concepts of ecosystem energetics, top-down vs. bottom-up controls, and scaling laws might be integrated to provide both a theoretical framework that reduces the apparent complexity of landscape disturbance and a window into its underlying mechanisms.

McKenzie and Kennedy (Chap. 2) review quantitative scaling relationships in fire regimes and describe how they can be used to discern controls operating at different scales. They review the basis for scaling laws in fire-size distributions, fire frequency, and fire hazard. These authors also use scaling laws to illuminate the spatial autocorrelation structure in fire-history data, which in turn reveals the dominant drivers of historical fire occurrence and extent.

In Chap. 3, Moritz, Hessburg, and Povak focus on scaling laws that describe fire size distributions and show how the spatial domain over which these scaling laws obtain is linked to dominant scales of regulation. They further present ideas about how self-organized ecosystem dynamics play out at these characteristic "landscape scales", possibly building or enhancing landscape resilience.

Section II attends to one of the most important drivers of landscape fire dynamics: climate. Fire climatology references spatial scales broader than the usual domain of landscape ecology and is the subject of these two chapters. Gedalof (Chap. 4) reviews fire climatology with an emphasis on broad spatial patterns of climate drivers of fire and how they interact with biome-scale vegetation across North America. He invokes the idea of top-down vs. bottom-up controls on landscape fire, introduced in Chaps. 1–3, as they apply at regional to continental scales.

In Chap. 5, Littell and Gwozdz develop statistical fire-climate models at a finer spatial scale in the Pacific Northwest, USA. They introduce the idea of seasonal water-balance deficit as an overarching control of fire extent at regional scales and present ideas for scaling climate-fire models down to landscapes while maintaining the water-balance mechanism as a control.

Section III focuses on the ecological consequences of landscape fire dynamics. In Chap. 6, Smithwick reviews the interactions of fire with the biogeochemistry of ecosystems, using the well studied Greater Yellowstone Ecosystem as an example of the lessons learned about biogeochemical resilience. Whereas most fire-effects research looks at species, populations, and communities, Smithwick discusses the relatively unexplored idea that ecosystem functions such as decomposition and nutrient cycling are important contributors to resilience in the face of disturbance.

Swetnam, Falk, Hessl, and Farris (Chap. 7) provide an overview of methods for reconstructing historical fire perimeters from fire-scar records (which are essentially point data) as a tool for understanding the landscape spatial patterns of unmanaged fire. They review methods of interpolation, comparing both accuracy and assumptions implicit in a variety of methods. They then give a prospectus of the application of spatial reconstruction to both contemporary and future fire management.

In Chap. 8, Keeley, Franklin, and D'Antonio use the large and biologically rich state of California, USA, as a geographic template for examining the interplay of fire, climate, invasive species, and human populations. California's forests, shrublands, and grasslands, along with other Mediterranean ecosystems, are some of the world's most diverse with respect to species composition, landforms, and land use. Ecosystem dynamics in this region are analogously complex and provide a challenging arena for understand landscape fire dynamics in the face of extensive invasion by persistent non-native species.

Cushman, Wasserman, and McGarigal (Chap. 9) examine potential consequences of landscape fire dynamics for wildlife habitat in a Rocky Mountain landscape in northern Idaho, USA. They report a simulation experiment on the relative effects of climate change vs. management alternatives on habitat for two wildlife species with contrasting life-history traits. Their work poses the very relevant question of whether even fairly aggressive management can be effective given expected future changes in climate.

Our focus on the relatively uninhabited lands of western North America in no way obviates the need to consider the human dimension of the landscape ecology of fire in a contemporary context. Section IV provides two perspectives on fire management in the future. In Chap. 10, Peterson, Halofsky, and Johnson discuss fire management opportunities on landscapes that are moderately to intensively managed. They present both a technical overview of fire and fuels management, with implications for ecosystem function in future climate, and a review of adaptation strategies from a consensus of land managers.

By contrast, Miller, Abatzoglou, Syphard, and Brown (Chap. 11) look at fire management in areas protected as wilderness across the western United States. Acknowledging that fire regimes and their management do not exist in isolation from exogeneous forces of change, they explore how the future context of wilderness fire management might change with two future trends: increasing temperatures leading to more episodes of extreme fire weather, and increasing housing densities leading to greater risk and greater incidence of human-caused fires in wilderness areas. Using two contrasting examples, they discuss how the challenge to meet fire-management objectives could intensify in many wilderness areas.

A single book cannot cover the entire field of landscape fire ecology. Consequently, we have eschewed coverage of some topics that might be central to a broad survey of the field but have been well covered in other recent publications. For example, we do not review landscape fire simulation models or remote sensing of fire characteristics. Similarly, we do not provide surveys of the use of landscape metrics in the description of fire pattern and dynamics, or of spatial considerations in sampling designs in fire ecology. Instead, we focus on new and emerging ideas about the landscape ecology of fire that are not well covered in the existing literature. We hope that the chapters in this book stretch familiar concepts, touch upon new ideas and directions, and present a range of perspectives for the study of landscape fire ecology. We encourage the reader to use this volume as a complement to existing published work.

Seattle, WA   Donald McKenzie
Missoula, MT   Carol Miller
Tuscon, AZ   Donald A. Falk

Acknowledgments

We thank all 27 contributing authors for their creative work and their patience with the long editorial process. Each chapter was reviewed by at least two experts in the field and we thank the following referees for their diligence and their timely reviews: Greg Aplet, Steve Archer, Natasha Carr, Jeanne Chambers, Mike Flannigan, Gregg Garfin, Hong He, Becky Kerns, Marc Parisien, Scott Stephens, Phil van Mantgem, Jan van Wagtendonk, Peter Weisberg, and anonymous referees.

We are especially grateful to two colleagues who contributed their expertise to improving the finished product. Robert Norheim, geospatial analyst and cartographer with the School of Forest Resources, University of Washington, produced all the maps in the book to a consistent cartographic standard. Ellen Eberhardt, technical information specialist with the Fire and Environmental Research Applications team, US Forest Service, compiled and edited all the chapters and graphics into a publication-ready document.

We gratefully acknowledge funding from the Pacific Northwest Research Station and the Rocky Mountain Research Station, US Forest Service; and the School of Natural Resources and the Environment and the Laboratory of Tree-Ring Research, University of Arizona, which supported this project from concept to completion.

We thank our students, colleagues, friends, and families for their support during the period in which we were engaged producing this book.

Seattle, WA   Donald McKenzie
Missoula, MT   Carol Miller
Tuscon, AZ   Donald A. Falk

Contents

Contributors

John Abatzoglou
Department of Geography, University of Idaho, Moscow,
ID 83844-2130, USA
jabatzoglou@uidaho.edu

Timothy Brown
Desert Research Institute, 2215 Raggio Parkway, Reno,
NV 89512-1095, USA
tim.brown@dri.edu

Samuel A. Cushman
Rocky Mountain Research Station, U.S. Forest Service, Missoula,
MT 59801-5801, USA
scushman@fs.fed.us

Carla D'Antonio
Environmental Studies Program, University of California Santa Barbara,
Santa Barbara, CA 93106-4160, USA
dantonio@lifesci.ucsb.edu

Donald A. Falk
School of Natural Resources and the Environment, University of Arizona,
Tucson, AZ 85721-0001, USA
dafalk@u.arizona.edu

Calvin Farris
National Park Service, Pacific West Region, PO Box 1713,
Klamath Falls, OR 97601-0096, USA
calvin_farris@nps.gov

Janet Franklin
Schools of Geographical Sciences and Urban Planning and Life Sciences,
Arizona State University, Tempe, AZ 85287-5302, USA
janet.franklin@asu.edu

Ze'ev Gedalof
Department of Geography, University of Guelph, Guelph,
ON N1G 2W1, Canada
zgedalof@uoguelph.ca

Richard B. Gwozdz
School of Forest Resources, University of Washington, Seattle,
WA 98195-2100, USA
rgwozdz@uw.edu

Jessica E. Halofsky
School of Forest Resources, University of Washington, Seattle,
WA 98195-2100, USA
jhalo@uw.edu

Paul F. Hessburg
Wenatchee Forestry Sciences Laboratory, Pacific Northwest Research Station,
U.S. Forest Service, Wenatchee, WA 98801-1229, USA
phessburg@fs.fed.us

Amy E. Hessl
Department of Geology and Geography, West Virginia University,
Morgantown, WV 26506-6300, USA
amy.hessl@mail.wvu.edu

Morris C. Johnson
Pacific Wildland Fire Sciences Laboratory, Pacific Northwest Research Station,
U.S. Forest Service, Seattle, WA 98103-8600, USA
mcjohnson@fs.fed.us

Robert E. Keane
Missoula Fire Sciences Laboratory, Rocky Mountain Research Station,
U.S. Forest Service, Missoula, MT 59808-9361, USA
rkeane@fs.fed.us

Jon E. Keeley
Western Ecological Research Center, U.S. Geological Survey,
Sequoia National Park, 47050 Generals Hwy, Three Rivers,
CA 93271, USA
jon_keeley@usgs.gov

Maureen C. Kennedy
School of Forest Resources, University of Washington,
Seattle, WA 98195-2100, USA
mkenn@uw.edu

Jeremy S. Littell
CSES Climate Impacts Group, University of Washington, Seattle,
WA 98195_5672, USA
jlittell@uw.edu

Kevin McGarigal
Department of Natural Resources Conservation, University of Massachusetts,
Amherst, MA 01003-9285, USA
mcgarigalk@nrc.umass.edu

Donald McKenzie
Pacific Wildland Fire Sciences Laboratory, U.S. Forest Service,
400 N 34th St., Ste. 201, Seattle,
WA 98103-8600, USA
dmck@u.washington.edu

Carol Miller
Aldo Leopold Wilderness Research Institute, Rocky Mountain Research Station,
U.S. Forest Service, 790 E. Beckwith Ave., Missoula, MT 59801, USA
cmiller04@fs.fed.us

Max A. Moritz
Department of Environmental Science Policy and Management,
Division of Ecosystem Sciences, University of California Berkeley,
Berkeley, CA 94720, USA
mmoritz@berkeley.edu

David L. Peterson
Pacific Wildland Fire Sciences Laboratory, Pacific Northwest Research Station,
U.S. Forest Service, Seattle, WA 98103-8600, USA
peterson@fs.fed.us

Nicholas A. Povak
Wenatchee Forestry Sciences Laboratory, Pacific Northwest Research Station,
U.S. Forest Service, Wenatchee, WA 98801-1229, USA
npovak@fs.fed.us

Erica A.H. Smithwick
Department of Geography & Intercollege Graduate Program in Ecology,
The Pennsylvania State University, 318 Walker Building, University Park,
PA 16802-5011, USA
smithwick@psu.edu

Tyson Swetnam
School of Natural Resources and the Environment, University of Arizona,
Tucson, AZ 85721-0001, USA
tswetnam@gmail.com

Alexandra D. Syphard
Conservation Biology Institute, 10423 Sierra Vista Ave., La Mesa,
CA 91941-4385, USA
asyphard@consbio.org

Tzeidl N. Wasserman
School of Forestry, Northern Arizona University,
200 E. Pine Knoll Drive Flagstaff, AZ 86011-0001
tnw23@nau.edu

Part I
Concepts and Theory

Chapter 1
Toward a Theory of Landscape Fire

Donald McKenzie, Carol Miller, and Donald A. Falk

1.1 Introduction

Landscape ecology is the study of relationships between spatial pattern and ecological process (Turner 1989; Turner et al. 2001). It is the subfield of ecology that requires an explicit spatial context, in contrast to ecosystem, community, or population ecology (Allen and Hoekstra 1992). One major theme in landscape ecology is how natural disturbances both create and respond to landscape pattern (Watt 1947; Pickett and White 1985; Turner and Romme 1994). Landscape disturbance has been defined *ad nauseum*, but here we focus on its punctuated nature, in that the rates of disturbance propagation are not always coupled with those of other ecological processes that operate more continuously in space and time. Disturbance can therefore change landscape pattern abruptly, and large severe disturbances can be a dominant structuring force on landscapes (Romme et al. 1998).

Fire is a natural disturbance that is nearly ubiquitous in terrestrial ecosystems (Fig. 1.1). Because fire is fundamentally oxidation of biomass, the capacity to burn exists virtually wherever vegetation grows. Occurring naturally in almost every terrestrial biome, fire and its interactions with ecosystems enable the study of landscape pattern and process under a wide range of climates and geophysical templates (Bowman et al. 2009).

Fire represents one of the closest couplings in nature of abiotic and biotic forces (Chap. 6). Fires are frequent, severe, and widespread enough in multiple regions and ecosystems to have served as a selective evolutionary force, engendering adaptive responses across a variety of plant and animal taxa (Bond and Midgley 1995; Hutto 1995; Bond and van Wilgen 1996; Schwilk 2003). Conveniently, the combustion process itself does not undergo evolutionary change. In that way it is unlike insects

D. McKenzie (✉)
Pacific Wildland Fire Sciences Laboratory, U.S. Forest Service,
400 N 34th St., Ste. 201, Seattle, WA 98103-8600, USA
e-mail: dmck@u.washington.edu

D. McKenzie et al. (eds.), *The Landscape Ecology of Fire*, Ecological Studies 213,
DOI 10.1007/978-94-007-0301-8_1, © Springer Science+Business Media B.V. 2011

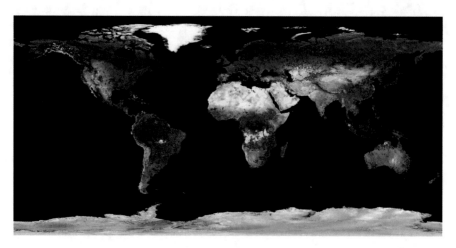

Fig. 1.1 Global compilation of MODIS fire detections between 19 and 28 June 2004 (Image courtesy of MODIS Rapid Response System http://rapidfire.sci.gsfc.nasa.gov/firemaps/)

responsible for outbreaks, which evolve (and co-evolve) with host species over millennia (Royama 1984; Logan and Powell 2001). Fire as a physical and chemical process is fundamentally the same today that it was millions of years ago, and arguably will be the same a million years from now, although its behavior and effects on landscapes change with the development of ecosystems and vegetation.

Starting from simple triggers (lightning, striking a match), fire on landscapes develops into a complex spatio-temporal process both driven and regulated by abiotic and biotic factors (Johnson 1992; Johnson and Miyanishi 2001; van Wagtendonk 2006). Fire behavior and fire effects reflect the relative strengths of multiple drivers, interacting at variable scales of space and time (Table 1.1). At fine scales (10^{-1}–10^1 m^2), fire spread and intensity are conditioned by properties of fuel (mass, availability, spatial arrangement, and moisture), ignition (type, intensity, frequency, and spatial distribution), and ambient weather (air temperature, wind speed, and humidity). As a fire spreads over larger spatial scales (10^1–10^3 m^2) other factors gain in importance, particularly topographic variation (aspect, slope, and slope position). As a result of these interactions, a fire can cover 5,000 ha or more in a day, or smolder and creep through ground fuels for months.

The spatial and temporal scales of fire are intuitively observable and comprehensible by humans, although reconciling them quantitatively with the spatiotemporal domain of "normal" ecosystem processes introduces profound challenges, chiefly because of the different rates and scales at which processes occur. Fire can reset landscape processes and their spatial pattern, often across community and watershed boundaries, thereby forcing managers to take a landscape perspective. Planning at scales that are too fine will fail to account for disturbances that arise outside small management units; planning at scales that are too coarse, such as regional scales, will not account for local patterns of spatial and temporal variability

Table 1.1 Spatiotemporal properties of fire regimes and drivers of fire behavior and effects. Drivers act on means, variances, and extremes of properties (Adapted from Falk et al. (2007))

	Climate, weather	Vegetation, fuels	Topography, landform
Temporal distribution			
Frequency or fire interval	Ignition availability and flammability; wind, humidity, and temperature patterns; fuel moisture	Vegetation productivity, postfire recovery and fuel buildup	Interaction of fire size with fuel availability; topographic barriers to fire spread
Duration	Drought or days without rain; frontal and synoptic climatic dynamics	Fuel biomass, condition, size distribution, connectivity; consumption rates	Topographic controls on rate of spread; fire spread barriers; rain shadows
Seasonality	Seasonal progression and length of fire season; effects on fuel phenology	Fuels phenology: green up, curing, and leaf fall	Topographic effects on fuel types, moisture, and phenology
Spatial distribution			
Extent	Local and synoptic weather control of ignition and fire spread	Vegetation (fuels) abundance and connectivity	Topographic influences on fire spread; fire compartments
Pattern (patch size, aggregation, contagion)	Orographic and frontal atmospheric instability, wind vectors, spatial distribution of ignitions	Spatial pattern of landscape fuel types (fuel mosaic)	Topographic influences on fire spread and spatial distribution of fuel types and condition
Intensity and severity	Microclimate and weather influences on spatial patterns of fuel moisture and abundance	Vegetation (fuel) mass, density, life-history traits, configuration; vertical and horizontal connectivity of surface and canopy layers	Slope and aspect interactions with local microclimate and weather

and are in danger of applying one-size-fits-all solutions (Chap. 10). Likewise, although fires occur as "events" over time spans of days to months, the postfire ecosystem response can unfold over decades to centuries. Landscape ecology provides a template for the analysis of both fire behavior and fire effects, and as a discipline offers the concepts and tools for understanding fire across scales (Turner et al. 2001; Falk et al. 2007).

A central concern in landscape ecology is the feedback that can exist between landscape pattern and ecological processes (White 1987; Turner 1989). In the case of fire, the mechanisms for this pattern-process dynamic are reasonably well understood at the fine scales for which fire behavior models were built

(Johnson and Miyanishi 2001; Linn et al. 2006), albeit not always quantified accurately enough for reliable landscape predictions (Keane and Finney 2003; Cushman et al. 2007). As fire opens canopies, causes differential mortality, consumes standing biomass, affects watershed hydrology and soils, and prepares seedbeds, it acts as a powerful agent of landscape pattern formation. At the same time, however, the spread and behavior of fire depend explicitly on some of those very same landscape attributes, such as the distribution, type, age, and condition of vegetation. The spatial and temporal distributions of biomass and moisture influence the spread of fire, inhibiting the spread of fire where biomass is too scarce or too wet, and allowing fire to spread only where conditions are favorable to combustion. Fire is therefore a *contagious* disturbance (Peterson 2002), in that its intensity depends explicitly on interactions with the landscape.

The feedback between fire and landscape pattern is strong and ecosystem-specific, and provides a perfect illustration in nature of the interaction of pattern and process. Over time this pattern-process interaction creates *landscape memory*, a legacy of past disturbance events and intervening processes (Peterson 2002). This memory can be spatially sparse, but temporally rich, as with a spatial pattern of fire-scarred trees (Kellogg et al. 2008), or the converse, as with a landscape pattern of age classes and structural types (Hessburg and Agee 2005). Landscape memory extends to the less visible but no less important functional properties of ecosystems, such as biogeochemical processes (Chap. 6).

Fire effects illustrate this interaction of pattern and process. Fire consumes biomass as it spreads, producing a patch mosaic of burned areas on the landscape, whose heterogeneity reflects the combined effects of the spatial patterns of fuels, topographic variation, and microscale variation in fire weather. Burned areas produce characteristic patterns of spatial variability in severity and patch sizes. This tendency is the basis for the widespread use of remote sensing and geographic information systems (GIS) to quantify and evaluate fire as a patch-generating landscape process.

Remotely derived imagery has revolutionized the field of burn severity mapping, especially by greatly improving the precision and accuracy of characterizations of postfire environments (MTBS 2009). Both qualitative and quantitative metrics of burn severity can be derived from satellite imagery based on reflected and emitted electromagnetic radiation (Miller and Yool 2002; Holden et al. 2005; Key and Benson 2006). Although most burn severity work to date has used just two spectral bands from LANDSAT images at 30-m resolution, multi-spectral and panchromatic data are increasingly available at multiple resolutions as fine as 1 m. Hyperspectral imaging (Merton 1999) and LiDAR (Lentile et al. 2006) also hold promise for more refined analysis of the three-dimensional structure of postfire landscapes.

A recently burned landscape is striking to look at. Spatial patterns of burn severity are often very heterogeneous, even within fires assumed to be stand-replacing (Fig. 1.2). Indices abound to quantify and interpret landscape spatial pattern (McGarigal et al. 2002; Peterson 2002), and have been used widely to understand spatial patterns specifically with respect to fire (Romme 1982; Turner et al. 1994). Our interest here, however, lies specifically in the processes that both generate and are controlled by that spatial pattern. For example, patterns of burn severity and the spatiotemporal structure of fire-scar records emerge from the cumulative effects of individual events and their interactions, but how these dynamic interactions play out over larger spatial and temporal scales is less well understood. A framework is needed for connecting these events and interactions that is conceptually and computationally feasible at the scales of landscapes. In this chapter we propose a theoretical framework that reduces the apparent complexity of ecosystem processes associated with fire. A full development of this theory would entail a formal structure for landscape fire dynamics and quantitative models for individual transformations of its elements (*sensu* West et al. 2009). Here we are content with suggesting a way of thinking about landscape fire that "streamlines" its complexity to a level that is tractable for both research and management.

1.2 An Energetic Framework for Understanding Landscape Fire

Earth system processes reflect the distribution of energy across scales of space and time (Pielou 2001). The climate system, for example, is a direct manifestation of the flows of energy near the Earth's surface, including the uplift of equatorial air masses and major convection processes such as Hadley cells and atmospheric circulation, all of which redistribute incoming solar energy. Ocean circulation is likewise driven by system energetics, which are evident in three dimensions between deep and surface waters across thermohaline gradients and major quasi-periodic ocean-atmosphere couplings (El Niño Southern Oscillation, Pacific Decadal Oscillation, Atlantic Multidecadal Oscillation, North Atlantic Oscillation). Earth's fluxes of energy drive biogeochemical cycles that connect flows of materials and energy within and among ecosystems. Biogeochemical cycles, such as those of carbon and nitrogen, link the biotic and abiotic domains and reflect feedbacks between biological and non-biological components of the Earth system. Ecosystem ecologist H. T. Odum (1983) observed that biogeochemical cycles can be considered a form of energy flow at all scales, and that other ecological processes such as succession and productivity can be viewed as expressions of organized energetics.

The ecosystem energy perspective offers a general framework for understanding landscape fire as a biophysical process. Fire redistributes energy, and in doing so, can dramatically transform landscape pattern. Here we outline a framework for understanding the landscape ecology of fire from an energetic perspective. In this energy—regulation—scale (ERS) framework we view fire as an ecosystem process that can be understood by examining how energy is transformed and redistributed, subject to regulation, across scales. We seek metrics associated with both energy and regulation that will be building blocks for a fully quantitative theory. The term *regulation* is intended in a broad heuristic sense, and is not intended to imply or be parallel to any genetic or molecular mechanism.

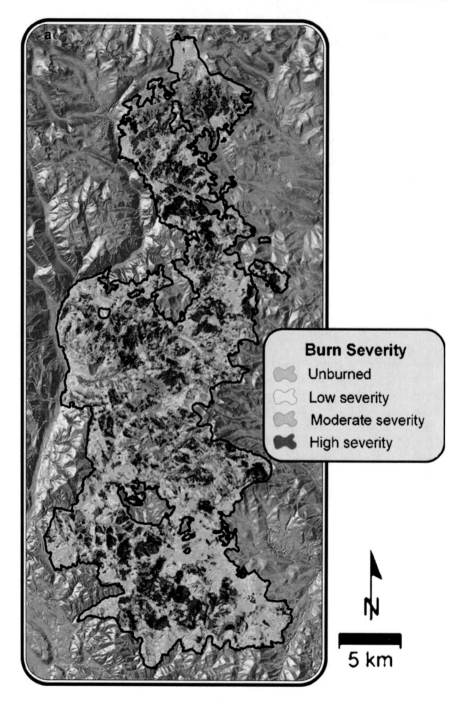

Fig. 1.2 (**a**) Fire-severity classes on the 2006 Tripod Complex Fire in northcentral Washington, USA. Fire severity classes are identified from LANDSAT imagery using the algorithm of Key and Benson (2006). (**b**) Photos demonstrate low-mixed severity as crown scorch (*above*), and mixed severity as juxtaposed high- and low-severity patches (*below*). Fire-severity data are from the Monitoring Trends in Burn Severity (MTBS) project. http://www.mtbs.gov. Accessed 1 November, 2009 (Photos courtesy of C. Lyons-Tinsley)

1. Energy. Incoming solar energy is the ultimate basis for plant growth and thus the fuels involved in combustion. Solar energy is also the basis for atmospheric circulation and the weather that influences moisture conditions of fuels and fire behavior. Vertical energy transfer in the atmosphere generates lightning, the primary non-human source of ignitions. The preconditions for fire are thus related inextricably to energy sources and fluxes.

2. Regulation. Ecosystems are subject to controls that affect the energy flux rates important to landscape fire. Forests store energy (fuel) as living and dead biomass aboveground and in soils, and the time it takes to accumulate a storehouse of biomass that will burn is subject to biotic and abiotic controls on growth and decomposition that vary across ecosystems (Aber and Melillo 1991). The energy fluxes associated with the combustion process itself are facilitated or constrained by atmospheric humidity, temperature, and air-mass movement (weather). Topography works in a similar fashion with landscapes having regions of low resistance to fire spread (e.g., steep slope gradients in the direction of wind) or high resistance (cliffs, lakes, persistent fuel breaks). Indeed all three elements of the traditional "fire triangle"—fuels, weather, and topography—can be interpreted as ecosystem components involved in regulating the flow of energy across a landscape (Table 1.2).

3. Scale. Flows of energy and mass (stored energy) are concentrated at characteristic scales of space and time (Holling 1992). For example, the main regulators of combustion at the space and time scales of millimeters and seconds (combustible fuel mass and moisture, a heat input source, and sufficient oxygen to sustain combustion) are different from those that regulate fire occurrence at subcontinental and decadal scales (interannual to decadal variation in winter precipitation, spring and summer temperature and humidity, prior fire history and regrowth of flammable biomass). Between these two ends of the scaling "gradient", fire dynamics play out across landscapes, in ways that are more complex and heterogeneous, and less tractable to analyze.

Within this "ERS" framework, we can recast the standard pattern-process polarity in landscape ecology (Turner et al. 2001) by examining energy in landscape fire. Following basic physics, we partition energy into potential and kinetic energy. Potential energy (PE) is stored mostly in biomass, in the form of molecular bond energy. Increases in biomass (productivity) are affected by kinetic energy (KE) in the form of photosynthetically active radiation (PAR), and regulated by levels of soil and foliar moisture. The potential energy in biomass is transformed rapidly into kinetic energy during a fire. Heat flux (radiative, convective, conductive) is basic to the physics of fire spread. The spatial interplay of heat flux with the connectivity of potential energy in fuels manifests as contagion on the landscape. Rates and directions of fire spread are determined by the interaction of heat flux, generated by the transformation of potential energy in fuels and driven by fire weather, with landscape pattern (*regulation*), producing the observed complex spatial patterns of landscape fire.

Table 1.2 Some important energetic and regulatory functions of elements of the "fire triangle" that are particularly relevant to landscape fire. Energy can be in kinetic (KE) or potential (PE) form. Energy storage and regulation of energy fluxes in landscape fire involve myriad ecosystem components

Fire triangle component	Energy sources and fluxes	Regulation of energy conversion
Weather and climate	Solar energy is the primary KE input, driving temperature and precipitation patterns that provide preconditions for ignition	Fuel moisture and fuel temperature affect the rate of PE→KE conversion, regulating ignition of fuels, fire intensity, and fire spread
	KE is distributed to ecosystems via circulation (wind, convection, and turbulence) contributing to fire spread	Energy regulation in the climate system is expressed in temporal and spatial patterns of precipitation, temperature, seasonality, and ocean-atmosphere teleconnections
Fuels and vegetation	Photosynthetic plants convert solar energy to PE in the form of chemical-bond energy in biomass	Abundance, compactness, and arrangement of fuels affect ignition, heat-transfer rates, and fire spread
	PE is stored on the landscape, measured as living and dead biomass and productivity. During combustion, these energy pools become sources of energy (KE) redistributed to the system	Tree density and canopy cover affect regulation by fuel moisture and temperature. Rates of postfire plant growth and decomposition influence how often fires occur
Topography and landform	N/A (By themselves they do not provide nor convert energy)	Slope steepness affects heat-transfer rates and fire spread
		Solar incidence varies with aspect, affecting fuel moisture and fuel temperature, and thus the ignition of fuels, fire intensity, and fire spread
		Shape of terrain and topographic barriers influence connectivity and the spatial pattern of fire spread

Energy fluxes associated with physiological processes of photosynthesis and respiration, and the ecosystem level processes of growth and decomposition involved in succession, proceed at very different rates from the energy fluxes associated with fire. The heat transfer in fire spread is pulsed, whereas the fluxes in growth and decomposition are more or less continuous, albeit time-varying. Fire therefore represents a dramatic and relatively instantaneous transformation of potential energy to kinetic energy, in contrast to the slower transformations associated with stand dynamics, which ultimately convert the kinetic energy from the sun into potential energy stored as biomass (Fig. 1.3).

Interactions among energy fluxes, and their cumulative effects over time, are evident in feedbacks to the process of landscape fire. These feedbacks can be negative, where fire is self-limiting, or positive, where fire is self-reinforcing. Fire as a landscape process is governed by available biomass, terrain properties that influence combustion, and meteorological variables that affect ignition, wind speed, temperature, and humidity. As a fire occurs, it effects a transformation of biomass (as potential energy) into thermal (kinetic) energy, which is then redistributed within and beyond the site. This transformation drives fire effects, including redistribution of organic and inorganic compounds (in foliage and soil) and water. The postfire environment integrates the legacy of the prefire landscape and the energy transformation from fire behavior to generate a new landscape on which stored energy has been redistributed. In this way, fire behavior, fire effects, and postfire ecosystem changes combine to create landscapes with unique self-regulating properties (Fig. 1.4).

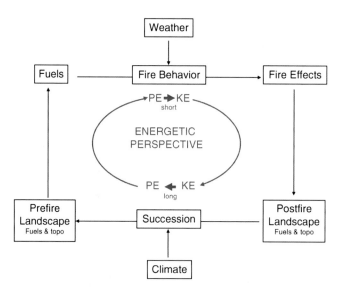

Fig. 1.3 The familiar landscape fire cycle is shown in black. Elements in boxes are things fire scientists (*top portion*) and landscape ecologists (*bottom portion*) are accustomed to measuring or modeling. In red is the energetic perspective. Short pulses of potential to kinetic energy (KE) occur during a fire, and kinetic energy is transformed into potential energy (PE) over long periods of time by plants. The spatial pattern of PE is continually being redistributed, subject to regulatory controls

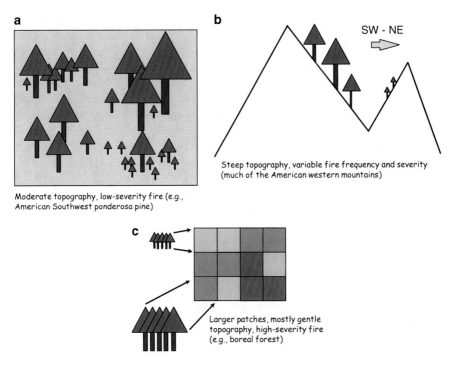

Fig. 1.4 Examples of energetic vs. regulatory emphasis in dynamics of self-limiting properties of landscape fire. (**a**) In moderate topography, fires may not carry through an entire area depending on the connectivity of fuels and the characteristic scale of variability in potential energy (correlation length). (**b**) The physical template (steep topography) regulates the energetic dynamics by introducing physical barriers that create resistance to fire spread. In theory, one could have the same correlation length in these two systems, with different dynamic underpinnings. (**c**) In a very different system subject to *top-down* controls (climate), correlation length is much larger, reflected in patch-scale variation in age classes

1.2.1   Self-Limiting Properties of Landscape Fire

The behavior and spread of fire on a landscape depend in part on current conditions (e.g., today's weather), and in part on the legacy of past fire events and subsequent ecosystem processes (e.g., the mosaic of flammable vegetation). By definition, in an ERS framework each fire—each combustion event—alters the distribution of stored energy in the form of fuels to create a new postfire environment. In prescribed surface fires, fire intensity is controlled such that consumption is limited to herbaceous and dead woody fuels, whereas canopy consumption can approach 100% of foliage and even small branches in a high-energy crown fire (Stocks et al. 2004). How long the legacy of this redistribution of stored energy persists, and the extent to which the landscape fuel mosaic resembles the pre-fire mosaic, depend on many factors, including the type of vegetation, fire intensity (heat output per unit time and space), fuel conditions (e.g., moisture content) at the time of the fire, and the productivity of the site, which governs how quickly vegetation can regrow.

Each fire alters the conditions for the next fire in the same location. Fire managers know well that the intensity and rate of spread often moderate when a fire spreads into a recently burned area. Indeed, such understanding is the basis for the widespread application of prescribed fire and wildland fire use (Mitchell et al. 2009). Behavior of wildfires burning under all but the most severe weather conditions moderates when fuel conditions are altered by thinning or prescribed fire (Agee and Skinner 2005; Finney et al. 2005; Maleki et al. 2007).

A similar self-limiting dynamic can also be seen in unmanaged landscapes. For example, in a study in the central Sierra Nevada, Collins et al. (2009) found that under all but the most extreme conditions, the spread of a fire slows when it burns into recently burned areas, with the most noticeable effects arising when the previous fire occurred less than 20 years ago. Similar self-regulating landscape properties have also been inferred in pre-management historical fire regimes (Taylor and Skinner 2003; Scholl and Taylor 2010). In this way, any one fire exerts a negative-feedback regulatory influence on the subsequent fire event, with varying periods of persistence. As this dynamic is ramified across many patches on the landscape, the result is self-regulation, which may be a fundamental property of fire as an ecosystem process (Chap. 3). From the energetic perspective, these self-limiting interactions might be viewed as an equilibrium—if an uneasy one—regulated by cycles of conversion between potential and kinetic (thermal) energy (Fig. 1.3).

1.2.2   Self-Reinforcing Properties of Landscape Fire

Another kind of landscape regulation also occurs, the self-reinforcing case. The clearest example of this is the tendency of many vegetation types—grasslands, ponderosa pine, chaparral, and lodgepole pine forest—to create fire regimes that favor their perpetuation and expansion. This occurs because dominant species create the physical environment and fuel complex that govern the fire regime, and in turn the fire regime reinforces a competitive hierarchy that favors these species (Rowe 1983; Agee 1993). For instance, the architecture of lodgepole pine (*Pinus contorta*) forests in the interior West—dense stands of trees with high canopy connectivity—tends to favor crown fire propagation, which kills most trees, giving an advantage to cohort reproduction by lodgepole due to its evolved capacity for serotiny (FEIS 2009).

Similarly, the open stand structure of many southwestern ponderosa pine (*P. ponderosa*) stands creates an open layer of surface fuels and grasses that carries relatively low intensity surface fires, killing seedlings and maintaining an open forest structure while generally causing relatively little or no mortality among canopy trees (Allen et al. 2002). Many grassland ecosystems have self-reinforcing fire regimes, with cured grasses providing fuel for fast-moving fires that burn off cured foliage and kill seedlings of woody species, while little heat penetrates to the apical meristem of the grasses, which has evolved to survive precisely such events (Brown and Smith 2000).

Whereas landscapes that are controlled by the self-limiting dynamic occupy a basin of attraction, under some conditions "escape" from this basin occurs, and the system moves into a new dynamic space (Gunderson and Holling 2001). Escape from an attractor may arise from stochastic rare events, including forcing by exogenous factors. For example, repeated fires at unusually short intervals may inhibit the recovery of certain plant species, allowing colonization by new species and a shift in the successional trajectory (Keeley et al. 1981; Suding et al. 2004). Weather conditions that promote an unusually severe or extensive fire, such as extended droughts, can also alter successional patterns. If the new vegetation is more flammable, slower growing, or more or less susceptible to a local insect or pathogen, the shift in the disturbance/succession dynamic may be sufficient to move the landscape to a stable state in a new basin of attraction (Chap. 8).

Climate change may accelerate these shifts to new basins of attraction, as disturbances such as fire change landscapes abruptly. Coupled with other complicating factors like invasions, landscape self-regulation can become chaotic. For example, climate-driven changes in fire extent, severity, or frequency, in conjunction with an invasive species such as cheatgrass (*Bromus tectorum*), buffelgrass (*Pennisetum ciliare*), or less prolific annuals, can quickly reset the connectivity of a fire-prone landscape such that species composition and spatial structure accelerate away from the previous attractor into a very different system (Zedler et al. 1983; Fischer et al. 1996; Esque et al. 2006). Typically, such landscapes will exhibit more spatial homogeneity and simple structure—in the worst-case (so far) scenario, vast areas covered by invasive annuals in which there was formerly a mosaic of longer-lived shrubs and discontinuous fine fuels. These novel systems can be impressively resistant to change, however, as reflected in the difficulty of returning an invaded grassland to its pre-invasion composition. Part of the reason is that the new system includes a strong element of self-reinforcement in its new configuration. For example, desert grasslands that have been invaded by Old World grasses have greater fine fuel mass and continuity than the pre-invasion community; this new fuel complex promotes fire spread, which eliminates fire-sensitive native species while favoring the pyrophilic invaders (Zouhar et al. 2008; Stevens and Falk 2009).

1.2.3   Top-down Vs. Bottom-up Controls

Energetic inputs and their regulation can be top-down or bottom-up, depending on the scale of spatial heterogeneity at which they act. For example, solar radiation, whether used to fix carbon between fires or to heat and dry fuels during a fire, is a top-down KE input. This energetic input is then subjected to further top-down regulation by locally homogeneous spatial fields of humidity, atmospheric pressure, temperatures, and precipitation. Fuels (stored PE) also become a source of thermal (kinetic) energy during a fire (Table 1.2). At finer scales, varying fireline intensity or flame length are associated both with fine-scale heterogeneity of fuels (spatial patterns of bottom-up inputs of PE), and with bottom-up regulation (e.g. by fuel moisture and topographic control of fire spread) of the PE→KE conversion associated with spatial variation in topography or fuel abundance at finer scales. Topographic barriers to fire spread shape and limit the size of individual fires, by creating spatial variation in flux rates, and over time produce spatiotemporal patterns of fire history of varying complexity (Kellogg et al. 2008). In general, variables with coarser resolution than these spatio-temporal patterns are associated with top-down controls, whether energetic or regulatory, whereas variables with finer resolution than this energy transfer are bottom-up controls.

In the language of pattern and process, energy flux represents process in landscape fire ecology, whereas regulation associated with the spatial distribution of energy represents landscape pattern. An obvious example of the latter is the spatial distribution of fuels (potential energy). Ideally we should be able to both quantify and predict landscape pattern change by measuring the relative strength of top-down vs. bottom up regulatory controls. For example, a dominance of top-down energy or regulation will homogenize and coarsen landscape pattern, whereas a dominance of bottom-up components will induce more complex (heterogeneous) spatial patterns to emerge. The spatial scale at which fire is "expressed" on the landscape is intermediate between the scales of variation of top-down vs. bottom-up components.

The expression of energy and regulation changes across scales, as some processes act cumulatively and others change qualitatively. For example, the energy transformed in the combustion process is a measurable physical property that is additive as a fire spreads, with output rates (e.g., $J\ s^{-1}\ m^2$) varying with external drivers and regulatory constraints such as fuel moisture and slope steepness. In contrast, topographic regulation across the landscape (e.g., ridges and valleys, barriers vs. corridors) changes combustion conditions and fire behavior in coherent spatial patterns correlated with aggregate patterns of slope and aspect. Similarly, with fuels, the expression of spatial heterogeneity changes from variation at fine scales (e.g., packing ratio) to larger-scale variation in landscape connectivity that influences fire shapes, sizes, and duration.

1.2.4 Landscapes and the Middle-Number Domain

The top-down and bottom-up organization implicit in the ERS framework might suggest that hierarchy theory could be a useful framework for studying landscape fire. Hierarchies are proposed to evolve in open dissipative systems, such as landscapes, establishing a regulatory structure (O'Neill et al. 1986). To our knowledge, however, hierarchy theory has not been applied successfully to landscape fire or similar landscape disturbances. We believe that the contagious and mercurial nature of fire, expressed as rapid temporal fluxes that greatly exceed the rates of other energy fluxes at both fine and coarse scales, confounds a hierarchical approach to the landscape ecology of fire. What works well for trophic structure in ecosystems, which can be studied over time scales of days to years, breaks down under the "metabolic" rates associated with fire: velocities can vary by orders of magnitude and temporal pulses of fire effects are far shorter than successional recovery. As such, fire is a "perturbing transitivity" (Salthe 1991) that melts hierarchical structure. Furthermore, hierarchy theory posits that ecosystem function is "driven" (forced) from lower hierarchical levels (finer scales) and constrained by upper levels (coarse scales). In our view, drivers (energy) and constraints (regulation) can issue from both coarser (top-down) and finer (bottom-up) scales than the level of interest, i.e., the landscape.

At the broadest scales, we can model fire occurrence and extent with aggregate statistics (e.g., Chap. 5; Littell et al. 2009) and capture meaningful information about fire regimes. Broad-scale regulators such as climate or derived variables such as water deficit can explain much of the variance in flux rates that manifest as regional area burned (Chaps. 4 and 5). At fine scales, fire's interactions with individual ecological objects (e.g., trees) are fairly straightforward to quantify. For example, individual tree mortality is closely associated with fireline intensity and flame length (energy flux) and tree resistance (e.g., bark thickness as a flux resistor) (Ryan and Reinhardt 1988). At both ends of the spectrum, both the energetic and regulatory components can be identified.

It is the intermediate scales that are problematic in the study of fire because of the interaction of bottom-up and top-down regulation. Recall that we have characterized a contagious disturbance as one whose properties depend on its interactions with landscape elements (Peterson 2002). The spatial heterogeneity of these interactions confounds attempts to predict fire area (or more importantly, fire severity) from spatially homogeneous top-down controls (e.g., weather), while also propagating and exacerbating estimation errors for many properties of fires that are computable at fine scales (Rastetter et al. 1992; McKenzie et al. 1996; Keane and Finney 2003). Fire as a contagious disturbance is thus inherently a multi-scale process.

This "modal" domain of fire, influenced by top-down and bottom-up controls on energy fluxes, which we refer to as the "landscape", is a middle-number system (O'Neill et al. 1986) with respect to ecological objects we can observe (growing trees, fuel transects, pixels, fire scars, animals—Fig. 1.5). We hypothesize here (and elsewhere—see Chap. 2) that in disturbance-prone landscapes, the physical limits to the extent of contagious disturbance coincide with the upper end of the middle-number domain. This is roughly equivalent to the spatial extent of the largest fires and the time frame of the fire cycle. At spatial scales much larger than the largest fires, and at time frames longer than the characteristic fire cycle, aggregate statistics suffice to characterize fire regimes. Indeed, for the purpose of understanding fire we define "landscape" as the spatial scale at which these middle-number relationships converge.

Ideally, analyses in the middle-number domain will be suitable for application of the ERS paradigm, if we can identify two thresholds. At the fine end of the gradient (near the origin in Fig. 1.5), what energetic and regulatory functions (Table 1.2) are in play up until a threshold at which spatial pattern starts to matter, where spatial contagion becomes a player in ecosystem dynamics? At the coarser

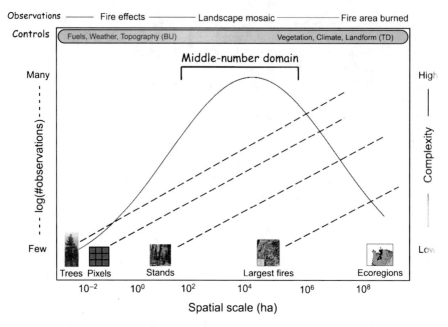

Fig. 1.5 Spatial scaling domain of landscape fire. Landscape fire regimes occupy the middle number domain for objects of analysis—trees, stands, pixels, etc. A middle-number domain is "in between" the finer scales at which the number of observations and computations on them are still analytically tractable and the coarser scales at which aggregate statistics can explain sufficient proportions of system variability for meaningful inferences. This is the spatial domain of maximum complexity (O'Neill et al. 1986), where *bottom-up* (BU) and *top-down* (TD) controls converge. The Gaussian-like curve represents a mean of many processes whose individual "complexity curves" may be less regular, e.g., perhaps even monotonic or bimodal

end of the scale gradient, what are the energetic and regulatory components effecting the breakdown of contagion, such that top-down controls are in effect and simple aggregate statistics like means and variances suffice to capture variation in process and pattern? Between these thresholds, we would further see some measure of how contagion changes across scale, as we have with more traditional properties of fire regimes such as fire frequency (Falk et al. 2007). To that end, we need to move from an *ad hoc* definition of contagion in relation to disturbance (see above) and establish a metric, or set of related metrics, to quantify it in a meaningful way. A spatially correlated physical process such as heat flux would be a good candidate for a covariate; i.e., the neighborhood effects of heat flux should modulate the strength of contagion, along with spatial variation in potential energy (fuel).

Perhaps the most elegant solution to quantifying how contagion changes across scale would involve deriving a scaling law linked to other sources of variability. For example, self-similar topography, if sufficiently dissected to produce bottom-up controls on energy flux (fire spread), produces scaling laws in fire-regime properties

(Kellogg et al. 2008; Chap. 2). Characteristic scales, or correlation lengths, of bottom-up controls (Fig. 1.4) might determine at what spatial scale contagion must be produced solely by external (top-down) drivers such as fire weather unconstrained by topography or connectivity of fuels. In this reduced case, limits to contagion would depend only on the spatial extent of extreme weather events (combined with available fuel), which are known to drive the largest wildfires.

To motivate the ERS framework as a potential solution to the middle-number problem in landscape fire, we therefore need to demonstrate how explicit scaling laws can bridge the gap between simple means and variances, i.e. aggregate statistics that work at fine or broad scales, and the complexity of middle-number systems that varies across scales in non-obvious ways. Specifically, we need to specify scaling of energy and regulation in a way that reduces the dimensionality, and potential for error propagation, through calculations on middle-number data.

We take it as axiomatic that energy and regulation covary across scales of space and time. Scaling laws represent stochastic processes that have been codified from multiple realizations across spatial and temporal scales (Lertzman et al. 1998). They also preserve the total information in a system better than aggregate statistics. For example, historical fire regimes comprise multiple realizations of individual events, whose landscape memory is in fire-size distributions and the vegetation mosaic (Chap. 3) or time-series of fire scars (Kennedy and McKenzie 2010). We may not be able to accurately reconstruct each individual realization (Chap. 7; Hessl et al. 2007), but we can back-engineer elements of the stochastic process (McKenzie et al. 2006; Kellogg et al. 2008) from the scaling relations, preferably in units of energy and regulation.

Falk et al. (2007) showed analytically how fire-regime information can be preserved in a scaling relation for fire frequency—the interval-area (IA) relation. Modeling across the middle-number domain under the ERS framework would explore analogous scaling patterns involving the more mechanistic "primitives" of fire regimes associated with the classic "fire triangle" (Table 1.2). Both energy-like and regulating elements are subject to scaling relations, and something akin to a covariance structure across scale is quantified, using these elements separately or in combination. An example of the latter would be the potential energy in a weighted combination of slope aspect, fuel loading, and packing ratio (*sensu* Rothermel 1972) represented in variograms.

We reiterate that the details of this theory are yet to be specified. Given such a theoretical framework, we need then to develop landscape experiments—probably simulation experiments—that not only test inferences but also demonstrate their tractability for quantifying landscape disturbance in the middle-number domain. Following this, we should attempt to "track" ERS components through ecosystem processes, beginning with the energetics themselves, from quantification of productivity and biomass pools (KE→PE and storage time) to heat-release dynamics (PE→KE). A later state of development could translate these energy fluxes to spatial fire-effects information, such as burn severity matrices, postfire patch characteristics, and other changes at scales useful to management.

1.3   Some Implications

Real improvements in landscape ecology theory will eventually be reflected in improved management of landscapes. The urgency for optimizing landscape management is heightened today with changes in climate, land use, and disturbance regimes affecting landscapes at ever broader scales (Chaps. 10 and 11). The translation of theory and science into appropriate and realistic management is always imperfect (Schmoldt et al. 1999). How does an energetic framework for landscape disturbance inform and improve management?

First, there are profound examples of what can happen when we ignore these principles of energetics, regulation, and scale. A basic principle of ecosystem energetics is that energy is not permanently stored in biomass; at some point, that biomass must either decompose or combust, releasing energy. By ignoring this, suppression policies throughout the twentieth century led to accumulated potential energy in biomass, and we should not have been surprised by the extent and severity of late twentieth-century wildfires. The expectation that every fire could be suppressed became less and less realistic as the potential energy in ecosystems grew, such that now the synergy of increased kinetic energy in a warming climate with abundant fuel has jeopardized ecosystems as sustainable sources of goods and services (Peters et al. 2004; Baron et al. 2009; Joyce et al. 2009).

We have also been caught by surprise when we have ignored some powerful regulators and what can change when they are no longer in force. For example, before the establishment of exotic vegetation such as cheatgrass and buffelgrass (Fischer et al. 1996; Esque et al. 2006), fire extent in arid rangelands was limited by the patchiness of flammable vegetation. This "regulation" maintained spatially heterogeneous landscapes with a concomitant diversity of habitat for species such as sage grouse (Fischer et al. 1996). Cheatgrass and other similar invasives have taken the energy-regulation dynamic out of equilibrium in arid rangelands, leading to a self-reinforcing pattern of change in the fire regimes (Chap. 8).

Attention to the principles of scaling can improve the focus and spatial resolution of management. For example, understanding how patch structure changes across scale is important for designing management plans, reserves, etc. (Baker 1989; Fahrig 1992; Parody and Milne 2004). Much energy has gone into documenting the importance of scale in landscape ecology (Peterson and Parker 1998; Turner et al. 2001; Wu et al. 2006), but considering the interplay of energy and regulation can be particularly cogent.

As an example, consider the tradeoff between maximizing C sequestration (in forests) and maintaining resilient landscapes under future fire regimes in a warming climate (North et al. 2009). Fuel treatments, and other practices in managed landscapes such as reduced planting densities, may remove biomass and release C but (if surface fuels are removed) can reduce the extent and intensity of subsequent fires (Peterson and Johnson 2007; Chap. 10). How does the ERS framework inform our choices about where, when, and how much? Fuel management is only as effective as the top-down drivers of fire will let it be. How effective will our fuel treatments be under drier more extreme fire weather (Chaps. 4 and 5)? How much of the landscape needs to be "treated" (in ERS terms, the spatial pattern of PE altered) to reduce the spread and growth of a fire? There are also temporal-scaling issues. For example, biomass (PE) accumulates at different rates in different ecosystems. How often do treatments need to be done to be effective?

We suggest that quantifying the potential energy stored in fuels, the strength of regulation (topographic complexity and fuel connectivity), and the kinetic energy associated with fire weather could provide valuable information for optimizing C sequestration over a chosen temporal domain. Alternatively one could jointly optimize C sequestration and other landscape metrics of interest (Kennedy et al. 2008), using inputs of energy and regulation. Identifying thresholds beyond which regulation breaks down (e.g., multiple megafires in the same year such as the Hayman, Rodeo-Chediski, and Biscuit fires of 2002) would also be essential (Chap. 5). With limited resources for active management of landscapes, a parsimonious model, such as we seek to enable with the ERS framework, could be a valuable tool to optimize the effectiveness of management.

1.4   Conclusions

We have proposed a theoretical model of landscape fire grounded in the interactions between energy fluxes and pools, and their controls, or "regulators", across spatial and temporal scales. If successful, an ERS framework could help identify the nature and strength of top-down vs. bottom-up controls on landscape fire, and help to solve two classic problems: the pulsed nature of fire *cf.* most ecosystem processes, and the middle-number problem, which can make landscape-scale analyses intractable at worst and fraught with uncertainty at best. A quantitative theory would need to compare favorably with existing paradigms in reproducing observed structures and processes on landscapes, while providing parsimony in both analysis and computation that could reduce uncertainty and increase the scope, both spatial and temporal, of inference. We return to this idea in Chap. 12, in which we suggest specifically how the analyses throughout this book might be transformed by an energetic perspective.

References

Aber, J.D., and J.M. Melillo. 1991. *Terrestrial ecosystems*. Philadelphia: Saunders.

Agee, J.K. 1993. *Fire ecology of Pacific Northwest forests*. New York: Island Press.

Agee, J.K., and C.N. Skinner. 2005. Basic principles of forest fuel reduction treatments. *Forest Ecology & Management* 211: 83–96.

Allen, T.F.H., and T.W. Hoekstra. 1992. *Toward a unified ecology*. New York: Columbia University Press.

Allen, C.D., M. Savage, D.A. Falk, K.F. Suckling, T.W. Swetnam, T. Schulke, P.B. Stacey, P. Morgan, M. Hoffman, and J. Klingel. 2002. Ecological restoration of southwestern ponderosa pine ecosystems: A broad perspective. *Ecological Applications* 12: 1418–1433.

Baker, W.L. 1989. Landscape ecology and nature reserve design in the boundary waters canoe area. *Ecology* 70: 23–35.

Baron, J.S., L. Gunderson, C.D. Allen, E. Fleishman, D. McKenzie, L.A. Meyerson, J. Oropeza, and N. Stephenson. 2009. Options for national parks and reserves for adapting to climate change. *Environmental Management* 44: 1033–1042.

Bond, W.J., and J.J. Midgley. 1995. Kill thy neighbour: An individualistic argument for the evolution of flammability. *Oikos* 73: 79–85.

Bond, W.J., and B.W. van Wilgen. 1996. *Fire and plants*. London: Chapman & Hall.

Bowman, D.M.J.S., J.K. Balch, P. Artaxo, et al. 2009. Fire as an earth system process. *Science* 324: 481–484.

Brown, J.K., and J.K. Smith, eds. 2000. *Wildland fire in ecosystems: Effects of fire on flora*. General Technical Report RMRS-GTR-42-vol. 2. Ogden: U.S. Forest Service.

Collins, B.M., J.D. Miller, A.E. Thode, M. Kelly, J.W. van Wagtendonk, and S.L. Stephens. 2009. Interactions among wildland fires in a long-established Sierra Nevada natural fire area. *Ecosystems* 12: 114–128.

Cushman, S.A., D. McKenzie, D.L. Peterson, J.S. Littell, and K.S. McKelvey. 2007. *Research agenda for integrated landscape modeling*. General Technical Report RMRS-GTR-194. Fort Collins: U.S. Forest Service.

Esque, T.C., C.R. Schwalbe, J.A. Lissow, D.F. Haines, D. Foster, and M. Garnett. 2006. Buffelgrass fuel loads in Saguaro National Park, Arizona, increase fire danger and threaten native species. *Park Science* 24: 33–37.

Fahrig, L. 1992. Relative importance of spatial and temporal scales in a patchy environment. *Theoretical Population Biology* 41: 300–314.

Falk, D.A., C.M. Miller, D. McKenzie, and A.E. Black. 2007. Cross–scale analysis of fire regimes. *Ecosystems* 10: 809–823.

FEIS (Fire Effects Information System). 2009. http://www.fs.fed.us/database/feis/plants/tree/pinconl/all.html. Accessed 8 Oct 2009.

Finney, M.A., C.W. McHugh, and I.C. Grenfell. 2005. Stand and landscape effects of prescribed burning on two Arizona wildfires. *Canadian Journal of Forest Research* 35: 1714–1722.

Fischer, R.A., K.P. Reese, and J.W. Connelly. 1996. An investigation on fire effects within xeric sage grouse brood habitat. *Journal or Range Management* 49: 194–198.

Gunderson, L.H., and C.S. Holling. 2001. *Panarchy: Understanding transformations in systems of humans and nature*. Washington: Island Press.

Hessburg, P.F., and J.K. Agee. 2005. Dry forests and wildland fires of the inland Northwest USA: contrasting the landscape ecology of the pre–settlement and modern eras. *Forest Ecology and Management* 211: 117–139.

Hessl, A.E., J. Miller, J. Kernan, and D. McKenzie. 2007. Mapping wildfire boundaries from binary point data: Comparing approaches. *Professional Geographer* 59: 87–104.

Holden, Z.A., A.M.S. Smith, P. Morgan, M.G. Rollins, and P.E. Gessler. 2005. Evaluation of novel thermally enhanced spectral indices for mapping fire perimeters and comparisons with fire atlas data. *International Journal of Remote Sensing* 26: 4801–4808.

Holling, C.S. 1992. Cross–scale morphology, geometry, and dynamics of ecosystems. *Ecological Monographs* 62: 447–502.

Hutto, R.L. 1995. Composition of bird communities following stand-replacement fires in northern Rocky Mountain (U.S.A.) conifer forests. *Conservation Biology* 9: 1041–1058.

Johnson, E.A. 1992. *Fire and vegetation dynamics: Studies from the North American boreal forest*. Cambridge: Cambridge University Press.

Johnson, E.A., and K. Miyanishi. 2001. *Forest fires: Behavior and ecological effects*. San Diego: Academic.

Joyce, L.A., G.M. Blate, S.G. McNulty, C.I. Millar, S. Moser, R.P. Neilson, and D.L. Peterson. 2009. Managing for multiple resources under climate change: National forests. *Environmental Management* 44: 1022–1032.

Keane, R.E., and M.A. Finney. 2003. The simulation of landscape fire, climate, and ecosystem dynamics. In *Fire and climatic change in temperate ecosystems of the Western America*, eds. T.T. Veblen, W.L. Baker, G. Montenegro, and T.W. Swetnam, 32–68. New York: Springer.

Keeley, S.C., J.E. Keeley, S.M. Hutchinson, and A.W. Johnson. 1981. Post-fire succession of the herbaceous flora in southern California chaparral. *Ecology* 62: 1608–1621.

Kellogg, L.-K.B., D. McKenzie, D.L. Peterson, and A.E. Hessl. 2008. Spatial models for inferring topographic controls on low-severity fire in the eastern cascade range of Washington, USA. *Landscape Ecology* 23: 227–240.

Kennedy, M.C., and D. McKenzie. 2010. Using a stochastic model and cross-scale analysis to evaluate controls on historical low-severity fire regimes. Landscape Ecology 25:1561–1573.

Kennedy, M.C., E.D. Ford, P. Singleton, M. Finney, and J.K. Agee. 2008. Informed multi-objective decision-making in environmental management using Pareto optimality. *Journal of Applied Ecology* 45: 181–192.

Key, C.H., and N.C. Benson. 2006. Landscape assessment: Sampling and analysis methods. In *FIREMON: Fire effects monitoring and inventory system*, eds. D.C. Lutes, R.E. Keane, J.F. Caratti, C.H. Key, N.C. Benson, S. Sutherland, and L.J. Gangi. General Technical Report RMRS-GTR-164-CD. Ogden: U.S. Forest Service.

Lentile, L.B., Z.A. Holden, A.M.S. Smith, M.J. Falkowski, A.T. Hudak, P. Morgan, S.A. Lewis, P.E. Gessler, and N.C. Benson. 2006. Remote sensing techniques to assess active fire characteristics and post-fire effects. *International Journal of Wildland Fire* 15: 319–345.

Lertzman, K., J. Fall, and B. Dorner. 1998. Three kinds of heterogeneity in fire regimes: At the crossroads of fire history and landscape ecology. *Northwest Science* 72: 4–23.

Linn, R., J. Winterkamp, J. Canfield, J. Sauer, J. Colman, J. Reisner, C. Edminster, F. Pimont, J. Dupey, and P. Cunningham. 2006. Versatility of FIRETEC, a physics-based wildfire model. *Forest Ecology and Management* 234(Suppl 1): S94.

Littell, J.S., D. McKenzie, D.L. Peterson, and A.L. Westerling. 2009. Climate and wildfire area burned in western U.S. ecoprovinces, 1916–2003. *Ecological Applications* 19: 1003–1021.

Logan, J.A., and J.A. Powell. 2001. Ghost forests, global warming, and the mountain pine beetle. *American Entomologist* 47: 160–172.

Maleki, S., C., Skinner, and M. Ritchie. 2007. Tested by fire: the Cone Fire and the lessons of an accidental experiment. Science Perspectives PSW-SP-008. Albany: U.S. Forest Service.

McGarigal, K., Cushman, S.A., Neel, M.C., and E. Ene. 2002. FRAGSTATS: Spatial pattern analysis program for categorical maps. http://www.umass.edu/landeco/research/fragstats/frag-stats.html. Accessed 25 Jan 2010.

McKenzie, D., D.L. Peterson, and E. Alvarado. 1996. Extrapolation problems in modeling fire effects at large spatial scales: A review. *International Journal of Wildland Fire* 6: 65–76.

McKenzie, D., A.E. Hessl, and L.-K.B. Kellogg. 2006. Using neutral models to identify constraints on low-severity fire regimes. *Landscape Ecology* 21: 139–152.

Merton, R.N. 1999. Multi-temporal analysis of community scale vegetation stress with imaging spectroscopy. Ph.D. thesis, University of Auckland, Aukland.

Miller, J.D., and S.R. Yool. 2002. Mapping forest post-fire canopy consumption in several overstory types using multi-temporal Landsat TM and ETM data. *Remote Sensing of Environment* 82: 481–496.

Mitchell, S.R., M.E. Harmon, and K.E.B. O'Connell. 2009. Forest fuel reduction alters fire severity and long-term carbon storage in three Pacific Northwest ecosystems. *Ecological Applications* 19: 643–646.

MTBS (Monitoring Trends in Burn Severity). 2009. http://mtbs.gov/. Accessed 1 Nov 2009.

North, M., M. Hurteau, and J. Innes. 2009. Fire suppression and fuels treatment effects on mixed-conifer carbon stocks and emissions. *Ecological Applications* 19: 1385–1396.

Odum, H.T. 1983. *Systems ecology: an introduction*. Chichester: Wiley.

O'Neill, R.V., D.L. deAngelis, J.B. Waide, and T.F.H. Allen. 1986. *A hierarchical concept of ecosystems*. Princeton: Princeton University Press.

Parody, J.M., and B.T. Milne. 2004. Implications of rescaling rules for multi-scaled habitat models. *Landscape Ecology* 19: 691–701.

Peters, D.P.C., R.A. Pielke Sr., B.T. Bestelmeyer, C.D. Allen, S. Munson-McGeed, and K.M. Havstad. 2004. Cross-scale interactions, nonlinearities, and forecasting catastrophic events. *Proceedings of the National Academy of Sciences* 101:15130–15135.

Peterson, G.D. 2002. Contagious disturbance, ecological memory, and the emergence of landscape pattern. *Ecosystems* 5: 329–338.

Peterson, D.L., and M.C. Johnson. 2007. Science-based strategic planning for hazardous fuel treatment. *Fire Management Today* 67: 13–18.

Peterson, D.L., and V.T. Parker, eds. 1998. *Ecological scale: theory and applications.* New York: Columbia University Press.

Pickett, S.T., and P.S. White. 1985. *The ecology of natural disturbance and patch dynamics.* New York: Academic.

Pielou, E.C. 2001. *The energy of nature.* Chicago: University of Chicago Press.

Rastetter, E.B., A.W. King, B.J. Cosby, G.M. Hornberger, R.V. O'Neill, and J.E. Hobbie. 1992. Aggregating finescale ecological knowledge to model coarser scale attributes of ecosystems. *Ecological Applications* 2: 55–70.

Romme, W.H. 1982. Fire and landscape diversity in subalpine forests of Yellowstone National Park. *Ecological Monographs* 52: 119–221.

Romme, W.H., E.H. Everham, L.E. Frelich, and R.E. Sparks. 1998. Are large infrequent disturbances qualitatively different from small frequent disturbances? *Ecosystems* 1: 524–534.

Rothermel, R.C. 1972. *A mathematical model for predicting fire spread in wildland fuels.* Research Paper INT-115. Ogden: U.S. Forest Service.

Rowe, J.S. 1983. Concepts of fire effects on plant individuals and species. In *The role of fire in northern circumpolar ecosystems,* eds. R.W. Wein and D.A. MacLean, 135–154. New York: Wiley.

Royama, T. 1984. Population dynamics of the spruce budworm *Choristoneura fumiferana*. *Ecological Monographs* 54: 429–462.

Ryan, C.K., and E.D. Reinhardt. 1988. Predicting postfire mortality of seven western conifers. *Canadian Journal of Forest Research* 18: 1291–1297.

Salthe, S.N. 1991. Two forms of hierarchy theory in western discourses. *International Journal of General Systems* 18: 251–264.

Schmoldt, D.L., D.L. Peterson, R.E. Keane, J.M. Lenihan, D. McKenzie, D.R. Weise, and D.V. Sandberg. 1999. *Assessing the effects of fire disturbance on ecosystems: A scientific agenda for research and management.* General Technical Report PNW-GTR-455. Portland: U.S. Forest Service.

Scholl, A.E., and A.H. Taylor. 2010. Fire regimes, forest change, and self-organization in an old-growth mixed-conifer forest, Yosemite National Park, USA. *Ecological Applications* 20(2): 362–380.

Schwilk, D.W. 2003. Flammability is a niche construction trait: canopy architecture affects fire intensity. *American Naturalist* 162: 725–733.

Stevens, J., and D.A. Falk. 2009. Can buffelgrass (*Pennisetum ciliare* (L.) Link) invasions be controlled in the American Southwest? Using invasion ecology theory to explain buffelgrass success and develop comprehensive restoration and management. *Ecological Restoration* 24: 417–427.

Stocks, B.J., M.E. Alexander, and R.A. Lanoville. 2004. Overview of the international crown fire modelling experiment (ICFME). *Canadian Journal of Forest Research* 34: 1543–1547.

Suding, K.N., K.L. Gross, and G.R. Houseman. 2004. Alternative states and positive feedbacks in restoration ecology. *Trends in Ecology and Evolution* 19: 46–53.

Taylor, A.H., and C.N. Skinner. 2003. Spatial patterns and controls on historical fire regimes and forest structure in the Klamath Mountains. *Ecological Applications* 13: 704–719.

Turner, M.G. 1989. Landscape ecology: The effect of pattern on process. *Annual Review of Ecology and Systematics* 20: 171–197.

Turner, M.G., and W.H. Romme. 1994. Landscape dynamics in crown fire ecosystems. *Landscape Ecology* 9: 59–77.

Turner, M.G., W.H. Hargrove, R.H. Gardner, and W.H. Romme. 1994. Effects of fire on landscape heterogeneity in Yellowstone National Park, Wyoming. *Journal of Vegetation Science* 5: 731–742.

Turner, M.G., R.H. Gardner, and R.V. O'Neill. 2001. *Landscape ecology in theory and practice: Pattern and process.* New York: Springer.

van Wagtendonk, J.W. 2006. Fire as a physical process. In *Fire in California's ecosystems*, eds. N.G. Sugihara, J.W. van Wagtendonk, J. Fites-Kaufman, K.E. Shaffer, and A.E. Thode, 38–57. Berkeley: University of California Press.

Watt, A.S. 1947. Pattern and process in the plant community. *Journal of Ecology* 35: 1–22.

West, G.B., Enquist, B.J., and J.H. Brown. 2009. A general quantitative theory of forest structure and dynamics. *Proceedings of the National Academy of Sciences* 106: 7040–7045.

White, P.S. 1987. Natural disturbance, patch dynamics, and landscape pattern in natural areas. *Natural Areas Journal* 7: 14–22.

Wu, J., K.B. Jones, H. Li, and O.L. Loucks. 2006. *Scaling and uncertainty analysis in ecology.* Dordrecht: Springer.

Zedler, P.H., C.R. Gautier, and G.S. McMaster. 1983. Vegetation change in response to extreme events: the effect of a short interval between fires in California chaparral and coastal scrub. *Ecology* 64: 809–818.

Zouhar, K., Smith, J.K., Sutherland, S., and M.L. Brooks. 2008. *Wildland fire in ecosystems: Fire and nonnative invasive plants.* General Technical Report RMRS-GTR-42-v6. Ogden: U.S. Forest Service.

Chapter 2
Scaling Laws and Complexity in Fire Regimes

Donald McKenzie and Maureen C. Kennedy

2.1 Introduction

Use of scaling terminology and concepts in ecology evolved rapidly from rare occurrences in the early 1980s to a central idea by the early 1990s (Allen and Hoekstra 1992; Levin 1992; Peterson and Parker 1998). In landscape ecology, use of "scale" frequently connotes explicitly spatial considerations (Dungan et al. 2002), notably *grain* and *extent*. More generally though, scaling refers to the systematic change of some biological variable with time, space, mass, or energy. Schneider (2001) further specifies *ecological scaling* sensu Calder (1983) and Peters (1983) as "the use of power laws that scale a variable (e.g., respiration) to body size, usually according to a nonintegral exponent" while noting that this is one of many equally common technical definitions. He further notes that "the concept of scale is evolving from verbal expression to quantitative expression" (p. 545), and will continue to do so as mathematical theory matures along with quantitative methods for extrapolating across scales. In what follows, we operate mainly with this quoted definition, noting that other variables can replace "body size", but we also use such expressions as "small scales" and "large scales" somewhat loosely where we expect confusion to be minimal. We examine the idea of contagious disturbance, how it influences our cross-scale understanding of landscape processes, leading to explicit quantitative relationships we call *scaling laws*. We look at four types of scaling laws in fire regimes and present a detailed example of one type, associated with correlated spatial patterns of fire occurrence. We conclude briefly

D. McKenzie (✉)
Pacific Wildland Fire Sciences Laboratory, Pacific Northwest Research Station,
U.S. Forest Service, 400 N 34th St, Ste. 201, Seattle, WA 98103-8600, USA
e-mail: dmck@u.washington.edu

D. McKenzie et al. (eds.), *The Landscape Ecology of Fire*, Ecological Studies 213,
DOI 10.1007/978-94-007-0301-8_2, © Springer Science+Business Media B.V. 2011

with thoughts on the implications of scaling laws in fire regimes for ecological processes and landscape memory.

Landscape ecology differs from ecosystem, community, and population ecology in that it must always be spatially explicit (Allen and Hoekstra 1992), thereby coupling scaling analysis with spatial metrics. For example, characteristic scales of analysis such as the stand, watershed, landscape, and region are associated with both dimensional spatial quantities (e.g., perimeter, area, elevational range) and dimensionless ones (e.g., perimeter/area ratio, fractal dimension). Similarly, properties of landscapes such as patch size distributions are also associated with spatial metrics. The tangible physical dimensions of landscapes obviate the often circuitous methods required to define and quantify scales in communities or ecosystems.

2.2   Scale and Contagious Disturbance

A contagious disturbance is one that spreads across a landscape over time, and whose intensity depends explicitly on interactions with the landscape (Peterson 2002). Some natural hazards (Cello and Malamud 2006), such as wildfires, are therefore contagious, whereas others, such as hurricanes, are not, even though their propagation may still produce distinctive spatial patterns. By the same criterion, biotic processes can be contagious (e.g., disease epidemics, insect outbreaks, grazing) or not (e.g., clearcutting). Contagion has two components: momentum (also see energy, Chap. 1) and connectivity. Together they create the aforementioned interaction between process and landscape. For an infectious disease—the best-known contagious process—a sneeze can provide momentum, while the density of nearby people provides connectivity. For fire, momentum is provided by fire weather via its effects on fireline intensity and heat transfer, whereas connectivity is provided by the spatial pattern and abundance of fuels.

Momentum and connectivity covary in a contagious disturbance process such as fire. Increases in momentum generally increase connectivity, and changes in connectivity can be abrupt when the number of patches susceptible to fire reaches a percolation threshold (Stauffer and Aharony 1994; Loehle 2004). For example, Gwozdz and McKenzie (unpublished data) found that decreasing humidity across a mountain watershed (momentum provided by fire weather) can abruptly change the connectivity of fuels when the percentage of the landscape susceptible to fire spread crosses a percolation threshold.

Interactions between momentum and connectivity may appear to be scale-dependent in that they yield qualitative changes in the behavior of landscape disturbances when viewed at different scales, even though the mechanisms of contagion per se do not change across scales. For example, the physical mechanisms of heat transfer remain the same across scales, and fire spread does depend on local connectivity of fuels, but estimates of connectivity across landscapes are sensitive to spatial resolution (Parody and Milne 2004).

2.3 Extrapolating Across Scales

Much study has gone into understanding how spatial processes change across scales (Levin 1992; Wu 1999; Miller et al. 2004; Habeeb et al. 2005). Scale extrapolation is universally seen to be obligatory, because detailed measurements are often only available at fine spatial scales (McKenzie et al. 1996), but also difficult. Given a set of observations at coarse scales, however, it is important to understand the distinction between *average behavior* of fine-scale processes and the *emergent behavior* (Milne 1998; Levin 2005) of a system. Emergent behavior "appears when a number of simple entities (agents) operate in an environment, forming more complex behaviors as a collective".[1] In the first case, the principal difficulty in extrapolation is error propagation, producing biased estimates of the average or expected behavior at broad scales because of the cumulative error from summing or averaging many calculations (Rastetter et al. 1992; McKenzie et al. 1996). In the second case, the difficulty is more profound, in that one must identify scales in space and time at which qualitative changes in behavior occur.

Some qualitative models can partition scale axes in tractable ways. For example, Simard (1991) developed a classification of processes associated with wildland fire and its management that spanned many orders of magnitude on space and time axes. This "taxonomy" of wildland fire, though not derived quantitatively from data, was enough to build a logical connection to the National Fire Danger Rating System (NFDRS—Cohen and Deeming 1985) that was of practical use (Simard 1991). Nevertheless, the limitations of such models are clear, in that qualitative changes in system behavior and key variables are established *a priori*. In order to relate processes quantitatively across scales, whether one is interested in average behavior or emergent behavior, a tractable theoretical framework is needed.

Scaling laws are quantitative relationships between or among variables, with one axis (usually X) often being either space or time. Many scaling laws are bivariate and linear or log-linear, and are developed from statistical models, theoretical models, or both. Most commonly they are based on frequency distributions or cumulative distributions wherein variables, objects, or events with smaller values occur more frequently than those with larger values. The simplest scaling law is a *power law*, for which a histogram in log-log space of the frequency distribution follows a straight line (Zipf 1949, as cited in Newman 2005). Following Newman (2005), let $p(x)\, dx$ be the proportion of a variable with values between x and dx. For histograms that are straight lines in log-log space, $\ln p(x) = -\alpha \ln x + c$, where α and c are constants (Newman 2005). Exponentiating both sides and defining $C = \exp(c)$, we have the standard power law formulation

$$p(x) = Cx^{-\alpha} \tag{2.1}$$

---

[1] Wikipedia contributors, "Emergence," Wikipedia, the Free Encyclopedia, http://en.wikipedia.org/wiki/Emergence. Accessed 25 Jan 2010.

The parameter of interest is the slope α (always negative for frequency distributions), whereas C serves as a normalization constant such that p(x) sums to 1 (Newman 2005). In the case of a frequency distribution, where Y values in a histogram are counts, C can be rescaled in order to compare slopes among distributions. Power-law relationships are often fit statistically by various binning methods, with subsequent regression of bin averages on event size, but more complicated maximum-likelihood methods may be more robust (White et al. 2008; Chap. 3).

Newman (2005) gives 12 examples of quantities in natural, technical, and social systems that are thought to follow power laws over at least some part of their range. His diverse examples include intensities of wars (Roberts and Turcotte 1998), magnitude of earthquakes (National Geophysical Data Center 2010), citations of scientific papers (Redner 1998), and web hits (Adamic and Huberman 2000). Newman (2005) specifically excludes fire size distributions, while admitting that they might follow power laws over portions of their ranges. Current opinion is divided among those who would globally assign power laws to fire-size distributions (Minnich 1983; Bak et al. 1990; Malamud et al. 1998, 2005; Turcotte et al. 2002; Ricotta 2003) and those who would attribute them only to portions of distributions or rule them out altogether in favor of alternatives (Cumming 2001; Reed and McKelvey 2002; Clauset et al. 2007; Chap. 3).

2.4   Scaling Laws and Fire Regimes

Wildfires affect ecosystems across a range of scales in space and time, and controls on fire regimes change across scales. The attributes of individual fires are spatially and temporally variable, and the concept of *fire regimes* has evolved to characterize aggregate properties such as frequency, severity, seasonality, or area affected per unit time. These aggregate properties are often reduced to metrics such as means and variances, thereby simplifying much of the complexity of fire by focusing on a single scale and obscuring ecologically important cross-scale interactions (Falk et al. 2007).

Scaling laws can deconstruct aggregate statistics of fire regimes in two ways: via frequency distributions that exhibit scaling laws, or by examining the scale dependence of individual metrics. Fire-size distributions are an example of the first, in that frequency distributions of fire sizes often follow power laws over at least portions of their ranges (Malamud et al. 1998, 2005; Turcotte et al. 2002; Moritz et al. 2005; Millington et al. 2006). Fire frequency, fire hazard, and spatial patterns of fire occurrence in fire history data are examples of the second, in that these statistics often change systematically and predictably across the spatial scale of measurement (Moritz 2003; McKenzie et al. 2006a; Falk et al. 2007; Kellogg et al. 2008). Here we briefly discuss both the scaling *patterns* that have been found within each of these four metrics of fire regimes (size, frequency, hazard, spatial pattern) and the more problematic attribution of *mechanisms* responsible for the scaling patterns.

2.4.1 Fire Size Distributions

Power laws have been statistically fit to fire size distributions from simulation models and empirical data at many scales, from virtual raster landscapes generated by the "Forest Fire Model" (Bak et al. 1990) to historical wildfire sizes throughout the continental United States (Malamud et al. 2005). Not all scaling relationships found in fire-size distributions are power laws. For example, Cumming (2001) found that a truncated exponential distribution, which defines an upper bound to fire size, had the best fit to data from boreal mixedwood forests in Canada. Reed and McKelvey (2002) suggest that the power law serves as an appropriate *null model*, but that additional parameters in a "competing hazards" model improved the fit to empirical data at regional scales. Ricotta (2003) suggests that power law exponents can change with spatial scale, based on hierarchical fractal properties of landscapes, providing a rejoinder to detractors of the power-law paradigm. An excellent review of this topic, with discussion, is found in Millington et al. (2006). These authors state, and we concur, that the value of discerning power-law behavior, or alternative, more complex nonlinear functions, would increase greatly if the ecological mechanisms driving such behavior could be identified (West et al. 1997; Brown et al. 2002).

Two mechanisms in particular have been proposed to explain power-law behavior in fire-size distributions. Self-organized criticality (SOC—Bak et al. 1988) refers to an emergent state of natural phenomena whereby a system (be it physical, biological, or socioeconomic) evolves to a state of equilibrium characterized by variable event sizes, each of which resets the system in proportion to event magnitude. In theory, the frequency distribution of events will approach a power law because the recovery time from "resetting" varies with event magnitude. SOC has been associated mainly with physical systems, particularly natural hazards such as earthquakes and landslides (Cello and Malamud 2006), but its attribution to power laws in fire regimes has typically been only at small scales (Malamud et al. 1998) or inferred from small-scale behavior (Song et al. 2001).

In contrast to SOC, highly optimized tolerance (HOT) emphasizes structured internal configurations of systems that involve tradeoffs in robustness (Carlson and Doyle 2002; Moritz et al. 2005), rather than the emergent outcomes of stochastic though correlated events as in SOC. For example, a HOT model that can be applied to wildfires is the probability-loss ratio (PLR) model (Doyle and Carlson 2000; Moritz et al. 2005), a probabilistic model of tradeoffs between resources (e.g., some ecosystem function in natural systems or efforts to protect timber in managed systems) and losses (e.g., from fire). Solving the PLR model analytically produces a frequency distribution of expected fire sizes that follows a power law (Moritz et al. 2005). HOT provides a theoretical framework for examining ecosystem resilience in response to fire events (Chap. 3).

2.4.2 Fire Frequency

The terms *fire frequency* and *fire-return interval* (FRI) are part of the currency of ecosystem management. Fire frequency is often compared among different geographic regions and between the current and historical periods. For example, considerable FRI data exist across the western United States (NOAA 2010), which can be compared and used to build regional models of fire frequency (McKenzie et al. 2000). Both comparisons and model-building assume that all FRI data points represent a *composite fire return interval* (CFRI)—the average time between fires that are observed within a sample area, but the likelihood of detecting a fire event clearly increases as the search area is expanded. FRIs are inherently scale-dependent, despite sophisticated methods for unbiased estimation of fire-free intervals (Reed and Johnson 2004).

Scaling laws in fire frequency thus quantify the relationship between the area examined for evidence of fire and the estimated fire return interval. This *interval-area* relationship (IA—Falk et al. 2007) appears in low-severity fire regimes producing fire-scars on surviving trees, mixed-severity fire regimes where fire perimeters are estimated, and raster simulation models that produce a range of fire severities and fire sizes (Falk 2004; McKenzie et al. 2006a; Falk et al. 2007). In each case, the IA can be fit to a power law, whose slope (exponent) captures other aggregate properties of the fire regime (Fig. 2.1). For example, larger mean fire sizes produce less negative slopes, because small-area samples are more likely to detect large fires than small fires. Simulations suggest that greater variance in fire size, given equal means, also produces less negative slopes, for reasons that are presently unclear (see Falk et al. 2007 for details).

In theory, then, the intercept in log-log space of the IA relationship reflects the mean point fire-return interval (sample area=0 in the case of a point, or the area of the minimum mapping unit otherwise), providing a "location" parameter to the scaling law (Falk et al. 2007). Also in theory, the exponents in the IA relationship could be derived from the properties of fire-size distributions, possibly means and variances alone, although extreme values (rare large fires) make this difficult. This connection to fire size is useful because predictive modeling of fire sizes, though subject to substantial uncertainty, is less problematic than predicting fire frequency (McKenzie et al. 2000; Littell et al. 2009). Further work is necessary, though, to connect the IA relationship to estimates of fire sizes, or fire-size distributions.

Another metric of fire frequency, the *fire cycle*, or *natural fire rotation*, refers, on a particular landscape, to the time it takes to burn an area equal to that landscape. The fire cycle is presumably independent of spatial scale if the sample landscape is much larger than the largest fire recorded within it (Agee 1993), but calculating it

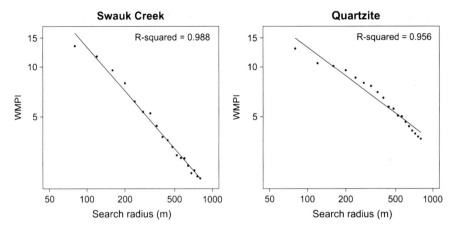

Fig. 2.1 Interval-area (IA) relationships (power laws) in log-log space for two watersheds in eastern Washington. WMPI = Weibull median probability interval. The more negative slope in Swauk Creek is a result of smaller fire sizes and more frequent fire occurrence than in Quartzite. Quartzite displays a minor but noticeable (concave down) departure from linearity (Redrawn and rescaled from McKenzie et al. (2006a))

depends on accurate estimates of the sizes of every fire in the sample. This is a difficult task in historical low-severity fire regimes, in which most fire-frequency work has been done (Hessl et al. 2007; Chap. 7). Furthermore, Reed (2006) showed that the mathematical equivalence between the fire cycle and the mean point FRI only holds if all fires are the same size, limiting the usefulness of the fire cycle as a metric of fire frequency.

2.4.3 Fire Hazard

Fire hazard in fire-history research quantifies the instantaneous probability of fire, and is derivable from distribution functions of the exponential family (e.g., negative exponential and Weibull) associated with the fire cycle (stand-replacing fire—Johnson and Gutsell 1994) and the distribution of fire-free intervals (fire-scar records—McKenzie et al. 2006a). The *hazard function* may be constant over time, reflecting a memory-free system in which current events do not depend on past events, and producing exponential age class distributions of patches in stand-replacing fire regimes (Johnson and Gutsell 1994). In contrast, an increasing hazard of fire over time (or decreasing, but this is rarely seen in fire regimes) reflects a causative factor, i.e. the growth of vegetation and buildup of fuel that facilitates fire spread. This increasing hazard is represented mathematically by a shape parameter in the Weibull distribution that is significantly greater than 1 (if this parameter is 1 the distribution reduces to the negative exponential—Evans et al. 2000). Moritz (2003) observes, however, that the ecological significance of the shape parameter covaries with the scale parameter, representing, with fire, the mean fire-free interval. For long fire-free intervals, shape parameters ≤ 2 represent fire hazard that increases negligibly over time (Moritz 2003).

When the hazard function changes with spatial scale, it reflects changing controls on fire occurrence. McKenzie et al. (2006a) and Moritz (2003) identified patterns in hazard functions that were associated with the relative strength of transient controls on fire occurrence and fire spread. In low-severity fire regimes in dry forests of eastern Washington state, USA, McKenzie et al. (2006a) sampled composite fire records at different spatial scales to examine the scale dependence of fire frequency and fire hazard. At small sampling scales, hazard functions were significantly greater than 1 (increasing hazard over time), particularly in watersheds with complex topography, but declined monotonically with increasing sampling scale (Fig. 2.2). McKenzie et al. (2006a) suggest that fire hazard on eastern Washington landscapes increases over time at spatial scales associated with a characteristic size of historical fires, reflecting the effects of fuel buildup within burned areas.

In high-severity fire regimes of shrublands in southern California, USA, Moritz (2003) found no scale dependence in the hazard function except for one landscape whose location and topography protected it from extreme fire weather (Fig. 2.3). Fire hazard increased in response to the increasing flammability of fuels over time. Over most of the region, however, fuel age-classes burned with equal likelihood, because almost all large fires occurred during extreme fire weather, providing sufficient inertia to overcome the patchiness of fuels and rendering the hazard function essentially constant. In both these examples, then, scaling laws in fire hazard were

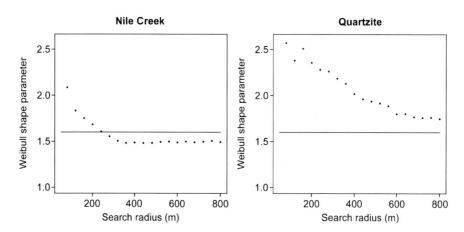

Fig 2.2 The Weibull shape parameter decreases with scale of sampling in two watersheds in eastern Washington. WMPI = Weibull median probability interval. Horizontal line marks the value (1.6) at the 95% upper confidence bound for testing whether the parameter is different from 1.0—meaning no increasing hazard over time. Fires were larger and less frequent in Quartzite than in Swauk Creek, so a shape parameter significantly greater than 1.0 may still be negligible ecologically, because shape and scale parameters co-vary (Moritz 2003 and Fig. 2.3) (Redrawn from McKenzie et al. (2006a))

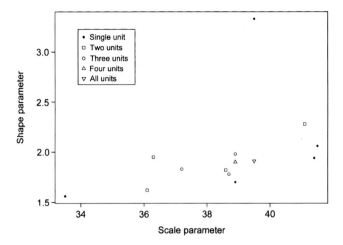

Fig. 2.3 Hazard function scale and shape parameters sampled at different scales in high-severity fire regimes in shrublands of southern California. The single point in the upper right represents one sample at the finest spatial scale that was protected from extreme fire weather and shows significantly increasing hazard over time. The positive covariance of the two parameters widens confidence intervals on significance tests of the shape parameter's difference from 1.0, *sensu* McKenzie et al. (2006a) and Moritz (2003), such that even values ≈ 2.0 may not indicate increasing fire hazard with time (Redrawn from Moritz (2003))

apparent only when controls were "bottom-up" (Kellogg et al. 2008, Chaps. 1 and 3), i.e., produced by interactions between fine-scale process (the buildup of fuels over time) and landscape pattern (topography and the spatial variability in fuel loadings), and where extreme fire weather was uncommon.

2.4.4   *Correlated Spatial Patterns*

We emphasized earlier that a key property of landscape fire is contagion. The relative connectivity of landscapes with respect to fire spread and the momentum provided by fire intensity and fire weather jointly affect the probability that two locations will experience the same fire event. If this probability attenuates systematically with distance, it can in theory be represented by a scaling law related to contagion.

The cumulative effect of these probabilities over time can be seen clearly as the similarity between two locations of the time series of years recording fire. In low-severity fire regimes, this similarity is measured between two recorder trees (point fire records) or area samples (composite fire records). Kellogg et al. (2008) compiled these time series for every recorder tree in each of seven watersheds in Washington state, USA. They used a classical ecological distance measure, the Jaccard distance (closely related to the Sørensen's distance [see below]—Legendre and Legendre 1998), to compare pairs of recorder trees at different geographic distances, generating scatterplots analogous to empirical variograms (hereafter *SD variograms*). Spherical variogram models, and power-law functions, were fit to these aggregate data for each watershed (McKenzie et al. 2006b; Kellogg et al. 2008; and the example below). Both types of models had better explanatory power in more topographically complex watersheds.

2.4.5   *Mechanisms*

Power laws abound in nature and society, but to date explicit mechanisms that produce them, and the parameters associated with their variability, have been difficult to identify. Purely stochastic processes can produce power laws (Reed 2001; Brown et al. 2002; Solow 2005), as can general dimensional relationships among variables, the most familiar being Euclidean geometric scaling (Brown et al. 2002). Brown et al. (2002) suggest that when scaling exponents in power laws (α in Eq. 2.1) take on a limited or unexpected range of values they are more likely to have arisen from underlying mechanisms. Examples of this are in organismic biology, where the fractal structure of networks and exchange surfaces clearly leads to allometric relationships (West et al. 1997, 1999, 2002) and in ecosystems in which there are strong feedbacks between biotic and hydrologic processes (Scanlon et al. 2007; Sole 2007).

How might we identify the mechanisms behind scaling laws in fire regimes? We propose two general criteria, based on our overview above, as hypotheses to be tested. Criterion #1 suggests how mechanisms produce scaling laws, whereas criterion #2 provides necessary conditions for scaling laws in fire regimes to be linked to driving mechanisms.

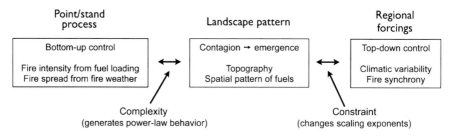

Fig. 2.4 Scaling laws in fire regimes are expected when bottom-up controls predominate and they interact strongly with landscape elements. For the contagious process of fire, fine-scale mechanisms provide momentum and topography and spatial pattern of fuels control connectivity (see text for discussion of contagion). In contrast, top-down controls (climate) increase fire size and therefore fire synchrony on landscapes where they are dominant, e.g., with gentle topography or continuous fuels. This favors irregular frequency distributions and lessens the scale dependence of fire frequency, hazard functions, and spatial patterns

1. Bottom-up controls are in effect: Drawing on O'Neill et al. (1986), we propose a hierarchical view of fire regimes that focuses interest on landscape scales (Fig. 2.4). Mechanisms at a finer scale below drive fire propagation, and interactions between process (fire spread) and pattern (topography and fuels) generate complex spatial patterns. When landscape spatial complexity is sufficient, fire spread and fuel consumption produce the spatial patterns that are reflected in the IA relationship, the hazard function, and the SD variogram. Conversely to one paradigm of complexity theory that posits that simple generating rules can produce complex observable behavior, we therefore see that relatively simple aggregate properties of natural phenomena—scaling laws—are the result of complex interactions among driving mechanisms.

2. Contagion provides a linkage among observations: We submit that if events (fires) are separated by more distance in space or time than some limit of contagion, observed scaling laws cannot be reasonably linked to a driving mechanism. Mechanism requires "entanglement" (as in the quantum-mechanical sense). For example, both SOC and HOT, mentioned above, require that events within a domain influence each other, whether one event resets system properties in proportion to its magnitude (SOC) or multiple events interact as they propagate through a system (HOT). The range limit of contagion clearly changes as a function of variation in fine-scale drivers. As we said earlier (see also Chap. 1), increasing energy (momentum) effectively increases connectivity, e.g., when extreme fire weather overcomes barriers to fire spread that are associated with landscape heterogeneity (Turner and Romme 1994).

Criterion #2 does not preclude some mechanism for power-law behavior across continental-to-global scales; it just limits the hierarchical interpretation in criterion #1 to spatial scales at which contagion occurs. Other explanations for power laws in nature and society do exist, however, including the purely mathematical (Reed 2001; Solow 2005).

2.5 Example: Power Laws and Spatial Patterns in Low-Severity Fire Regimes

We now turn to an example, briefly alluded to above, from low-severity fire regimes of eastern Washington state, USA (Everett et al. 2000; Hessl et al. 2004, 2007; McKenzie et al. 2006a; Kellogg et al. 2008; Kennedy and McKenzie 2010). Detailed fire-history data were collected in seven watersheds east of the Cascade crest, along a southwest–northeast gradient (Fig. 2.5). In contrast to most fire history studies, exact locations of all recorder trees were identified, creating an unprecedented opportunity for fine-scale spatial analysis (McKenzie et al. 2006a; Hessl et al. 2007; Kellogg et al. 2008). For a detailed description of the data and methods, see Everett et al. (2000) or Hessl et al. (2004).

a

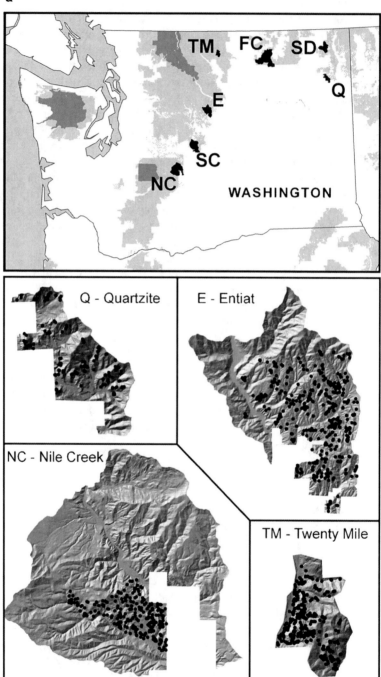

b

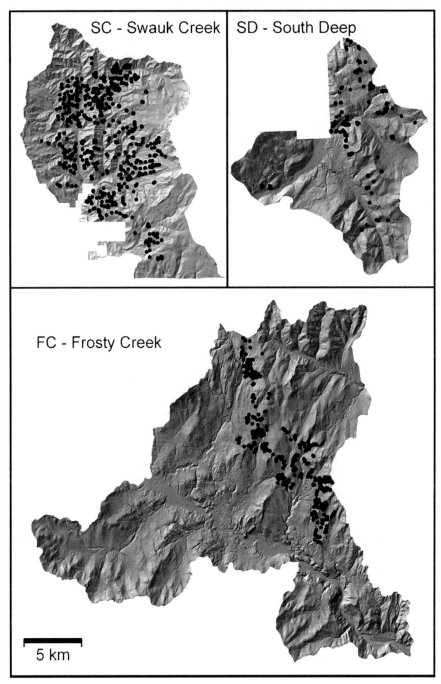

Fig. 2.5 Fire history study sites, east of the crest of the Cascade Mountains, Washington, USA. (**a**) Watershed locations. (**b**) Inserts that display hill shaded topography with dots representing the locations of recorder trees

Kellogg et al. (2008) fit the aforementioned empirical SD variograms to spherical models, in keeping with standard practice in geostatistics (Rossi et al. 1992), which uses variograms chiefly for spatial interpolation. Interpolation is generally only feasible with spherical, exponential, or Gaussian variogram models, due to certain mathematical conveniences (Isaaks and Srivastava 1989), but the spherical model in particular is a rather cumbersome artifact, with two separate equations applying to observations within or beyond the range (Kellogg et al. 2008). McKenzie et al. (2006b) examined the same empirical variograms in double logarithmic space and found that for some watersheds, the variograms seemed linear or nearly so, both graphically and when fit with linear regression. This suggested that power laws govern the correlated spatial pattern of fire histories. The observed pattern in these variograms was consistent across varying distance lags used to construct the variogram. We seek to test the hypothesis in criterion #1 (above) by trying to replicate the power-law behavior by controlling fine-scale processes (bottom-up control), using a neutral landscape model (Gardner and Urban 2007).

2.5.1   Neutral Model for Fire History

McKenzie et al. (2006a) developed a simple neutral fire history model to simulate recorder trees on landscapes that are scarred by fires of different sizes and frequencies. The purpose of the neutral model is to separate intrinsic stochastic processes from the effects of climate, fuel loadings, topography and management. We have enhanced the model to spread fires probabilistically on raster landscapes (Kennedy and McKenzie 2010; Fig. 2.6). The raster model produces 200-year fire histories on a neutral landscape, with homogenous topography and fuels. The raster landscape is initialized with a spatial point pattern of recorder trees; this pattern is simulated as a Poisson pattern of complete spatial randomness (CSR—Diggle 2003). A mean fire return interval (μ_{fri}) is specified for the whole "landscape", yielding a random number of fires (n_{fire}), drawn from a negative exponential distribution, within the 200-year fire history. For each fire, a random fire size is drawn from a gamma probability distribution (Evans et al. 2000) with the scale and shape parameters adjusted to produce a specified mean fire size (μ_{size}). For each fire in the fire history, an ignition point (pixel) is randomly assigned and the fire is spread until it reaches the randomly drawn fire size (i.e., area), or until all tests for fire spread fail in a given iteration. When a pixel is burned, each of the four immediate neighbors that are not yet burned is tested for fire spread against the spread probability (p_{burn}). After the neighbors are tested for fire spread, the burned pixel can no longer spread fire.

In a given fire, if a pixel is burned, then all trees located in that pixel are tested independently for scarring in the same time step. This is a simple probability test, with a specified scar probability (p_{scar}) that is uniform across all trees. This neutral model was produced in particular to evaluate whether the pattern in the observed SD variogram could be replicated by a simple stochastic model of fire spread, and

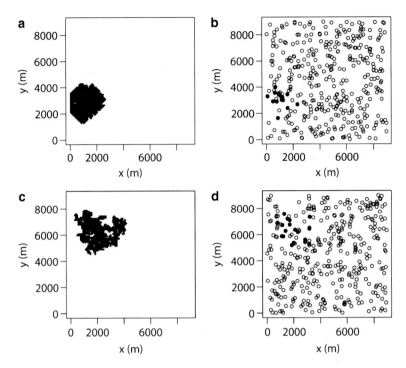

Fig. 2.6 Fire spread for (**a**) $p_{burn} = 0.75$ and (**c**) $p_{burn} = 0.50$. A complete spatial randomness (CSR) process generates recorder trees (points), with trees scarred by associated fire (black-filled points in **b** and **d**). A higher p_{burn} yields a more regular fire shape, although the difference in fire shape is difficult to discern visually in the scar pattern

to explain what differentiates variograms that appear linear in log-log space from those that do not. In order to satisfy the second goal, we considered whether the value of Sørensen's distance between two trees could be predicted by features of the neutral model.

2.5.2 Prediction of Sørensen's Distance

The Sørensen's distance can be analytically derived from conditional probabilities associated with fire spread and the scarring of recorder trees. Within the context of this neutral model, and under several assumptions verified by simulation, Kennedy and McKenzie (2010) found that the Sørensen's distance (SD) for a pair of trees a given distance apart is predicted by two features of the neutral model. The first is the probability a tree in a burned pixel is scarred (p_{scar}, which is spatially independent), which in the neutral model is constant across all recorder trees in the simulated landscape. The second model feature is the probability that two trees are both in a burned pixel in a given fire year (but not necessarily the same burned pixel). Specifically, for the pair of trees A and B, we calculate the probability that tree B is in a burned pixel (B_{fire}) given that tree A is in a burned pixel ($P(B_{fire}|A_{fire})$). For the stochastic model we consider the expected value of SD, and we found that it is predicted by (Kennedy and McKenzie 2010)

$$E(SD) = 1 - P\left(B_{fire} \mid A_{fire}\right)^* p_{scar} \qquad (2.2)$$

The probability the second tree is in a burned pixel given the first is in a burned pixel is not constant across pairs of trees, as it depends on the distance between the two trees, the fire size, and fire shape (Fig. 2.7).

As the distance between two trees approaches 0, then the conditional probability the second is in the fire given that the first is ($P(B_{fire}|A_{fire})$) approaches 1, and Eq. 2.2 reduces to

$$E(SD) = 1 - p_{scar} \qquad (2.3)$$

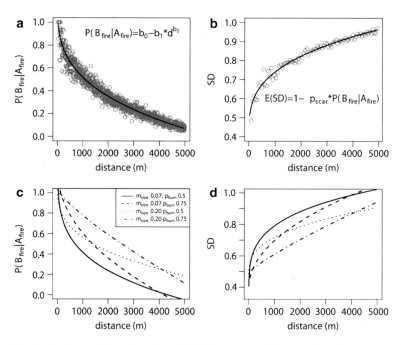

Fig. 2.7 Verification of the derivation of E(SD) via simulation and nonlinear regression. (**a**) $P(B_{fire}|A_{fire})$ with distance (*d*) predicted by 3-parameter model (neutral model μ_{size} =0.15 = 1500 pixels). (**b**) The fit to $P(B_{fire}|A_{fire})$ $\{\hat{b}_0, \hat{b}_1, \hat{b}_2\}$, with the p_{scar} set in the simulation (=0.5), used to predict E(SD) and compared to calculated SD variogram from the same simulation (i.e., Eq. 2.5). It fits well. (**c**) The relationship of $P(B_{fire}|A_{fire})$ with distance changes with mean fire size (μ_{size}) and fire shape as modified by the burn probability (p_{burn}); (**d**) these differences are also shown in changes to the shape of the SD variogram

Therefore, one can estimate the p_{scar} from an empirical SD variogram by the mean SD at the smallest distance bin. Simulations confirmed that the value of p_{scar} would be $\geq$ the mean value at the smallest distance bin.

We used a least-squares nonlinear regression algorithm in the R statistical program (nls; R Foundation 2003) to fit simulated $P(B_{fire}|A_{fire})$ against distance (up to half the maximum distance between simulated recorder trees—the same criterion used to evaluate SD), for three candidate functions (Kennedy and McKenzie 2010). The best fit with respect to an information–theoretic criterion (AIC) was found with a three-parameter function:

$$P\left(B_{fire} \mid A_{fire}\right) = b_0 - b_1 d^{b_2} \tag{2.4}$$

and, therefore,

$$E\left(SD\right) = 1 - p_{scar}\left(b_0 - b_1 d^{b_2}\right) \tag{2.5}$$

The coefficients $\{b_0, b_1, b_2\}$ thereby characterize the change in $P(B_{fire}|A_{fire})$ with distance, and consequently the change in SD with distance. The estimates of b_0, b_1 and b_2 in the neutral model change with increasing fire size, in a manner that depends on the shape of the fire (Fig. 2.7). Fire shape is closely associated with p_{burn}, with lower values of p_{burn} producing more irregular and complex shapes (Fig. 2.6). As the fire becomes larger and more regular, then the relationship between $P(B_{fire}|A_{fire})$ approaches a straight line with intercept b_0 and slope $-b_1$, i.e., b_2 gets closer to 1 (Fig. 2.7c; Table 2.1), and the slope (b_1) becomes less negative. In contrast, for irregularly shaped fires characteristic of $p_{burn} = 0.5$, the decline of $P(B_{fire}|A_{fire})$ remains nonlinear with estimates of b_2 well below 1 across a range of values for μ_{size} (Fig. 2.7c).

Note also that when $b_0 = 1/p_{scar}$, a power law describes the SD variogram, because we have:

$$E\left(SD\right) = p_{scar} b_1 d^{b_2}, \tag{2.6}$$

which is the power-law relationship presented in Eq. 2.1.

Recall that the relationship $P(B_{fire}|A_{fire})$ is independent of p_{scar}, and values of $\{b_0, b_1, b_2\}$ change with p_{burn} and μ_{size}. It is therefore possible to calibrate the values

Table 2.1 Parameter estimates for neutral model results with varying μ_{size} (0.07, 0.20) and p_{burn} (0.5, 0.75), and for the observed variograms (Twentymile, Swauk). Note that the coefficients b_1 are all negative, also indicated, for clarity, by the minus sign in Eq. 3.4

		b_0	b_1	b_2
μ_{size} 0.07	p_{burn} 0.50	1.430	−0.1990	0.235
	p_{burn} 0.75	1.240	−0.0247	0.469
μ_{size} 0.20	p_{burn} 0.50	1.060	−0.0437	0.351
	p_{burn} 0.75	1.030	−0.0010	0.805
Twentymile	p_{scar} 0.704	0.979	−0.0008	0.788
Swauk	p_{scar} 0.689	1.492	−0.2270	0.195

of μ_{size}, p_{burn} and p_{scar} to make $b_0 * p_{scar}$ arbitrarily close to 1, and thus manipulate simulated results to produce a power-law relationship in the SD variogram. In the neutral model this is a consequence of the mathematical relationships that we have found, yet the exercise of calibrating the parameters reveals under what conditions, as represented by μ_{size}, p_{burn} and p_{scar}, power laws should be expected. These can then be compared to the patterns observed in real landscapes, and indicate the ecological conditions under which power laws are produced.

The challenge, then, is to evaluate the relevance of the neutral model results for real landscapes insofar as the derived mathematical relationships are able to predict the patterns observed. We fit Eqs. 2.3, 2.5, and 2.6 to the SD variograms of real landscapes on a gradient of topographic complexity; first we estimate p_{scar} as the mean SD at the smallest distance bin in the observed SD variogram, then we fit Eq. 2.5 to the variogram in order to estimate the coefficients $\{b_0, b_1, b_2\}$. Here we compare the two watersheds from Kellogg et al. (2008) that are at opposite ends of this topographic gradient: Twentymile (least complex) and Swauk Creek (most complex). Coefficient estimates are in Table 2.1, and Fig. 2.8 shows the contrasting fits of the SD variograms from Twentymile and Swauk Creek in log-log space.

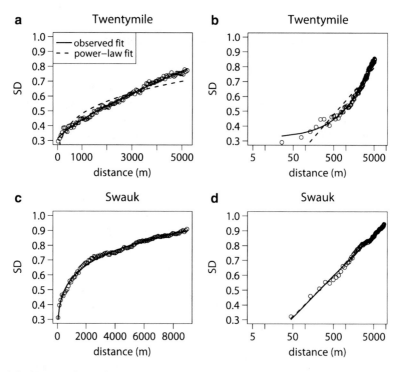

Fig. 2.8 Observed SD variograms for the least (Twentymile; **a,b**) and most (Swauk; **c,d**) topographically complex sites. Swauk increases more rapidly at smaller distances, and reaches a higher value. The Swauk fit is almost indistinguishable from the power-law prediction, with a small separation at the lowest distance bins

Clearly the relationship for Swauk Creek follows a power law (b_0 * p_{scar} = 1.492 * 0.689 = 1.028 ≈ 1; Eq. 2.6), whereas Twentymile does not (0.7 * 0.979 = 0.685).

These results suggest preliminary support for the hypothesis associated with Criterion #1 (above): Topographic complexity provides a bottom-up control on the spatial patterns of low-severity fire, producing relatively small fires and irregular fire shapes (SD increases more rapidly with distance, and reaches a higher peak, in Swauk Creek than Twentymile; Fig. 2.7). Neutral model runs with p_{burn} = 0.5 (irregular fire shapes; Fig. 2.6a) and relatively small mean fire sizes produced coefficient estimates similar to Swauk ({b_0, b_1, b_2}; Table 2.1) and SD variograms that followed power laws with p_{scar} near that estimated for Swauk. In contrast, neutral model runs with p_{burn} = 0.75 (regular fire shapes; Fig. 2.6c) and larger mean fire sizes produced coefficient estimates and SD variograms similar to those from Twentymile (Table 2.1).

What do we gain, then, by deconstructing these scaling laws via simulation; e.g., can we back-engineer a meaningful, preferably quantitative, description of fire regime properties that is relevant for landscape ecology and fire management? Certain combinations of the probability of scarring, the probability that a cell burns given that a neighboring cell has burned, and the mean fire size produce power-law behavior in an aggregate measure—the SD variogram—that represents the spatial autocorrelation structure of fire occurrence. For example, a low probability of scarring suggests variable fire severity at fine scales. A moderate likelihood of a cell's burning given that its neighbor has burned (i.e., p_{burn} = 0.5) suggest fine-scale controls on fire spread (topography and spatial heterogeneity of fuels). Mixed-severity fires subject to fine-scale landscape controls over time (decades to centuries) engender complex patterns that nonetheless produce simple mathematical structures (power laws). Further simulation modeling such as we describe here should illuminate what additional structures and scaling relationships can arise from the universe of complex interactions between the contagious process of fire and landscape controls.

2.6 Conclusions and Implications

Scaling laws in fire regimes are one aggregate representation of landscape controls on fire. Cross-scale patterns can reflect landscape memory (Peterson 2002). For example, fire-size distributions on landscapes small enough for fires to interact hold a memory of previous fires (Malamud et al. 1998; Collins et al. 2009), as do shape parameters of the hazard function on landscapes in which fuel buildup is necessary to sustain fire spread (Moritz 2003; McKenzie et al. 2006a). Scaling laws in our SD variograms hold a memory of all historical fires registered by recorder trees. We have conjectured above that scaling laws arise when bottom-up controls are in effect, but an additional possibility is that scaling relationships may be non-stationary over time, reflecting changes or anomalies in top-down controls, specifically climate (Falk et al. 2007). Mean fire size, fire frequency, and fire severity change with changes in climate and land use (Hessl et al. 2004; Hessburg and Agee 2005; Littell et al. 2009). A rapidly changing climate may at least change the parameters of scaling relationships, such as exponents in power laws derived from frequency distributions, and at most make them disappear altogether. Such behavior could indicate that a fire-prone landscape had crossed an important threshold (Pascual and Guichard 2005), with implications for ecosystem dynamics and management.

References

Adamic, L.A., and B.A. Huberman. 2000. The nature of markets in the World Wide Web. *Quarterly Journal of Electronic Commerce* 1: 512.

Agee, J.K. 1993. *Fire ecology of Pacific Northwest forests*. Washington: Island Press.

Allen, T.F.H., and T.W. Hoekstra. 1992. *Toward a unified ecology*. New York: Columbia University Press.

Bak, P., C. Tang, and K. Wiesenfeld. 1988. Self-organized criticality. *Physical Review A* 38: 364–374.

Bak, P.C., K. Chen, and C. Tang. 1990. A forest-fire model and some thoughts on turbulence. *Physics Letters A* 147: 297–300.

Brown, J.H., V.K. Gupta, B.L. Li, B.T. Milne, C. Restrepo, and G.B. West. 2002. The fractal nature of nature: Power laws, ecological complexity, and biodiversity. *Philosophical Transactions of the Royal Society B* 357: 619–626.

Calder, W.A. 1983. Ecological scaling: Mammals and birds. *Annual Review of Ecology and Systematics* 14: 213–230.

Carlson, J.M., and J. Doyle. 2002. Complexity and robustness. *Proceedings of the National Academy of Sciences* 99: 2538–2545.

Cello, G., and B.D. Malamud. eds. 2006. Fractal analysis for natural hazards. Special publication 261. London: Geological Society.

Clauset, A., C.R. Shalizi, and M.E.J. Newman. 2007. Power law distributions in empirical data. (http://arXiv:0706.1062v1).

Cohen, J.D., and J.E. Deeming. 1985. The national fire danger rating system: Basic equations. General Technical Report PSW-82. Berkeley: Forest Service.

Collins, B.M., J.D. Miller, A.E. Thode, M. Kelly, J.W. van Wagtendonk, and S.L. Stephens. 2009. Interactions among wildland fires in a long-established Sierra Nevada natural fire area. *Ecosystems* 12: 114–128.

Cumming, S.G. 2001. A parametric model of the fire-size distribution. *Canadian Journal of Forest Research* 31: 1297–1303.

Diggle, P.J. 2003. *Statistical analysis of spatial point patterns*, 2nd ed. London: Arnold.

Doyle, J., and J.M. Carlson. 2000. Power laws, HOT, and generalized source coding. *Physics Review Letters* 84: 5656–5659.

Dungan, J.L., J.N. Perry, M.R.T. Dale, P. Legendre, S. Citron-Pousty, M.-J. Fortin, A. Jakomulska, M. Miriti, and M.S. Rosenberg. 2002. A balanced view of scale in spatial statistical analysis. *Ecography* 25: 626–640.

Evans, M., N. Hastings, and B. Peacock. 2000. *Statistical distributions*, 3rd ed. New York: Wiley.

Everett, R.L., R. Schelhaas, D. Keenum, D. Spubeck, and P. Ohlson. 2000. Fire history in the ponderosa pine/Douglas-fir forests on the east slope of the Washington Cascades. *Forest Ecology and Management* 129: 207–225.

Falk, D.A. 2004. Scale dependence of probability models for fire intervals in a ponderosa pine ecosystem. Ph.D. dissertation. Tucson: University of Arizona.

Falk, D.A., C. Miller, D. McKenzie, and A.E. Black. 2007. Cross-scale analysis of fire regimes. *Ecosystems* 10: 809–823.

Gardner, R.H., and D.L. Urban. 2007. Neutral models for testing landscape hypotheses. *Landscape Ecology* 22: 15–29.

Gwozdz, R., and D. McKenzie. (unpublished data) Effects of topography, humidity, and model parameters on the spatial structure of simulated fine-fuel moisture. Seattle: Pacific Wildland Fire Sciences Lab, U.S. Forest Service (manuscript on file with: Don McKenzie).

Habeeb, R.L., J. Trebilco, S. Witherspoon, and C.R. Johnson. 2005. Determining natural scales of ecological systems. *Ecological Monographs* 75: 467–487.

Hessburg, P.F., and J.K. Agee. 2005. Dry forests and wildland fires of the inland Northwest USA: contrasting the landscape ecology of the pre–settlement and modern eras. *Forest Ecology and Management* 211: 117–139.

Hessl, A.E., D. McKenzie, and R. Schellhaas. 2004. Drought and Pacific decadal oscillation linked to fire occurrence in the inland Pacific Northwest. *Ecological Applications* 14: 425–442.

Hessl, A.E., J. Miller, J. Kernan, and D. McKenzie. 2007. Mapping wildfire boundaries from binary point data: Comparing approaches. *Professional Geographer* 59: 87–104.

Isaaks, E.H., and R.M. Srivastava. 1989. *An introduction to applied geostatistics*. New York: Oxford University Press.

Johnson, E.A., and S.L. Gutsell. 1994. Fire frequency models, methods, and interpretations. *Advances in Ecological Research* 25: 239–287.

Kellogg, L.-K.B., D. McKenzie, D.L. Peterson, and A.E. Hessl. 2008. Spatial models for inferring topographic controls on low-severity fire in the eastern Cascade Range of Washington, USA. *Landscape Ecology* 23: 227–240.

Kennedy, M.C., and D. McKenzie. 2010. Using a stochastic model and cross-scale analysis to evaluate controls on historical low-severity fire regimes. *Landscape Ecology*. doi:10.1007/s10980-010-9527-5.

Legendre, P., and L. Legendre. 1998. *Numerical ecology*, 2nd ed. Amsterdam: Elsevier Science B.V.

Levin, S.A. 1992. The problem of pattern and scale in ecology. *Ecology* 73: 1943–1967.

Levin, S.A. 2005. Self-organization and the emergence of complexity in ecological systems. *Bioscience* 55: 1075–1079.

Littell, J.S., D. McKenzie, D.L. Peterson, and A.L. Westerling. 2009. Climate and wildfire area burned in western U.S. ecoprovinces, 1916–2003. *Ecological Applications* 19: 1003–1021.

Loehle, C. 2004. Applying landscape principles to fire hazard reduction. *Forest Ecology and Management* 198: 261–267.

Malamud, B.D., G. Morein, and D.L. Turcotte. 1998. Forest fires: An example of self-organized critical behavior. *Science* 281: 1840–1842.

Malamud, B.D., Millington, J.D.A., and G.L.W. Perry. 2005. Characterizing wildfire regimes in the United States. *Proceedings of the National Academy of Sciences, USA* 102:4694–4699.

McKenzie, D. [N.d.]. Unpublished data. Seattle: Pacific Wildland Fire Sciences Lab (On file with: Don McKenzie).

McKenzie, D., D.L. Peterson, and E. Alvarado. 1996. Extrapolation problems in modeling fire effects at large spatial scales: A review. *International Journal of Wildland Fire* 6: 65–76.

McKenzie, D., D.L. Peterson, and J.K. Agee. 2000. Fire frequency in the Columbia River Basin: Building regional models from fire history data. *Ecological Applications* 10: 1497–1516.

McKenzie, D., A.E. Hessl, and Lara-Karena B. Kellogg. 2006a. Using neutral models to identify constraints on low-severity fire regimes. *Landscape Ecology* 21: 139–152.

McKenzie, D., L-K.B. Kellogg, D.A. Falk, C. Miller, and A.E. Black. 2006b. Scaling laws and fire-size distributions in historical low-severity fire regimes. *Geophysical Research Abstracts,* 8: 1607–7962/gra/EGU06–A–01436.

Miller, J.R., M.G. Turner, E.A.H. Smithwick, C.L. Dent, and E.H. Stanley. 2004. Spatial extrapolation: The science of predicting ecological patterns and processes. *Bioscience* 54: 310–320.

Millington, J.D.A., G.L.W. Perry, and B.D. Malamud. 2006. Models, data, and mechanisms: Quantifying wildfire regimes. In *Fractal analysis for natural hazards*, eds. G. Cello, and B.D. Malamud, 155–167. Special Publication 261. London: Geological Society.

Milne, B.T. 1998. Motivation and benefits of complex systems approaches in ecology. *Ecosystems* 1: 449–456.

Minnich, R.A. 1983. Fire mosaics in southern California and northern Baja California. *Science* 219: 1287–1294.

Moritz, M.A. 2003. Spatio-temporal analysis of controls of shrubland fire regimes: Age dependency and fire hazard. *Ecology* 84: 351–361.

Moritz, M.A., M.E. Morais, L.A. Summerell, J.M. Carlson, and J. Doyle. 2005. Wildfires, complexity, and highly optimized tolerance. *Proceedings of the National Academy of Sciences* 102: 17912–17917.

National Geophysical Data Center. 2010. NGDC: Natural hazards databases at NGDC. http://www.ngdc.noaa.gov/hazard/hazards.shtml. Accessed 25 Jan 2010.

NOAA. 2010. International multiproxy paleofire database. http://www.ncdc.noaa.gov/paleo/impd/paleofire.html. Accessed 25 Jan 2010.

Newman, M.E.J. 2005. Power laws, Pareto distributions, and Zipf's law. *Contemporary Physics* 46: 323–351.

O'Neill, R.V., D.L. deAngelis, J.B. Waide, and T.F.H. Allen. 1986. *A hierarchical concept of ecosystems.* Princeton: Princeton University Press.

Parody, J.M., and B.T. Milne. 2004. Implications of rescaling rules for multi-scaled habitat models. *Landscape Ecology* 19: 691–701.

Pascual, M., and F. Guichard. 2005. Criticality and disturbance in spatial ecological systems. *Trends in Ecology & Evolution* 20: 88–95.

Peters, R.H. 1983. *The ecological implications of body size.* Cambridge: Cambridge University Press.

Peterson, G.D. 2002. Contagious disturbance, ecological memory, and the emergence of landscape pattern. *Ecosystems* 5: 329–338.

Peterson, D.L., and V.T. Parker, eds. 1998. *Ecological scale: Theory and applications.* New York: Columbia University Press.

R Foundation. 2003. The R Project for statistical computing. http://www.r-project.org. Accessed 25 Jan 2010.

Rastetter, E.B., A.W. King, B.J. Cosby, G.M. Hornberger, R.V. O'Neill, and J.E. Hobbie. 1992. Aggregating finescale ecological knowledge to model coarser scale attributes of ecosystems. *Ecological Applications* 2: 55–70.

Redner, S. 1998. How popular is your paper? An empirical study of the citation distribution. *The European Physical Journal B* 4: 131–134.

Reed, W.J. 2001. The Pareto, Zipf, and other power laws. *Economics Letters* 74: 15–19.

Reed, W.J. 2006. A note on fire frequency concepts and definitions. *Canadian Journal of Forest Research* 36: 1884–1888.

Reed, W.J., and E.A. Johnson. 2004. Statistical methods for estimating historical fire frequency from multiple fire-scar data. *Canadian Journal of Forest Research* 34: 2306–2313.

Reed, W.J., and K.S. McKelvey. 2002. Power-law behaviour and parametric models for the size distribution of forest fires. *Ecological Modelling* 150: 239–254.

Ricotta, C. 2003. Fractal size distributions of wildfires in hierarchical landscapes: Natura facit saltus? *Comments on Theoretical Biology* 8: 93–101.

Roberts, D.C., and D.L. Turcotte. 1998. Fractality and self-organized criticality of wars. *Fractals* 6: 351–357.

Rossi, R.E., D.J. Mulla, A.G. Journel, and E.H. Franz. 1992. Geostatistical tools for modeling and interpreting spatial dependence. *Ecological Monographs* 62: 277–314.

Scanlon, T.M., K.K. Caylor, S.A. Levin, and I. Rodriguez-Iturbe. 2007. Positive feedbacks promote power-law clustering of Kalahari vegetation. *Nature* 449: 209–213.

Schneider, D.C. 2001. The rise of the concept of scale in ecology. *Bioscience* 51: 545–553.

Simard, A.J. 1991. Fire severity, changing scales, and how things hang together. *International Journal of Wildland Fire* 1: 23–34.

Sole, R. 2007. Scaling laws in the drier. *Nature* 447: 151–152.

Solow, A.R. 2005. Power laws without complexity. *Ecology Letters* 8: 361–363.

Song, W., F. Weicheng, W. Binghong, and Z. Jianjun. 2001. Self-organized criticality of forest fire in China. *Ecological Modelling* 145: 61–68.

Stauffer, D., and A. Aharony. 1994. *Introduction to percolation theory*, 2nd ed. London: Taylor and Francis.

Turcotte, D.L., B.D. Malamud, F. Guzzetti, and P. Reichenbach. 2002. Self-organization, the cascade model, and natural hazards. *Proceedings of the National Academy of Sciences* 99: 2530–2537.

Turner, M.G., and W.H. Romme. 1994. Landscape dynamics in crown fire ecosystems. *Landscape Ecology* 9: 59–77.

West, G.B., J.H. Brown, and B.J. Enquist. 1997. A general model for the origin of allometric scaling laws in biology. *Science* 276: 122–126.

West, G.B., J.H. Brown, and B.J. Enquist. 1999. A general model for the structure of plant vascular systems. *Nature* 400: 664–667.

West, G.B., W.H. Woodruff, and J.H. Brown. 2002. Allometric scaling of metabolic rate from molecules and mitochondria to cells and mammals. *Proceedings of the National Academy of Sciences* 99: 2473–2478.

White, E.P., B.J. Enquist, and J.L. Green. 2008. On estimating the exponent of power-law frequency distributions. *Ecology* 89: 905–912.

Wu, J. 1999. Hierarchy and scaling: Extrapolating information along a scaling ladder. *Canadian Journal of Remote Sensing* 25: 367–380.

Zipf, G.K. 1949. *Human behavior and the principle of least effort: An introduction to human ecology*. Reading: Addison-Wesley.

Chapter 3
Native Fire Regimes and Landscape Resilience

Max A. Moritz, Paul F. Hessburg, and Nicholas A. Povak

3.1 Introduction

First introduced by Holling (1973), the term "resilience" has been used widely in the ecological literature, but it is not always defined and is rarely quantified. Holling suggested that ecological resilience is the amount of disturbance that an ecosystem could withstand without changing self-organized processes and structures. His description suggests that resilience may be: (1) represented by an observable set of properties; (2) defined by measures of degree; and (3) related to system states and their (in)tolerance to reshaping, and that some properties of resilience may be quantifiable. We also see the idea of *fire resilience* in the literature (e.g., MacGillivray and Grime 1995; He and Mladenoff 1999; Díaz-Delgado et al. 2002; Brown et al. 2004; Pausas et al. 2004), but this term has different meanings in diverse contexts.

Despite disparate interpretations of resilience in the existing literature and of the role that fire may play, many agree that there is important linkage between naturally functioning fire regimes, the vegetation and terrain that fires move through, and the climate and weather that promote a fire ecology. This linkage manifests itself in fire-related plant traits (Bond and van Wilgen 1996), changes to landscape patterns, processes, and ecosystem functioning when fire is suppressed (Agee 1993; Hessburg et al. 2005), and potentially large changes as plant invaders alter native fire regimes and plant community structure (D'Antonio and Vitousek 1992).

One of the key challenges in defining properties of fire resilient landscapes is identifying mechanisms through which fire influences and reinforces landscape structure and functionality. What we observe on any single landscape is inevitably a mixture of both ecological interaction and adaptive response (Herrera 1992), providing only one snapshot in time and space. By choosing a relevant scale of

M.A. Moritz (✉)
Department of Environmental Science Policy and Management, Division of Ecosystem Sciences,
University of California Berkeley, Berkeley, CA 94720-001, USA
e-mail: mmoritz@berkeley.edu

D. McKenzie et al. (eds.), *The Landscape Ecology of Fire*, Ecological Studies 213,
DOI 10.1007/978-94-007-0301-8_3, © Springer Science+Business Media B.V. 2011

observation and similar biophysical settings, however, one may characterize a breadth of ecological structure and organization that is a function of interactions between species and processes operating within that scale (Peterson et al. 1998). Ultimately, positive and negative interactions between organisms, tradeoffs between exogenous and endogenous controls, and feedbacks between biotic and abiotic variables must all influence the patterns of vegetation within an ecosystem and the fire regime they co-create (Moritz et al. 2005).

The goal of this chapter is to examine mechanisms that might contribute to resilience in fire-prone ecosystems and their persistence in the face of ongoing natural disturbances and environmental variation. Our emphasis is on landscape patterns and processes, as opposed to finer spatial scales relevant to the fire ecology of a given species or patch. Because fire size distributions have been increasingly important to descriptions and explanations of ecosystem organization and structure, we examine datasets and examples from several different fire environments to look for consistent patterns among them. This was important to us because disturbances like fire have the greatest potential to restructure landscapes. Likewise, along with other factors, the living and dead structure of the landscape after fires provides endogenous feedback to future fire event and fire severity patterns (*sensu* Peterson 2002).

In particular, power law statistics have been used to characterize fire-event size distributions and what may control them, so we will examine the theories and methods related to this approach. Given that pattern and scale continue to present some of the most complex and interesting questions in ecology (Levin 1992), our intent is to shed new light on interactions between fire and its drivers at different scales. Without a better idea of how fire and ecosystem resilience are intertwined, their management and conservation may be impossible, as human influences and climatic changes continue to unfold.

3.2   Landscape Resilience

In order for an ecological system to persist and continue functioning in an environment with stochastic influences, it must be able to recover or rebound after disturbance. Intuitively, this is what many of us think of when the term resilience is used. Landscape resilience would thus apply to ecological persistence or a sort of meta-stability (*sensu* Wu and Loucks 1995) and continued functioning at a meso-scale, above that of vegetation patches and below that of physiographic province. Regardless of scale, a "ball-in-cup" model with one or more basins of attraction is often employed as a metaphor. This qualitative notion is somewhat vague, however, and there has been a profusion of literature and much confusion over terminology (e.g., Grimm and Wissel 1997) about what resilience actually means.

We introduce resilience concepts that are covered extensively in the edited volume of Gunderson and Holling (2002), due largely to members of the Resilience Alliance (*http://www.resiliencealliance.org*). In their parlance, there is *engineering resilience*, which focuses on system stability and the capacity to resist movement away from an equilibrium state, as well as the speed with which it can return to equilibrium. On the other hand, *ecosystem resilience* highlights non-equilibrium conditions and the ability of a system to absorb disturbance, changing and reorganizing to maintain structure and functionality over time. Holling and Gunderson (2002) draw a distinct contrast between engineering resilience (i.e., emphasizing efficiency, control, constancy, predictability) and ecosystem resilience (i.e., emphasizing persistence, adaptation, variability, unpredictability), although aspects of both would appear necessary for resilience in the face of ongoing fire events.

To consider landscape resilience, it is useful to scale these two concepts. Engineering resilience seems to emerge most clearly at relatively fine scales of space and time, under relatively more homogeneous conditions that arise from local, deterministic, and mostly bottom-up or endogenous controls. In contrast, ecosystem resilience emerges at meso- and broader scales, arising from a mix of bottom-up, top-down, and stochastic influences. *Cross-scale resilience*, a third concept put forth by Peterson et al. (1998), emphasizes the distribution of functional diversity within and across scales to allow, for example, regeneration after disturbances such as wildfire. To paraphrase Peterson et al. (1998), we will suggest that although most patterns and processes interact within system levels, there is also a certain amount of cross-talk or "leakage" between levels. This is especially true where processes and patterns reach their upper and lower bounds, or process domains, and where bounding is fuzzy or porous in nature. Wu and Loucks (1995) referred to this as loose vertical coupling: strongly coupled interconnections of patterns and processes within an observed level of organization, but cross-scale connections from the context (bottom-up) and constraint (top-down) levels in the fuzzy transition zones between levels. We illustrate an example of this later in the chapter.

Although processes and patterns at different scales are said to self-organize (Kauffman 1993), the origin and structure of relevant feedbacks and forcing factors are seldom quantified. How do these feedbacks and factors relate to landscape resilience? We know that species interactions with their local environments, disturbance regimes, and other ecological processes can lead to species sorting, structuring of communities, and ecological patterns of conditions to support them over moderate time scales (e.g., centuries to millennia, Moritz et al. 2005). However, species persistence and ecological functioning must also accommodate infrequent extreme events that may overwhelm bottom-up controls and any self-reinforcing feedbacks that may have developed in conjunction with more moderate disturbances. How ecosystems recover and continue functioning across the full distribution of events in a naturally functioning fire regime is therefore a key to landscape resilience.

3.3 Fire Regime Characterization

The fire regime (Gill 1975; Romme 1980) is a simplifying construct used throughout this book, so only a few of the relevant features are covered here. A conceptual framework for depicting controls on fire at different scales is presented in Fig. 3.1,

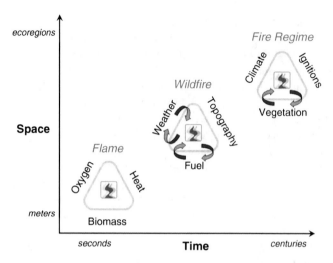

Fig. 3.1 Controls on fire at different scales of space and time. This framework adds a fire regime triangle (*upper right*) to the traditional triangles used to characterize combustion (*lower left*) and the fire environment (*middle*). Mechanisms most relevant to landscape resilience would tend to be operating at and in between the scales of wildfire and a fire regime. *Arrows* represent feedbacks between fire and the forces controlling fire at different scales

which combines the traditional "fire triangles" for combustion and wildfire with one for fire regime controls at the broadest scales. This framework was introduced in Moritz (1999) and developed in subsequent work (Davis and Moritz 2001; Moritz et al. 2005; Krawchuk et al. 2009; Parisien and Moritz 2009). Others, notably Martin and Sapsis (1992) and Bond and van Wilgen (1996), have also identified similar fire regime controls.

An excellent source for more general background on disturbance regimes is the edited volume of Pickett and White (1985). Despite the ongoing relevance of this early synthesis, there has been relatively little theoretical progress in disturbance ecology over the last 25 years. There have certainly been advances in understanding of individual ecological disturbances, such as fires, avalanches, debris flows, and floods, and a significant literature documents these insights. On the whole, however, still lacking are a general conceptual framework and body of quantitative methods that form the basis of disturbance ecology as a thriving research field in its own right (White and Jentsch 2001).

Because fire is a naturally recurring process, it can be statistically characterized by how often it occurs, when it occurs, the extent of area burned, and burn intensity. For example, one might be interested in the mean return interval of fire, some measure of interval variance, or the best fitting statistical distribution that describes the probabilities of all possible return intervals. An applied use of fire return interval distributions is to transform them into a measure of the "hazard of burning" (Johnson and Gutsell 1994). This hazard measure is used to quantify the degree to which fire probabilities change with time since the last fire (e.g., due to plant age or size effects, species composition, fuel density, or accumulation). This analytical approach originates largely in the forestry literature, but it has also been applied to the fire regime of chaparral shrublands across southern California (Moritz et al. 2004). Censored observations of fire return intervals (i.e., open-ended intervals that were at least age X when burned) are common in many fire datasets, and they can alter statistical outcomes and interpretations substantially (Polakow and Dunne 1999, 2001). Although return intervals are only one fire regime property, their characterization is an active area of research (Moritz et al. 2009).

Other fire regime parameters include fireline intensity (a measure of heat energy released), fire season (time of year), and fire size (area burned). Similar to fire interval data, the mean, variation, and statistical distributions of other fire regime parameters are often of interest. Note that all of these parameters refer to characteristics of fire itself, as opposed to the ecological effect of fire. The net ecological impact of fire—fire severity—is a function of several different fire regime parameters in an unlimited variety of combinations, and it may not manifest itself in vegetation or soils until well after a fire. In addition, ecological structure and function in one ecosystem may be highly sensitive to specific fire regime parameters, but much less to others (Romme et al. 1998). An ecologically severe fire in chaparral, for example, would be one that occurs soon after a preceding event (i.e., short fire interval), eliminating many of the native dominant species that require a decade or two to become sexually mature and contribute seeds to a soil seedbank (Zedler et al. 1983). An ecologically severe fire in a ponderosa pine forest might be one that occurs after a long fire-free interval, burning at a higher intensity than the dominant trees can survive (Agee 1993; Swetnam and Betancourt 1998).

In terms of the area affected, many fire regimes are dominated by the largest events (e.g., Strauss et al. 1989; Moritz 1997) and are sometimes said to have "heavy-tailed" fire size distributions (e.g., Malamud et al. 1998). These descriptive terms have statistical definitions, and their relevance in analysis of power law relations is discussed later in this chapter. Whether the largest fires are unusual and severe ecological events, however, depends on the ecosystem in question. Similar to the fire interval examples previously given, ecological resilience after a large fire hinges on how well species and communities inhabiting an ecosystem can regenerate, reorganize, and persist in the face of fires of varying size. The notion that there should be some natural range of variation for a fire regime has therefore become central to the management of fire-prone landscapes (Hessburg et al. 1999a; Landres et al. 1999; Swetnam et al. 1999). Later in this chapter, we discuss possible origins of that natural range of variation and why it is intuitive to think in these terms.

3.4   Fire Regime Variation and Resilience

One may aim to define the "native" fire regime of an ecosystem by quantifying variation in fire characteristics and connectivity of natural landscape states (e.g., before modern human activities) over some defined climatic period. This approach has been the basis of ecosystem management efforts (e.g., Hessburg et al. 1999a;

Landres et al. 1999) and large government agency projects, such as LANDFIRE (Schmidt et al. 2002) and the Interior Columbia River Basin Project (Hann et al. 1997). Quantifying native fire regimes for use in forest management is also the basis for "emulation forestry" (Perera and Buse 2004, and chapters therein). Ideally, all fire regime parameters described above would be factored into this approach, because each can be ecologically meaningful. The parameter emphasized most frequently tends to be fire return interval, which is often assumed to produce an associated fire intensity. This is a widely held assumption for many fire-prone ecosystems. However, longer fire free intervals do not always result in higher fire intensities. Likewise, short fire return intervals do not always result in lower fire intensities. Examples include ecosystems in which extreme fire weather events (a top-down influence) can overwhelm constraints that time since the last fire, recovery pathway, and fuel accumulation might otherwise pose (bottom-up influences). This tradeoff in controls applies for many chaparral shrublands of southern California (Moritz 2003) and a variety of coniferous crown fire dominated ecosystems (Turner and Romme 1994). There are also examples of ecosystems in which rarely burned stands display a decreasing probability of intense fire, such as the forests of the western Klamath Mountains in California described by Odion et al. (2004, 2009).

In addition to paying more attention to some parameters than others, use of the historical range of variation (HRV) in native fire regimes also requires that a particular period of relevant climate be chosen as a reference (e.g., Landres et al. 1999; Swetnam et al. 1999). Therefore, one can arrive at different estimates for a given parameter, simply by considering different periods. For example, restricting the temporal baseline to the Little Ice Age (~1,400–1,850) could give quite different estimates than if the Medieval Warm Period (~800–1,300) were included. Several reference periods, however, can be highly informative about the dynamics and interplay between the climate, land, and biotic systems, and that is the primary utility of historical ecology.

Rather than being a single snapshot of conditions in space and time, we suggest that the HRV should represent the broad envelope of realizations that can occur in a given landscape, considering a particular climate reference period. When the climate system changes, the envelope of realizations drifts to include new conditions, but is not likely re-invented. This is due to the potent effect of the historical ecology, which is the system memory; i.e., prior influences can determine, to a large but incomplete extent, future landscape or ecosystem trajectories, and the effects can last for centuries (Peterson et al. 1998; Peterson 2002). A thought experiment for estimating the HRV of any landscape is to consider the range of conditions that would occur were we able to rewind time in a particular climatic period a 1,000 times or more, all else being equal (Hessburg et al. 1999a, b; Nonaka and Spies 2005). In this light, the HRV is an emergent property of landscapes and ecosystems (Peterson 2002), derived from the same exogenous and endogenous forcing factors that shape their resilience. Any future range of variation (FRV) is then a consequence of the prior HRV, plus changes in exogenous and endogenous forcings, and the resulting range of conditions.

A related alternative approach is to identify a bounded range of fire regime variation, regardless of what the past has demonstrated, within which long-term persistence of ecosystem structure and function might be possible. As opposed to a focus on a central-tendency measure of mean fire return interval, the emphasis here is on avoiding ecological thresholds. This would seem to be at the heart of fire resilience, but it presupposes knowledge of the thresholds to avoid, the manner and rate of ecosystem shifting once thresholds are exceeded, and which fire regime parameters are most ecologically influential (Romme et al. 1998). Such knowledge is seldom available. Another unknown is whether thresholds themselves shift in a dynamic climatic future and how species, communities, and processes might respond. So, while conceptually important, a focus on thresholds may offer limited guidance (e.g., only for certain species) until much more is learned about ecosystem dynamics in general.

In the face of climatic change, discussion has also emerged about reinforcing ecological resilience (Millar et al. 2007; Moritz and Stephens 2008), as opposed to recreating or restoring more natural disturbance regimes. This is largely due to uncertainty in whether the last few centuries can indicate how ecosystems will respond to climates of future decades and the fire regimes that may accompany them. Even so, it is not time to toss away the historical range of variation concept or historical ecology. Understanding the mechanisms that have to date controlled landscape resilience is of central importance, and a marriage of the aforementioned ideas seems warranted.

3.5 Fences and Corridors

Landscape resilience in stochastic environments must involve a variety of species and processes at different scales, some of which are redundant and others that are overlapping, such that reorganization and persistence of ecological function are possible after disturbances (Peterson et al. 1998). In the case of fire, there must also be mechanisms that generate "fences and corridors" on the landscape—the patchiness of conditions that retard or facilitate progress of combustion—that fire has to negotiate at any given time. We propose that fire's fences and corridors, both metaphorically and in reality, are a key to landscape resilience.

In a completely homogeneous (and hypothetical) landscape, an extreme situation would be that all biomass burns every year, and at all scales, assuming the infrequent ignition at some locations. For all but a few species, this lack of fences and corridors for fire would clearly be intolerable to their persistence. It is heterogeneity across the landscape that allows for patchiness in space and time, for vegetation as well as fire, and thus persistence of diverse ecosystems. Even after very large and stand-replacing fires like those of Yellowstone in 1988, heterogeneity at the landscape scale is seen as key to resilience and regeneration (Schoennagel et al. 2008). Landscape heterogeneity, variation in fire regimes, and patchiness in fire effects all contribute to landscape pattern complexity and different types of refugia for post-fire regeneration.

Areas that are less likely to burn (fences) and more flammable swaths of landscape (corridors) influence fire patterns and are due to both biotic and abiotic factors. Some landscape patterns that either constrain or facilitate the spread of fire will be relatively static, while others will change with the seasons, and with time since the last fire. Certain climatic trends (e.g., protracted drought) and extreme fire weather episodes (e.g., hot, dry, and strong winds) can also temporarily reduce constraint on fire spread across the landscape. Over long enough time scales, feedbacks that occur between vegetation and fire eventually lead to vegetation patterns that are tolerant of – and often adapted to – the fire regime that exists there. Since these feedbacks are partially responsible for the frequencies and types of fires that are characteristic of a given region, they also reinforce the network of fences and corridors in a given ecosystem.

It seems self-evident that landscape heterogeneity should affect the rates and patterns of biomass consumption by fire. But does this heterogeneity have inherent structure? Is there any reason to suspect that the size distributions of fires should somehow be similar across ecosystems that have different inherent rates of primary productivity or types of topographic complexity? If so, this would imply that the ensemble of fences and corridors characteristic of one ecosystem can produce fire patterns that are somehow comparable to those from another ecosystem.

3.6 Fire Size Distributions and Power Laws

Theory and observation hold that certain systems exhibit self-organizing properties (Turcotte 1999). Under a broad range of conditions, event size distributions of landslides, earthquakes, floods, and some argue, forest fires exhibit this behavior (e.g., Malamud et al. 1998; Turcotte and Malamud 2004). Event-size distributions are described using a power-law relation (Pareto I distribution), which implies scale-invariance of event frequency-size distributions, and system self-reinforcement.

Power laws have been found in many fire size distributions (e.g., Malamud et al. 1998, 2004; Song et al. 2001; Carlson and Doyle 2002; Reed and McKelvey 2002; Moritz et al. 2005; Boer et al. 2008), although there is substantial disagreement about what this shared characteristic signifies. A distribution may include very large and unlikely events—the signature of being "heavy-tailed"—but this does not necessarily mean it displays a power law relation. Specifically, a fire size distribution is said to fit a power law relation with slope α if the probability P of a fire of size (l) is given by:

$$P(l) \approx l^{-\alpha} \tag{3.1}$$

Using a cumulative form of the data (e.g. rank-ordered by size, or the cumulative distribution function, CDF) avoids having to choose bin widths and other potentially subjective decisions related to model fitting (Malamud et al. 1998). A constant must be added to Eq. 3.1 to normalize units of P such that values range from 0 to 1 in the cumulative probability distribution. Plots of data are typically shown after log-log transforming both axes, so that the slope α provides a linear fit to the data. Moreover, a distribution of fire sizes may be heavy-tailed and not be purely power law in a log-log plot, if the probability does not decrease in a linear fashion as fire size increases over the entire range of the distribution. As we will show later, several closely related statistical distributions have heavy tails and do not show a linear fit at either end of the CDF, yet they display robust power law behavior across a middle range of fire sizes. In the simplest form, purely power law relations are synonymous with the single parameter Pareto I (P1) model (Newman 2005).

Because fire size distributions have exhibited power-law behavior, despite very different geographic locations and vegetation types, some have seen this as evidence of a common mechanism, and of self-organization (e.g., Malamud et al. 1998, 2004; Ricotta et al. 1999; Song et al. 2001). Observation of power law characteristics over a broad range of spatial scales has led to descriptions of these relations as scale-invariant; that is, relations apparently exist regardless of the scale of observation.

3.7 Theories on the Origin of Power Laws

One body of theory, called self-organized criticality (SOC), argues that such system behavior is a function of purely endogenous controls (Bak 1996; Turcotte 1999). This has been shown, for example, in simple sand pile and forest fire simulation models, which exhibit scale-invariance of event frequency-size distributions and apparent system self-reinforcement. Criticality is said to be driven by distinct events (e.g., landslides, fires, earthquakes). Above a "critical" threshold, rates of endogenous processes produce cascades of events and a range of event sizes fitting a power law (P1) distribution (Turcotte 1999; Turcotte and Malamud 2004; Malamud and Turcotte 2006).

When one examines the simulation logic behind the SOC fire model, it is clear that these experiments must reveal chiefly endogenous controls, due to the simulation approach and the modeling rules driving critical events (e.g., fuel regrowth rates). At the other end of the spectrum, one can imagine a system in which event-size distributions are completely driven by exogenous factors. In the case of wildfires, for example, Boer et al. (2008) have argued that the frequency of wind events is the sole structuring mechanism of several fire size distributions they examined. While their comparison of wind severity distributions and fire size distributions is compelling, the analysis itself required the specification of a vegetation-related parameter – an endogenous factor – to match the power law exponents of wind and fire events.

Given the many interactions across different scales that ultimately produce a fire regime (Fig. 3.1), it is almost inconceivable that a full range of fire sizes could be controlled by a single exogenous or endogenous factor. Indeed, Reed and McKelvey (2002) have shown that fire size distributions in different regions fit power laws under certain circumstances and that multiple influences should be involved. Across a range of fire sizes, the importance of factors driving fire spread is approximately equal to that of factors causing fires to go out (i.e., mathematically, a balanced extinguishment : growth ratio).

The findings of Reed and McKelvey (2002) imply a type of meta-stability, which may have profound implications if generally true. First, they confirm that both fences (extinguishment) and corridors (growth) are involved in structuring fire size distributions, implying a variety of endogenous and exogenous factors at play. Furthermore, this suggests an ongoing tradeoff in the influence of constraints vs. drivers of fire spread, from which *we should actually expect power law distributions of fire sizes to emerge*. Marked deviations from a power law distribution could thus indicate ecosystems in which forces facilitating the process of combustion are consistently overwhelming those constraining it (or vice versa). Such a skewed dynamic might reflect ecosystems going through a major transition (e.g., due to climate change) or the possibly loss of inherent resilience mechanisms. Although it is not obvious what the power law slope should be for a robust and functioning ecosystem, nor over how many orders of magnitude this should be observed, the findings of Reed and McKelvey (2002) suggest the importance of structured networks of fences and corridors on fire-prone landscapes, as well as an expectation for power law distributions in fire sizes.

The idea that there are multiple inherent constraints on fire size and that ecosystems become somewhat "tuned" to the local fire regime is central to the concept of Highly Optimized Tolerance (HOT) in fire-prone ecosystems (Carlson and Doyle 2002; Moritz et al. 2005). HOT also provides an explanation for the slopes of observed power laws. HOT is a conceptual framework for studying organization and structure in complex systems, and the clearest examples come from biology and engineering, where adaptation and control theory have direct application (Carlson and Doyle 1999, 2000; Doyle and Carlson 2000; Zhou and Carlson 2000; Robert et al. 2001; Zhou et al. 2002). The HOT framework is based on the assumption that complex systems of interacting components must be robust to environmental variation within some characteristic range. Otherwise complex systems would not be able to persist and function in fluctuating and uncertain environments. Being more finely tuned to a narrow spectrum of conditions – even if increasing performance or efficiency under these conditions – will ultimately make a system more susceptible to failure in circumstances outside the narrow range of variation. This tradeoff is at the heart of what it means to be "robust yet fragile" in the HOT framework (Carlson and Doyle 2002), and it may offer substantial insight into landscape resilience. Notably, there are also direct parallels between the concept of HRV in fire regimes and the degree of environmental variation to which complex systems must be resilient.

In addition to providing theory for how tradeoffs and feedbacks operate in complex systems, HOT also employs an analytical framework for optimizing these tradeoffs under uncertainty, and solutions relate directly to the dimensionality of the problem (Carlson and Doyle 2002). In the case of fire and a managed forest, a goal might be to arrange barriers to fire spread among forest stands such that one

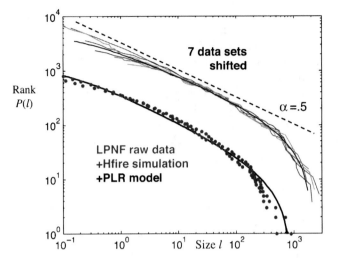

Fig. 3.2 Fire size statistics for a variety of fire datasets The lower curves include HRV fire size data for chaparral-dominated portions of Los Padres National Forest (*LPNF*) as well as for a simulation model (HFire) and the analytical model proposed by HOT (PLR). The vertical axis is the rank of the event size, while the horizontal axis is in km². The upper set of curves shows these datasets, plus 4 additional fire size catalogs from different regions of the world, rescaled to show their power law fit of slope −1/2 (i.e., exponent α = .5) over several orders of magnitude (Reprinted from Moritz et al. 2005)

minimizes the range of fire sizes observed in the system. Using linear (i.e., 1-dimensional) barriers, the analytical HOT solution to this problem leads to a size distribution of fires (~2-dimensional) that follows a power law. It has been shown that several real and modeled fire size datasets approximate a power law with slope −1/2, or 1 divided by the dimension of the events being minimized (Carlson and Doyle 2002; Moritz et al. 2005). Figure 3.2 shows a variety of fire datasets that have this characteristic shape and overall slope of −1/2 in their HRV of fire sizes.

3.8   Example Ecosystems

Although some have argued that power-law behavior should not necessarily be interpreted as evidence for ecological organization or inherent ecosystem structure (e.g., Reed and McKelvey 2002; Solow 2005), the consistent shape of many fire size datasets indicates an apparent "functional form" and is quite compelling. Furthermore, the power law slope of some of these distributions is that predicted by HOT, which would suggest a tendency in these systems toward minimizing the size range of disturbances. It is not clear, however, how HOT as a mechanism might accomplish this. How would tradeoffs in the influence of bottom-up (e.g., topography and vegetation) vs. top-down (e.g., fire weather and climate patterns) controls consistently generate a specific distribution of fire patterns under different combinations of environmental conditions? For HOT to apply in fire-prone ecosystems, one would expect consistencies between ensembles of fences and corridors for fire across ecosystems, as well as feedbacks that could at least partially create these generic structures.

In the remaining sections of this chapter, we further examine the origin, controls, and methods for identifying power law distributions in fire size data. A first example focuses on a crown-fire-adapted chaparral ecosystem, where fire severity essentially functions as a constant across all fire event sizes. In this example, we demonstrate application of HOT as a theoretical framework, which leads into several questions about fitting statistical distributions to fire size data and how to interpret the results. This is followed by a second example analyzing a variety of landscapes, including surface fire, crown fire, and mixed surface and crown-fire-adapted ecosystems, where fire-severity patterns vary considerably. The importance of rigorous statistical distribution fitting methods is also emphasized, as well as more mechanistic relations to topographic and physiographic controls on fire size distributions.

3.9   Fire Size Distributions in Chaparral Ecosystems

Our goal here is to demonstrate application of HOT to fire datasets to see how well they do or do not adhere to the distribution of fire sizes predicted by this framework. In particular, we aim to contrast regions that have varying degrees of similarity in fire regime controls, to determine if adjacent regions with different top-down influences still hold to HOT predictions.

Many fire size datasets show evidence of power law behavior over some mesoscale range (e.g., Fig. 3.2), with a "cutoff" at the upper event sizes (Burroughs and Tebbens 2001). A steepening of the slope in the largest fire-size range may correspond to some upper limit to the growth of fires in the study domain. Such an upper truncation could be caused by large fires stopping when they eventually reach landscape boundaries, such as adjacent oceans or deserts, ridgetops, or catchment boundaries. The upper limit could also be dictated by the duration of fire weather episodes (e.g., hot, dry, and strong wind events that typically last less than a week), which would constrain the final size of the largest events. A steepening of slope in the heavy tails of these distributions is therefore not a contradiction to HOT predictions; on the contrary, it is indicative of the scale of spatial controls operating in the creation of the largest patches.

One issue worth mentioning here is the choice of study domain size: How do we identify the most appropriate scale at which this type of analysis is to be performed? One could compile fire size data from a very large study area, which would contain many different ecosystems with quite different fire regimes. In that case, we might not expect evidence of a clear cutoff in the large fire size range, since many different upper limit boundaries are being mixed together in the dataset. Mixtures of fire regimes with different large event cutoffs could also lead to steeper power-law slopes over what would otherwise constitute the meso-scale range of the distribution (Doyle and Carlson 2000). Identifying the spatial limits of a region with a roughly homogeneous fire regime is thus an important and under-explored area of study.

In the smaller fire size range (the left tail of the fire size distribution), a shallower slope and the opposite tendency is often observed—i.e., relatively large increases in size between the probability of one fire and the next largest—up to a meso-scale range exhibiting power law behavior (e.g., see Fig. 3.2). One explanation for this flattening of slope could be that many of the events below the lower cutoff size are unrecorded, undetected, or undetectable, and their inclusion would steepen this portion of the distribution. Another explanation is that the interaction of factors driving and/or constraining the spread of smaller fires is basically different than that occurring across the meso-scale range displaying power law behavior, leading to a differing slope.

The analytical solution to the HOT model that minimizes average fire sizes (l) has the following cumulative form (referred to as the PLR or probability-loss-resource model, Moritz et al. (2005) and references therein), after including both the small (C) and large (L) event cutoffs:

$$P(l) \sim (C+L)^{-\alpha} - (C+L)^{-\alpha} \qquad (3.2)$$

Similar to Eq. 3.1 above, a constant is applied to the right-hand side of Eq. 3.2 to normalize units of probability P. The constant, the truncation parameters for the cutoffs, and the exponent can be chosen through an objective fitting algorithm (e.g., maximum likelihood), or values may be selected based on other criteria (e.g., smallest and largest events in record, hypothesized slope). Regardless of the slope or the mechanism in question, this lower- and upper-truncated power law function provides a simple tool for examining fire size data and the range over which power law behavior applies. In this example, we are not objectively fitting algorithms to determine parameter values; instead, we specify the cutoffs from the data themselves.

3.9.1 Exposed vs. Sheltered from Extreme Fire Weather

So far we have considered the meso-scale domain of fire regime controls to be that across which fire sizes display a power-law distribution, presumably structured by various feedbacks and forcing factors. If there are specific ensembles of fences and corridors characteristic of particular fire regimes – our hypothesized signature of landscape resilience – it is not yet obvious how broad-scale differences in top-down controls might alter fire size distributions from the "functional form" with power law slope of −1/2 described above.

One of the datasets examined in Moritz et al. (2005) and shown in Fig. 3.2 is for the combined chaparral-dominated shrublands of Los Padres National Forest (LPNF) in central coastal California. In analyzing the degree to which time since the last fire constrains subsequent burning probabilities, it has been shown that most shrublands of the region do not show a strong relationship between the age of fuels and the hazard of burning (Moritz 2003; Moritz et al. 2004). This is largely because these regions are routinely exposed to seasonal drought and Santa Ana wind episodes, which can drive fires through all age classes of vegetation. There is one region of LPNF, however, that is sheltered from Santa Ana winds and actually shows a moderate degree of age dependence in burning probabilities. Although the region near the town of Santa Barbara is subject to highly localized fire weather events known as "sundowner winds," the alignment of local mountain ranges appears to shelter the region from the more synoptic-scale Santa Ana winds that case massive fires in other parts of California (Moritz 2003; Moritz et al. 2004).

Disaggregating the fire data for LPNF into the Santa Barbara region and the adjacent Ventura region, we see in Fig. 3.3 that both distributions display quite similar shapes and hold closely to HOT predictions. These two regions vary markedly, however, in the amount of area burned in very large fires. The ten largest events, for example, comprise a total of ~95,000 and 213,000 ha burned in the Santa Barbara and Ventura regions, respectively (encircled in Fig. 3.3 and plotted in Fig. 3.4). Notably, the largest ten events account for the vast majority (~95%) of the difference in area burned by all fires shown for these regions.

Despite striking differences in conditions under which most of the area burns, the adjacent regions shown in Fig. 3.3 both appear to be good fits to a power law

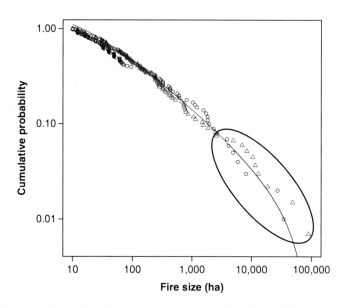

Fig. 3.3 Fire size distributions for subregions of Los Padres National Forest. Data includes fires > 10 ha and since 1950 for Santa Barbara (*black circles*) and Ventura (*red triangles*) regions, with largest 10 events encircled in lower *right*. *The black line* shows the HOT prediction of slope −1/2 (Eq. 3.2, $C = 10$ and $L = 100,000$)

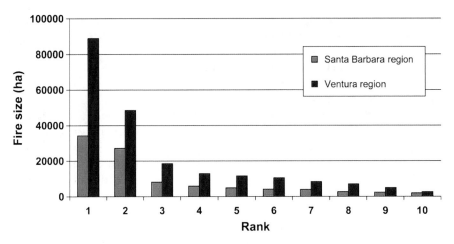

Fig. 3.4 Size comparison of 10 largest events for regions within the Los Padres National Forest

with a slope of −1/2. The fire size data for the Santa Barbara region are not as heavy-tailed as that of the Ventura region, but in a log-log plot, this difference is not as large a deviation as the total area burned would indicate. The majority of fire events occurring in the meso-scale range of the fire regime still exhibit a remarkably similar form in terms of fire-size distributions. This suggests somewhat similar types and scales of landscape heterogeneity between the regions examined. In other words, comparable ensembles of fences and corridors for fire spread may be encountered for most events in the fire regime of both regions.

3.9.2   Landscape Resilience in Chaparral

For the chaparral-dominated shrublands examined here, the interplay of endogenous and exogenous controls apparently maintains a specific structure in the fire size distributions, despite major differences in top-down fire weather types and frequencies. How does this relate to ecological resilience on a fire-prone landscape?

As noted earlier, an ecologically severe fire in chaparral would tend to involve short fire intervals in a given location. This is because several dominant chaparral plants have specialized life histories that for their persistence on the landscape require a seedbank to accumulate locally before the next fire. Thus, frequent fires can lead to the replacement of some of the native dominants by invasive annual species (Zedler et al. 1983). In and of themselves, large and intense fires are not ecologically severe events, as long as they are well separated in time. Maintaining this separation is at least partly dependent on ignition patterns, since more ignitions will increase the likelihood that a fire actually occurs under extreme fire weather conditions and be capable of burning through young and regenerating stands of vegetation. As this trend continues, larger and larger portions of the landscape become type-converted into highly flammable species that can support fire every year–a positive feedback that is known as the "grass/fire cycle" (D'Antonio and Vitousek 1992). Landscape resilience can thus be fundamentally altered, leading eventually to a new alternative state, if system sensitivities are challenged repeatedly and ecological thresholds are eventually crossed.

HOT provides a promising conceptual and analytical framework for understanding the ensemble of fences and corridors that structure fire patterns on landscapes subject to this natural disturbance. Admittedly, however, we have not rigorously demonstrated that the best-fitting slope for the truncated power law in Fig. 3.3 is actually $-1/2$. It is also possible that a different statistical distribution altogether may be a better fit to the fire size data we examined. Although chaparral ecosystems appear to have an inherent resiliency against infrequent large events in the tails of fire size distributions, the tradeoffs between constraints and drivers hypothesized in HOT have yet to be identified. Steps toward linking fire regimes to various endogenous and exogenous factors driving them would therefore include a more statistically rigorous approach to fitting fire-size data to statistical distributions, and direct evaluation of relations between endogenous and exogenous factors and the distributions themselves. We undertake these steps below.

3.10 Fire Size Distributions in Ecoregions of California

Much of the discussion of landscape and ecosystem resilience to date has been descriptive and theoretical in nature. Recently, however, several researchers have begun to take quantitative methods from laboratory simulation experiments and apply them to natural systems, as in the application of the HOT model to California chaparral just described. This is important on several levels, since it allows observation of natural systems that may be under purely endogenous (syn. fine scale, bottom-up), purely exogenous (syn. broad scale, top-down), or mixed controls (syn. meso-scale). Evidence for power-law relations among wildfire events has largely relied on the log-linear relationship of the frequency-size distributions of fires. Power laws have been suggested with satisfactory fits of ordinary least squares linear regression to log-log transformed, cumulative (CDF) or non-cumulative frequency-size distributions (Malamud et al. 1998). Distributions tend to be described using the one-parameter Pareto I distribution introduced earlier or some variation on it (e.g., the truncated form in Eq. 3.2). However, the intricacies of demonstrating a good power-law fit in the first place have received relatively little attention.

The underlying goals of this analysis were to objectively evaluate evidence for (1) power law behavior in the event size distributions of wildfires in California, and (2) potential top-down (exogenous) and bottom-up (endogenous) controls over the structure of these distributions. In ecological systems, we suspect that interactions among constraining and contextual influences (*sensu* Wu and Loucks 1995) may offer a fuller explanation for what drives system structure. We therefore attempt here to test for different forcing factors in a variety of fire-prone ecosystems, at the observation scale of ecoregions, for the State of California. Our objective was to provide quantitative evidence of both endogenous and exogenous forms of spatial control in natural systems, while also distinguishing their control domains.

We used an atlas of recorded fire event sizes in California for the period 1900–2007. Because fire records were spotty for the first half of the 20th century, we pared the atlas down to the period 1950–2007 to avoid the greatest potential bias in recording event-size distributions. We also note the likely incompleteness of the dataset for wildfires less than 40 ha occurring in forest or 120 ha in grass or shrubland habitats. These are threshold sizes when a fire start is considered a large wildfire incident, from a suppression standpoint. We assumed that most so-called large wildfire incidents were recorded, but that record-keeping of the smaller events was likely uneven due to their lesser operational importance.

An on-line geodatabase for the Bailey nested "ecoregions" was acquired (Bailey 1995, *http://www.fs.fed.us/rm/ecoregions/products/map-ecoregions-united-states/*), including spatial layers for the division, province, and section levels. We used the multi-level regionalization to determine whether the biogeoclimatic setting of the fires explained differences in event-size distributions, and at which scales of observation distributions showed the highest goodness-of-fit. Where ecoregions at one scale minimized variance in event-size distribution when compared with other scales of observation, this would be quantitative evidence of the approximate scale of top-down spatial control on event size distributions. To accomplish this, we stratified the California fire event atlas for the period 1950–2007 by the Bailey division, province, and section strata (Fig. 3.5). We then submitted the stratified fire-size distributions to the set of distribution fitting and goodness-of-fit techniques.

3.10.1 Distribution Fitting

Our first objective was to fit the HOT model to fire event sizes within Bailey's nested divisions, provinces, or sections across California (Fig. 3.5). We began by fitting the model using the constant slope of −0.5 (hereafter, the HOT_{2D} model), and then by using maximum likelihood estimation (MLE, Nash 1990) to find the best slope of the PLR model for the data (hereafter, the HOT_{MLE} model). As described earlier, the dimensionality of the HOT_{2D} model arises from the notion that fire event size increases as a function of a 2-dimensional spreading fire front with 1-dimensional perimeters of active fire spread or extinguishment. In essence, fire spread is constrained by polygons of fuel/non-fuel conditions, topography, and fire suppression (e.g., fences and corridors), the strength of which is moderated by climate and fire weather events. For each instance above (HOT_{2D} and HOT_{MLE}), fire size distributions were sequentially left-censored to find the range of patch sizes that best fit the distribution of the PLR models. We assessed PLR model goodness-of-fit (GOF) to the data using a bootstrapped version of the one-sample Kolmogrov-Smirnov (K-S) test through 2,500 iterations (Clauset et al. 2009). Acceptable model GOF

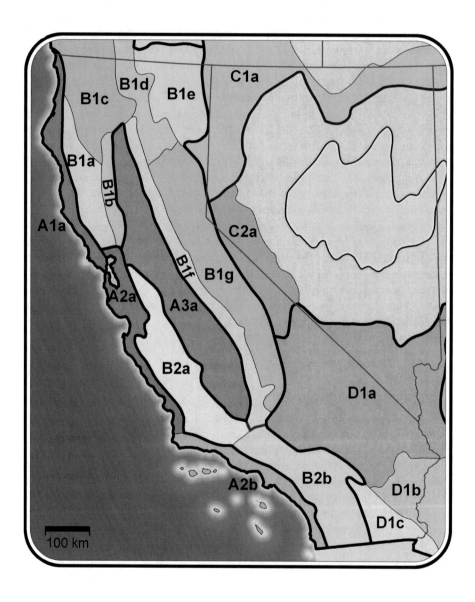

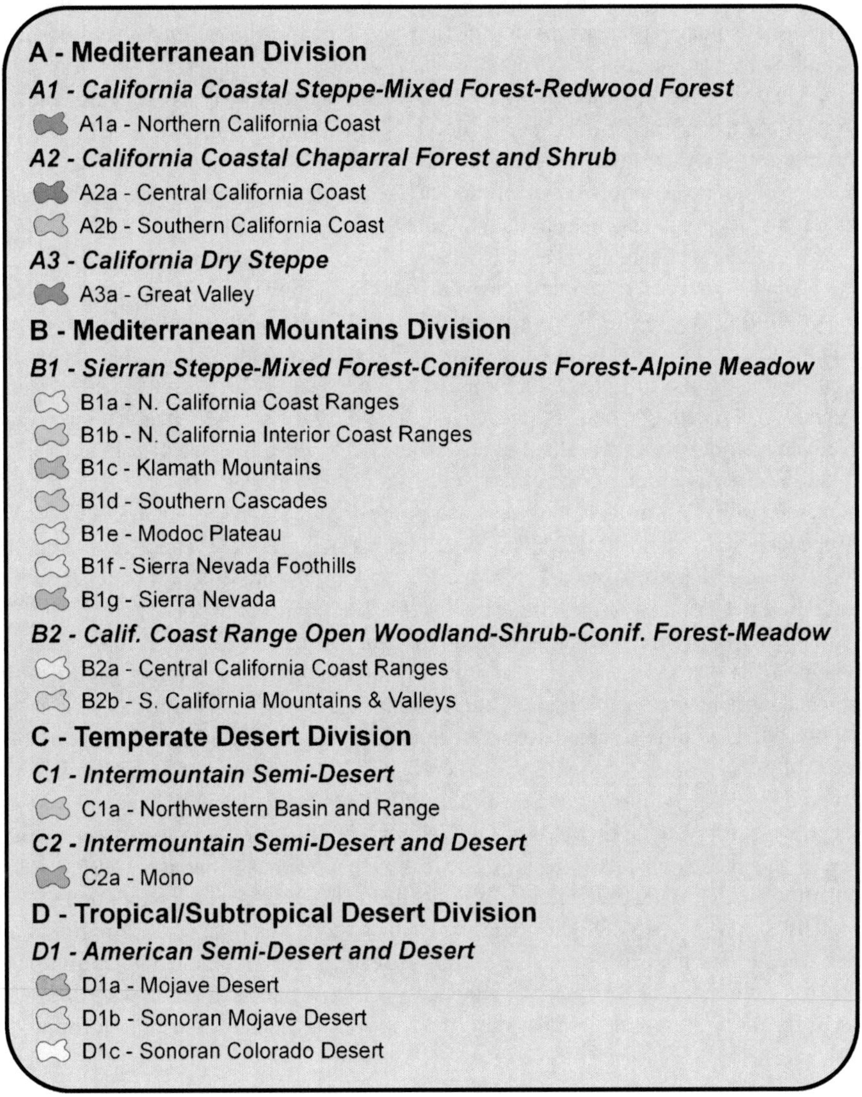

A - Mediterranean Division

A1 - California Coastal Steppe-Mixed Forest-Redwood Forest

A1a - Northern California Coast

A2 - California Coastal Chaparral Forest and Shrub

A2a - Central California Coast

A2b - Southern California Coast

A3 - California Dry Steppe

A3a - Great Valley

B - Mediterranean Mountains Division

B1 - Sierran Steppe-Mixed Forest-Coniferous Forest-Alpine Meadow

B1a - N. California Coast Ranges

B1b - N. California Interior Coast Ranges

B1c - Klamath Mountains

B1d - Southern Cascades

B1e - Modoc Plateau

B1f - Sierra Nevada Foothills

B1g - Sierra Nevada

B2 - Calif. Coast Range Open Woodland-Shrub-Conif. Forest-Meadow

B2a - Central California Coast Ranges

B2b - S. California Mountains & Valleys

C - Temperate Desert Division

C1 - Intermountain Semi-Desert

C1a - Northwestern Basin and Range

C2 - Intermountain Semi-Desert and Desert

C2a - Mono

D - Tropical/Subtropical Desert Division

D1 - American Semi-Desert and Desert

D1a - Mojave Desert

D1b - Sonoran Mojave Desert

D1c - Sonoran Colorado Desert

Fig. 3.5 Bailey's (1995) ecoregions within California. The analysis used all three levels of the classification: divisions (*single letters in caption*), regions (*letter + number*), and sections (*letter + number + lower-case letter*)

was indicated by $p > 0.10$, which indicates no significant difference between the data and the respective PLR model.

We also examined a variety of additional statistical distributions. Distribution fitting techniques in power law studies generally come in two flavors: (1) fitting ordinary least squares regressions to the log-log transform of either the empirical cumulative or non-cumulative frequency-size distributions (CDF), and (2) fitting the 1-parameter Pareto I distribution (P-I) using MLE, and assessing the fit of this model to the data using one of a variety of goodness-of-fit (GOF) tests (e.g., Chi-square, Kolmogrov-Smirnov tests). Several recent articles have convincingly argued for the latter method, because it is most appropriate for estimating the parameters of a Pareto model and its goodness-of-fit to data (White et al. 2008; Clauset et al. 2009).

We evaluated potential power-law behavior in three ways: (1) by fitting a variety of 2-, 3-, and 4-parameter complex Pareto models with known power law behavior, (2) directly fitting Pareto I (P1) and truncated Pareto I (TP1) models to fire-size distributions following the methods of Clauset et al. (2009), and (3) fitting broken-stick regression models to the inverse of the empirical CDF (Boer et al. 2008). Each of these methods has advantages and disadvantages (Table 3.1), and we used them to objectively evaluate the presence and scale(s) of power-law behavior in fire size distributions.

In the first assessment, we objectively fit a closely related family of Pareto and Generalized Beta II models to the inverse of log-log transformed empirical CDFs of fire event sizes using MLE. The distributions within the Generalized Beta II (GBII = Feller-Pareto, Arnold 1983) and Pareto families are 2–4 parameter models, including the Lomax (2P; = Pareto II), Inverse Lomax (2P), Fisk (2P; = Pareto III), Paralogistic (2P), Inverse Paralogistic (2P), Singh-Maddala (3P; = Pareto IV), and Dagum (3P) distributions. These models all have in common implied presence of power-law behavior in the middle and/or right tail of the distribution (Clark et al. 1999). MLE was performed using vector generalized linear models within the VGAM package in R version 2.9.1 (Yee 2006, 2008). To select the best model, we favored model parsimony and the minimum K-S test statistic. In the second assessment, we employed the methods of Clauset et al. (2009) to identify the lower boundary of the fire event sizes ($x.min$), above which power-law behavior most likely occurred. The third assessment involved fitting 1- or 2-break broken-stick regression models to the inverse CDFs to identify whether more than one scaling region was possible, as outlined by Boer et al. (2008). Scaling regions could indicate unique process domains and degrees of influence on fire event size. We assessed model GOF for the first two methods as described above under PLR distribution fitting.

3.10.2 Evaluating Top-down and Bottom-up Controls

For each of the three Bailey ecoregion levels, we evaluated the effect of top-down forcing by quantitatively comparing the fire event-size distributions among ecoregions. We used pairwise (two sample) K-S tests to determine the best stratification level for the data.

Table 3.1 Advantages and disadvantages of methods for determining the adequacy of power law model goodness-of-fit to fire event size distributions

Method	Advantages	Disadvantages
Fitting complex Pareto and GB II models with suspected power law tails to the entire distribution of patch-sizes (see Clark et al. 1999)	Can model the distribution of patch-sizes over entire range of observation using maximum-likelihood estimation (MLE) Can implement modified goodness-of-fit (GOF) tests to determine adequacy of model fits to the observed distributions Can compare model fit and parameter estimates within and among empirical distributions	Visually approximates the range of patch-sizes where power law behavior occurs Lack of model fit does not eliminate the possibility of power law behavior in the distributions Goodness-of-fit is dependent on the type or class of test used in analysis (e.g., KS, Chi-square, Anderson-Darling)
Fitting a 1-parameter Pareto I (power law) model to the right-tail of the distribution (see Clauset et al. 2009)	Fits power law model using MLE Adequacy of fit can be assessed using modified GOF tests Objectively determines scaling region in the right tail based on the Kolmogrov-Smironov (K-S) test statistic; most power law behavior in systems is known to exist in the right-tail (Clauset et al. 2009)	Location and GOF is dependent on the type of GOF test employed May miss the presence of a power law scaling region where model departures occur at the extreme end of the right-tail; these departures may be intuitively explained as upper physical or ecological limits on power law behavior Cannot identify multiple scaling regions in the data
Fitting a 1- or 2- parameter broken-stick model to identify scaling regions (see Boer et al. 2008)	Method is based on MLE and not on ordinary least-squares regression Can control the parameter number (breakpoints) in the model Can objectively determine lower and upper bounds on power law behavior, and identify multiple scaling regions, where present	GOF tests can be misleading as a good fit of the model to the data is not imperative in identifying scaling regions GOF tests will generally favor highly parameterized models Breaks may/may not be ecologically meaningful

To evaluate influence of bottom-up forcing, we evaluated patch size distributions of simple *aspect* (N or S) topographies derived from a 90-m digital elevation model (DEM). We also evaluated slope, curvature, and combined topographies but settled on aspect because it showed the best GOF when a left truncated Pareto-I model was fit to the aspect patch size data. Distribution fitting using MLE and GOF assessment for the topographic features followed the same methods used for the fire event-size distributions. We directly evaluated the influence of topography on fire event sizes by again using a pairwise K-S test on all event sizes and aspect patch sizes greater than the estimated *x.min* for the best fitting Pareto I model. To find the region of concordance between the aspect patch and fire event size distributions, we sequentially removed the patches from the right tail until a $p > 0.10$ was reached.

3.10.3 Characteristics of California Fires

Fire event sizes across California from 1950 to 2007 followed a distinctive pattern over most of the state, where small- to medium-sized fires were most common, and large fires >10,000 ha in size were relatively rare events. Fires ranged in size from 1 to 100,000 ha in size.

The greatest numbers of fires recorded were located in the Southern California Mountain Valley and Coast sections with 0.11 and 0.08 fires km^{-2}, respectively. Vegetation communities in this area are dominated by fire-adapted species, and physiognomies range from grasslands/shrublands and open hardwood woodlands in the foothills to ponderosa pine forests in lower-montane settings. Human population is also highest in these sections, with high concentration of anthropogenic ignitions. Fires of southern California are also influenced by Santa Ana (foehn) winds that have been linked with extreme fire behavior (Moritz 1997).

For most ecoregions, the 2- or 3-parameter GBII and Pareto models adequately fit the CDFs, based on a bootstrapped version of the K-S GOF test. These distributions all have in common the likely presence of an embedded power law region, suggesting that power law behavior is likely found above a certain minimum fire size. In the left tail of the fire event size distributions, where most of the fire events occurred (but represented the least part of wildfire affected area), there was evidence of a distinct change in slope at around $10^{1.5-2}$ ha, for most ecoregions, as can be seen in the empirical inverse CDF plots in Fig. 3.6. Factors accounting for this behavior may be: (1) fire reporting, recording, or mapping errors, (2) variable fire suppression efficacy, and (3) endogenous forcing. It is not possible to determine from the distributions alone which of these factors had the greatest influence on event sizes. However, highly dissimilar ecoregions, which vary in the amount of fire reporting errors and suppression efficacy, each followed this trend, indicating that endogenous factors may account for the lack of fit of the Pareto I model to the left-tails of the distributions.

3.10.4   Selecting an Optimal Ecoregion Scale

When attempting to detect direct evidence for controls on response variables, it is reasonable to first evaluate various regroupings of the data to observe those that are ecologically most intuitive and best minimize variance within the data. We used Bailey's ecoregion hierarchy to select an appropriate scale of observation for displaying potential top-down controls on fire event size distributions. Results of fitting various distributions to fire event sizes in Bailey's nested divisions, provinces, and sections showed that top-down ecoregional controls were best observed at the section level. At the division level 67% of the pairwise K-S test comparisons showed significant differences among ecoregions, while at the province and section

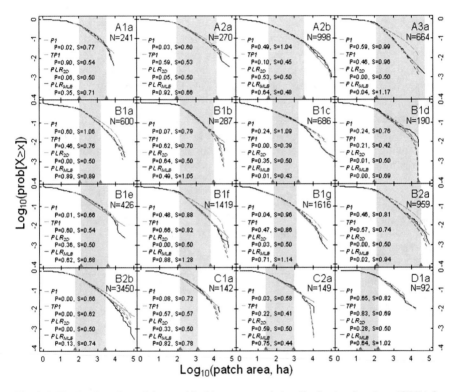

Fig. 3.6 The log-log plots of the empirical inverse cumulative distribution functions (CDFs) for event-size distributions of wildfires (> 1 ha) within Bailey's sections in California from 1950–2007. *Black* lines represent the empirical inverse CDF for fire patch-sizes. *Blue* lines represent the best-fit 1-parameter Pareto I (*P1*) distribution to the right-tail of the data, *orange dashed lines* represent the best truncated P1 fits, and the *green* and *red dashed lines* represent the best HOT_{2D} and HOT_{MLE} fits to the data, respectively. *Green triangles* represent the break-points for broken-stick regression models estimated by maximum-likelihood. Shaded areas represent a meso-scale domain where we theorize that endogenous and exogenous factors jointly influence the distribution of patch sizes

levels 79% and 88% of the comparisons were different, respectively. Thus, we report modeling results summarized to sections only.

3.10.5   Distribution Fits for California Fires

Results from our distribution fitting exercise support those from the earlier analysis we performed on a small chaparral-dominated region, and they provide evidence of HOT behavior for wildfires across much of California. Fourteen of 16 (88%) Bailey's sections showed support for HOT behavior; i.e., significant fits to the HOT_{2D} or HOT_{MLE} models, for fires larger than ~100 ha (Table 3.2).

Exceptions were the Great Valley and Central California Coast Range sections (Fig. 3.6), which we take up later.

The HOT_{2D} model fit seven of the 16 ecoregions (44%), based on the bootstrapped version of the K-S test. Sections located in desert (Mojave), semi-desert (NW Basin and Range, Mono), or chaparral (Southern California Coast) generally provided the best examples of the HOT_{2D} model (Table 3.2, Fig. 3.6). The Southern California Coast section represented the clearest example of a HOT_{2D} model, consistent with our earlier chaparral case study above, and this section includes the whole of that study area. The group of sections best explained by the HOT_{2D} model is dominated by fire-prone grassland or shrubland vegetation communities, all of which naturally have a high-severity or stand-replacement fire regime. Where fire

Table 3.2 Fit of the Pareto 1 (P1), truncated Pareto 1 (TP1), and HOT probability-loss-resource (PLR) models to the event size distributions of California wildfires > 100 ha for the period 1950–2007

Bailey's (1994) ecosection	N	P-1	TP1	PLR_{2D}	PLR_{MLE}
Northern California Coast	241	0.02	**0.90**	0.06	**0.35**
Central California Coast Ranges	270	0.03	**0.59**	0.05	**0.92**
Southern California Coast	998	**0.49**	0.10	**0.53**	**0.64**
Great Valley	664	**0.59**	0.46	0.00	0.04
Northern California Coast Ranges	600	**0.60**	0.46	0.00	**0.89**
Northern California Interior Coast Ranges	287	0.07	**0.62**	**0.64**	**0.49**
Klamath Mountains	686	**0.24**	0.00	**0.35**	0.01
Southern Cascades	190	**0.24**	**0.21**	0.01	0.00
Modoc Plateau	426	0.01	**0.60**	**0.36**	**0.62**
Sierra Nevada Foothills	1,419	**0.48**	**0.66**	0.00	**0.88**
Sierra Nevada	1,616	0.04	**0.47**	0.03	**0.71**
Central California Coast Ranges	959	**0.46**	**0.57**	0.00	0.02
Southern California Mountains and Valleys	3,450	0.00	0.00	0.00	**0.13**
Northwestern Basin and Range	142	0.08	**0.57**	**0.33**	**0.82**
Mono	149	0.03	**0.22**	**0.59**	**0.75**
Mojave Desert	92	**0.65**	**0.83**	**0.28**	**0.64**

Values in **bold** type face are significant ($p > 0.10$)

severity functions more or less as a constant, the HOT_{2D} model appears to most elegantly explain the origin of fire-size distributions. Thus, the map of grassland and shrubland landscapes functions as a mosaic of fuel/non-fuel patches resulting from prior disturbance and recovery, and event-sizes are driven by the magnitude and period of the climatic or weather influence during events. Similarly, fire event size distributions are highly relevant to understanding vegetation and disturbance patch dynamics, because fire-event and fire-severity patch-size distributions are more or less equivalent. Where fire severity is more variable, we theorize that fire event sizes are much less important. Rather, fire severity patch size distributions are likely the key.

The Klamath Section, which also fit the HOT_{2D} model, was a notable exception. The Klamath comprises roughly equal parts of rangeland and forest physiognomies (Bailey 1995). We hypothesize that the fire regime and forest type complexity of the Klamath should be further subdivided to better understand top-down and bottom-up controls on fire event-size distributions. The same is likely true for the Modoc Plateau Section (Table 3.2, Fig. 3.6).

Allowing for variable slopes, the HOT_{MLE} model fit 88% of fire event-size distributions at the section level, despite large ecological and geographical variation. Slope values for most sections were steeper than that of the HOT_{2D} model with the exception of the Klamath, Mono and Southern California Coast sections (Fig. 3.6). Where slopes are steeper than −0.5, the dimensionality of wildfires may be lower than that predicted by the HOT_{2D} model (Carlson and Doyle 1999). In California, this occurs in sections where relatively higher spatial complexity of topography, forest and rangeland types, structural conditions, climatic influences, and fire regimes is apparent. Falk et al. (2007) hypothesize that these relations might be expected. For example, they suggest that climatic anomalies that magnify weather extremes or lengthen fire seasons may lead to more variability in the distribution of fire sizes and larger maximums, which would tend to flatten the slope of the fire size distribution. In contrast, highly dissected topographies would tend to retard fire growth under non-extreme fire weather conditions, thereby reducing the largest fire sizes, which would tend to steepen the slope of the fire size distribution. (Carlson and Doyle 1999).

Doyle and Carlson (2000) posit that "landscapes which naturally break forests into regions of fractal dimension lower than 2 [slope is < −0.50] *would* have steeper [sloped] power laws by definition." With few exceptions, our results confirm that observation. A simple one-dimensional model, such as a network or flow route of linear features, would show a slope of around 1. In montane forests, winds during fires tend to be directional and wind flow is routed and concentrated by topography. Perhaps HOT model slopes tending towards 1 reflect a primary influence of fire flow routing in event size distribution. An important area of near-term research is unraveling the ecological meaning of differing slope values and their causal connections.

Several additional models (i.e., P1 and TP1) fit to all but one of the Bailey's sections; the Southern Mountain Valley Section. Similar to the HOT model GOF, these models fit best to fires larger than ~100 ha (Table 3.2). The P1 model fit best to the Great Valley (A3a), Mojave (D1a), Southern Cascades (B1d), Sierra Foothills (B1f), and Central Coast Ranges (B2a). For most other sections, the largest event sizes in these sections were smaller than those predicted by a pure power-law fit. This effect may be caused by physical constraint on the size distribution of aspect patches (and perhaps curvature and slope patches) imposed by geomorphic processes of an ecoregion. The best fitting section to the P1 model was the Great Valley. In the Great Valley, topography is flat to rolling, climatic influence is relatively more constant, and the land is highly parceled, owing to spatially continuous development and agriculture. As a consequence, we observe mostly anthropogenic and endogenous controls on wildfire spread. The Klamath and Southern California Coast sections shared a significant fit of the P1 to only the largest fire sizes, indicating that the largest fires in these sections might be under different controls than smaller fires.

3.11   The Meso-Scale Process Domain and a Role for Topography

We theorized that fire event sizes are controlled by different processes operating at different spatial scales (Fig. 3.9). For example, at fine scales ($<10^2$ ha), endogenous factors such as the spatial patterns of micro-topography and environment, stand dynamics and successional processes, and endemic insect and pathogen disturbances may affect fire size, regardless of human influence. At broad scales ($>10^4$ ha), exogenous factors may contribute to large and very large fire sizes, regardless of human influence (Fig. 3.9). Broad-scale controls might include climatic events such as multi-year droughts, multi-decadal climatic oscillations such as the PDO and ENSO (Heyerdahl et al. 2002; Hessl et al. 2004; Schoennagel et al. 2005), and gradient or foehn winds (Moritz 1997). At meso-scales ($\sim 10^2 - 10^4$ ha), however, fire-event sizes are influenced by a mixture of both endogenous and exogenous controlling factors (Turner 1989), and human influence can be most effective in influencing the distribution of medium to large fire sizes (Fig. 3.9).

We used broken-stick regression analysis to identify a possible meso-scale domain where exogenous and endogenous influences were both at work. Evidence from this analysis confirmed a meso-scale process domain likely exists between about 10^2 and 10^4 ha. These results, combined with the distribution modeling, indicate the presence of different scaling regions and process domains (fine, meso, and broad). Different forcings on fire-size distributions act independently (within their domain) and interact (on the edges of their domain) to control the distribution of wildfire event sizes.

Power-law behavior in a variety of earth systems has been studied and described extensively (Hergarten 2002, and references therein). Results from our analysis of topographies in California sections showed strong power law behavior in the distribution of north and south (N/S) aspect polygons. The left-censoring technique, which finds the minimum patch-size where the power law model best fits the data,

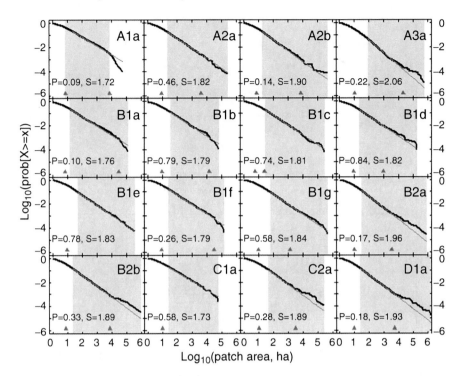

Fig. 3.7 The log-log plots of the empirical inverse cumulative distribution functions for size distributions of aspect (N/S) patches (> 1 ha) within Bailey's sections in California. *Black* lines represent the aspect patch-size data; *blue lines* represent the best-fitting 1-parameter Pareto I (*P1*) distribution to the *right*-tail of the data from the estimated lower cut-off (*vertical dotted line*). *Green triangles* represent the break-points for broken-stick regression models (*BSRS*) estimated by maximum-likelihood. Shaded areas represent the meso-scale domain as predicted by BSRs. *P-values* ≥ 0.10 indicate acceptable fits of the data to the P1 distribution; *S-values* indicate the slope of the best fitting P1 model

consistently identified power-law behavior for the distribution of patches $>10^2$ ha (Fig. 3.7). The scaling parameters for the Pareto I distributions of aspect patch sizes were generally slightly steeper than for the fire-size distributions, averaging ~ 1.85–1.9, depending on the Bailey's level.

Because the topographic and fire event-size patches were analyzed on equal logarithmic scales, they were directly comparable, and these results indicated that simple aspect N/S likely provides bottom-up controls on fire-size distributions (Fig. 3.8). In Fig. 3.8, for the meso-scale domains of the aspect and fire event size distributions shown, there was no significant difference between the two models, suggesting that meso-scale topographies may be entraining event size distributions in the same size range. We also evaluated slope and curvature topographies and combinations, and these showed significant control relations, but aspect produced the strongest apparent bottom-up spatial control.

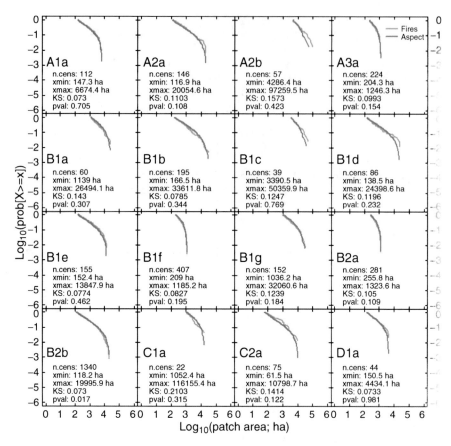

Fig. 3.8 The log-log plots of the statistical concordance between the empirical inverse cumulative distribution functions (*CDFs*) of aspect patch (N/S, >1 ha) and fire event size distributions with Bailey's sections in California. Over the regions shown, there is no significant difference between the two models (two-samples K-S test). The value of *n.cens* is the number of observations censored from the original value of N (see Fig. 3.6) for a given section

Results from our analysis suggest that both top-down geoclimatic and bottom-up topographic factors interact to control the distribution of fire event sizes in California and constrain the scales at which power law behavior is observed (Fig. 3.9). Topographic features such as aspect and slope (results not shown) have been shown to produce a myriad of effects on ecological patterns and processes at fine to meso-scales. Our results suggest that aspect may play an important role in controlling fire size distributions. This landscape effect of topography was observed over a large and diverse California landscape as seen by the similarities in fire-size distributions of sections. While these distributions shared features in common, there were also differences in fire size distributions among ecoregions (Fig. 3.6), in the best-fitting models, and in model parameters. Furthermore, individual section models provided consistently better P1 and TP1 fits than did the pooled sections model.

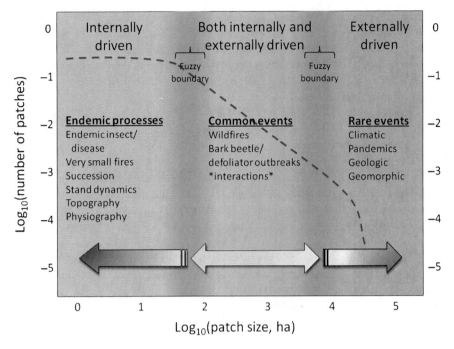

Fig. 3.9 Conceptual diagram of the spatial controls on fire event sizes. In the *left* tail, endogenous factors primarily influence event sizes. Analogously, in the *right* tail, exogenous factors primarily influence event sizes. In a meso-scale region, both endogenous and exogenous factors influence fire event sizes. The meso-scale domain has fuzzy boundaries that change as a consequence of the strength of endogenous and exogenous factors over time

These results suggest that top-down biophysical controls also have distinguishable effects on fire sizes.

3.12   From Whence Come the Distributions?

All of the previously described frameworks contribute to our understanding of landscape resilience, but none fully explains causal mechanisms behind apparently self-organized structure in California wildfires. In this work, we have used alternative ways of looking at California wildfire systems, in the same manner that one might look at some aspect of the world through different colored eye glasses. From what we have discovered thus far, we propose that landscape resilience in these ecosystems stems from the ongoing re-structuring of vegetation conditions by wildfires, controlled spatially and temporally from above by broad gradients of regional and subregional geology, geomorphology, and climate influence. From below, the underlying template of topography appears to cause relatively strong entrainment. Fires themselves are also advanced or retarded by the timing, severity, and spatial extent of prior disturbances, and their subsequent recovery regime. The net result is an ever-shifting template of fences and corridors, a structured heterogeneity of biotic and abiotic conditions that gives rise to the future mosaic of disturbed and recovering patches. This structured heterogeneity is the resistance surface that advances or retards the penetration of processes, especially disturbance processes, at all points in space or time.

Our observations are consistent with that proposed by Peterson (2002), who suggested that systems have an "ecological memory," whereby past disturbance and recovery create a patchwork mosaic of resistances to penetration by processes. Peterson found that ecological memory was a significant driver of the structure and organization in modeled systems. We provide evidence here for a structure to that ecological memory in natural ecosystems. Specifically, we find that controls on disturbance spread have identifiable scale-dependent domains, with endogenous factors likely dominating at finer scales (fires $<10^2$ ha) and exogenous controls dominating at broad scales (fires $>10^4$ ha).

Organization in natural systems is not static, but instead varies within a broad envelope of potential conditions, a function of the timing, severity, extent, and type of previous disturbances, the topographic setting, and climatic context. During relatively constant climatic conditions, this envelope develops the appearance of stationarity, while not being truly stationary. Under various climatic forcings, the envelope shifts as a function of the strength and duration of the climatic shifting. Nevertheless, within the meso-scale range of fire sizes ($\sim10^2$–10^4 ha), tradeoffs between various factors controlling fire spread – the shifting surface of fences and corridors that fire must repeatedly negotiate – lead to a relatively predictable power-law distribution of fire sizes. Topography, in particular, appears to play a previously under-appreciated role in generating the heterogeneity important to resilience in many fire-prone ecosystems. Therefore, even as the envelope of potential conditions shifts in a given region, topography remains (more or less) static, partially mediating climatic shifts and providing a template for ecological memory.

Power law relations are scale-invariant, meaning that the shape of patch-size distributions of the system in question is constant regardless of the scale of observation. This is an important relationship in SOC theory, because the mechanisms behind the organization of these hypothetical systems are solely dependent on the internal fuel configuration of the system and the frequency of ignitions. This unrealistic degree of internal regulation simply cannot apply in real fire-prone ecosystems. Similarly, few fire regimes could be entirely driven by exogenous forcings (e.g., wind events). In contrast to such one-sided views, HOT theory predicts that both bottom-up and top-down influences create fire regimes and ecosystems that are resilient to the degree of environmental variation experienced there. This is directly analogous to the HRV associated with a given ecosystem, which is commonly seen as important to resilience. HOT also predicts scale invariance and power law slopes of roughly $-1/2$, a theoretical result of minimizing the spread of two-dimensional events. Our observations show that scale invariance in natural systems does indeed occur over the meso-scale region of fire event sizes across California landscapes.

Perhaps we should not be at all surprised to find power-law distributions in fire sizes, as long as there is an approximate balance between fire growth and fire extinguishment (Reed and McKelvey 2002). It is conceivable that this will someday be intrinsically linked to ecosystem metabolism, with fire acting as fast respiration as biomass accumulates. Regardless, it is remarkable that so many regions in California have fire regimes that approximate a power-law distribution of fire sizes. Many of these regions appear to display a somewhat steeper slope than the $-1/2$ predicted by HOT, which may result from entrainment by the steeper size distribution of topographic characteristics. Fire in these regions could also be interpreted as following more of a network flow (i.e., closer to one-dimensional), since most are mixed-severity systems in which topography more strongly affects the paths of fires and intensities at which they burn.

We posit that in many fire-prone systems in California, the very large and rare fires play an important role in resetting the mosaic to different degrees, thereby affecting future fire spread, succession, and recovery processes. In most Bailey's sections the largest 10–15% of fires affected two-thirds to three-quarters of the section area. The largest and rarest fires must therefore be an integral part of the entrainment of future fires. [In forests, we suspect that it is the heterogeneous mosaic of fire severity patches within the area contributed by the largest fire events and the large fire event areas themselves that contribute to the entrainment of future fires.] Because of the inherent heterogeneity embedded across meso-scale landscapes, large fires are not necessarily associated with the "unraveling" of ecosystems. When their frequency exceeds that of HRV, however, such increases can exceed the resilience capacity of the system. Conditions for these rare events are mediated both externally by climatic factors and internally by the level of contagion inherent to the system at the time of disturbance. Anthropogenic forces also play a role: As humans increase the availability of ignitions during the most critical climatic conditions, the frequency of large events can approach levels that are destabilizing and far outside HRV.

3.13   Concluding Thoughts

The foregoing observations have implications for restoring a more naturally resilient fire ecology to fire-prone landscapes. If the meso-scale domain of event sizes provides an organizational structure for future fire sizes, then intentional human influences, whether positive or negative, will likely have the greatest impact in that domain. Large and rare events will occur regardless of human intervention, because they are under broad-scale spatial controls. Similarly, small events will occur regardless of human intervention, because they are governed by endemic fine-scale properties of ecosystems. However, in the middle domain, where endogenous and exogenous factors interact, humans may be able to rescale disturbance event size distributions by manipulating the patterns, conditions, and sizes of the patches that make up the fences and corridors that influence disturbance. For example, during typical wildfires in many forests (non-extreme event conditions), the flaming front is highly responsive to spatial mosaics of canopy and surface fuels (Finney et al. 2007). Conditions conducive to maintaining surface fires in forests typically yield surface fired patches and low- or mixed-severity effects. Those conducive to crown fire yield canopy fires and a preponderance of stand replacement. Likewise, root diseases, dwarf mistletoes, bark beetles, and defoliators are highly host specialized. Mosaics of differing actual vegetation cover type, stand age, density, size, and clumpiness represent varying degrees of resistance to spread of these processes. In the meso-scale range of patch sizes (~ 50 – 5,000 ha), restoration tactics appropriate to the natural fire ecology of the ecosystem will likely have the greatest effect on system meta-stability (Wu and Loucks 1995) and resilience.

It is yet unclear in fire-prone ecosystems exactly how factors other than topography might be involved in the production of specific power law fire size distributions across the meso-scale, although ecologically relevant mechanisms have been suggested (e.g., Moritz et al. 2005). It is likely that variation in fire severity patterns is really at the heart of landscape resilience, with fire event size distributions acting as a surrogate measure in most systems. We suggest that ecosystem-wide heterogeneity in fire effects, as well as landscape resilience that is associated with them, emerge primarily from patterns and processes operating at the meso-scale (Fig. 3.9). Linking distributions of fire severity patches to the HRV in fire sizes, frequencies, seasons, and intensities is an area ripe for a significant amount of new research.

How realistic are fire growth and fire extinguishment (Reed and McKelvey 2002) in generating power law distributions in fire sizes? It is intuitive that there should be some inherent limits on the growth and eventual sizes of typical fire events (Moritz et al. 2005), but how does this occur? In terms of ecosystem energetics, there must be a kind of see-saw balancing act between combustion of biomass and the primary productivity of a landscape. For a simple landscape in which fire is the only "consumer" of vegetation (e.g., Bond and van Wilgen 1996; Bond et al. 2005), one might expect rates of burning to roughly equal rates of biomass accumulation, when examined over broad scales of space and time. We already know that such relationships exist at the global scale, where fire activity shows strong links to net primary productivity patterns (Krawchuk et al. 2009). We do not yet understand all of the precise mechanisms behind the generation of fences and corridors that influence fire's dance across space and time. Pursuit of these questions and others should improve our ability to understand and further quantify the elusive nature of resilience, its variations, and its evolutionary mechanisms.

References

Agee, J.K. 1993. *Fire ecology of Pacific Northwest forests*. Washington: Island Press.
Arnold, B.C. 1983. *Pareto distributions*. Fairland: International Co-operative Publishing House.
Bailey, R.G. 1995. Descriptions of the ecoregions of the United States. 2nd ed. Misc. Publ. No. 1391, Map scale 1:7,500,000, U.S. Forest Service.

Bak, P. 1996. *How nature works: The science of self-organized criticality.* New York: Copernicus.

Boer, M.M., R.J. Sadler, R.A. Bradstock, A.M. Gill, and P.F. Grierson. 2008. Spatial scale invariance of southern Australian forest fires mirrors the scaling behaviour of fire-driving weather events. *Landscape Ecology* 23: 899–913.

Bond, W.J., and B.W. van Wilgen. 1996. *Fire and plants.* London: Chapman & Hall.

Bond, W.J., F.I. Woodward, and G.F. Midgley. 2005. The global distribution of ecosystems in a world without fire. *The New Phytologist* 165: 525–538.

Brown, R.T., J.K. Agee, and J.F. Franklin. 2004. Forest restoration and fire: Principles in the context of place. *Conservation Biology* 18: 903–912.

Burroughs, S.M., and S.F. Tebbens. 2001. Upper-truncated power laws in natural systems. *Pure and Applied Geophysics* 158: 741–757.

Carlson, J.M., and J. Doyle. 1999. Highly optimized tolerance: A mechanism for power laws in designed systems. *Physical Review E* 60: 1412–1427.

Carlson, J.M., and J. Doyle. 2000. Highly optimized tolerance: Robustness and design in complex systems. *Physical Review Letters* 84: 2529–2532.

Carlson, J.M., and J. Doyle. 2002. Complexity and robustness. *Proceedings of the National Academy of Science* 99(Suppl 1): 2538–2545.

Clark, R.M., S.J.D. Cox, and G.M. Laslett. 1999. *Geophysical Journal International* 136: 357–372.

Clauset, A., C. Shalizi, and M.E.J. Newman 2009. Power-Law Distributions in Empirical Data. *SIAM Review* 51(4): 661–703.

D'Antonio, C.M., and P.M. Vitousek. 1992. Biological invasions by exotic grasses, the grass/fire cycle, and global change. *Annual Review of Ecology and Systematics* 23: 63–87.

Davis, F.W., and M.A. Moritz. 2001. Mechanisms of disturbance. In *Encyclopedia of biodiversity*, ed. S. Levin, 153–160. New York: Academic.

Díaz-Delgado, R., F. Lloret, X. Pons, and J. Terradas. 2002. Satellite evidence of decreasing resilience in Mediterranean plant communities after recurrent wildfires. *Ecology* 83: 2293–2303.

Doyle, J., and J.M. Carlson. 2000. Power laws, highly optimized tolerance, and generalized source coding. *Physical Review Letters* 84: 5656–5659.

Falk, D.A., C. Miller, D. McKenzie, and A.E. Black. 2007. Cross-scale analysis of fire regimes. *Ecosystems* 10: 809–823.

Finney, M.A., R.C. Seli, C.W. McHugh, A.A. Ager, B. Bahro, and J.K. Agee. 2007. Simulation of long-term landscape-level fuel treatment effects on large wildfires. *International Journal of Wildland Fire* 16: 712–727.

Gill, A.M. 1975. Fire and the Australian flora: A review. *Australian Forestry* 38: 4–25.

Grimm, V., and C. Wissel. 1997. Babel, or the ecological stability discussions–an inventory and analysis of terminology and a guide for avoiding confusion. *Oecologia* 109: 323–334.

Gunderson, L.H., and C.S. Holling. 2002. *Panarchy: Understanding transformations in human and natural systems.* Washington: Island Press.

Hann, W.J., J.L. Jones, M.G. Karl, P.F. Hessburg, R.E. Keane, D.G. Long, J.P. Menakis, C.H. McNicoll, S.G. Leonard, R.A. Gravenmier, and B.G. Smith. 1997. An assessment of landscape dynamics of the Basin. In *An assessment of ecosystem components in the interior Columbia Basin and portions of the Klamath and Great Basins.* General Technical Report PNW-GTR-405, tech. eds. T.M. Quigley and S.J. Arbelbide. Portland: U.S. Forest Service.

He, H.S., and D.J. Mladenoff. 1999. Spatially explicit and stochastic simulation of forest-landscape fire disturbance and succession. *Ecology* 80: 81–90.

Hergarten, S. 2002. *Self-organized criticality in earth systems.* Germany: Springer.

Herrera, C.M. 1992. Historical effects and sorting processes as explanation for contemporary ecological patterns: Character syndromes in mediterranean woody plants. *The American Naturalist* 140: 421–446.

Hessburg, P.F., B.G. Smith, and R.B. Salter. 1999a. Detecting change in forest spatial patterns from reference conditions. *Ecological Applications* 9: 1232–1252.

Hessburg, P.F., B.G. Smith, and R.B. Salter. 1999b. *Using natural variation estimates to detect ecologically important change in forest spatial patterns: A case study of the eastern Washington Cascades.* Research Paper PNW-RP-514. Portland: U.S. Forest Service.

Hessburg, P.F., J.K. Agee, and J.F. Franklin. 2005. Dry mixed conifer forests and wildland fires of the inland Northwest: Contrasting the landscape ecology of the pre-settlement and modern eras. *Forest Ecology and Management* 211: 117–139.

Hessl, A.E., D. McKenzie, and R. Schellhaas. 2004. Drought and Pacific decadal oscillation linked to fire occurrence in the inland Pacific Northwest. *Ecological Applications* 14: 425–442.

Heyerdahl, E.K., L.B. Brubaker, and J.K. Agee. 2002. Annual and decadal climate forcing of historical fire regimes in the interior Pacific Northwest, USA. *Holocene* 12: 597–604.

Holling, C.S. 1973. Resilience and stability of ecological systems. *Annual Review of Ecology and Systematics* 4: 1–23.

Holling, C.S., and L.H. Gunderson. 2002. Resilience and adaptive cycles. In *Panarchy: understanding transformations in human and natural systems*, eds. C.S. Holling and L.H. Gunderson, 25–62. Washington: Island Books.

Johnson, E.A., and S.L. Gutsell. 1994. Fire frequency models, methods and interpretations. *Advances in Ecological Research* 25: 239–287.

Kauffman, S.A. 1993. *The origins of order: Self-organization and selection in evolution*. Oxford: Oxford University Press.

Krawchuk, M.A., M.A. Moritz, M.A. Parisien, J. Van Dorn, and K. Hayhoe. 2009. Global pyrogeography: The current and future distribution of wildfire. *PLoS ONE* 4(4), e5102. doi:10.1371/journal.pone.0005102.

Landres, P., P. Morgan, and F. Swanson. 1999. Overview of the use of natural variability in managing ecological systems. *Ecological Applications* 9: 1279–1288.

Levin, S.A. 1992. The problem of pattern and scale in ecology. *Ecology* 73: 1943–1967.

MacGillivray, C.W., and J.P. Grime. 1995. Testing predictions of the resistance and resilience of vegetation subjected to extreme events. *Functional Ecology* 9: 640–649.

Malamud, B.D., G. Morein, and D.L. Turcotte. 1998. Forest fires: An example of self-organized critical behavior. *Science* 281: 1840–1842.

Malamud, B.D., and D.L. Turcotte. 2006. The applicability of power law frequency statistics to floods. *Journal of Hydrology* 322: 168–180.

Malamud, B.D., J.D.A. Millington, and G.L.W. Perry. 2004. Characterizing wildfire regimes in the United States. *Proceedings of the National Academy of Sciences* 102: 4694–4699.

Martin, R.E., and D.B. Sapsis. 1992. Fires as agents of biodiversity: Pyrodiversity promotes biodiversity. In *Proceedings of the symposium on biodiversity of northwestern California*. Santa Rosa, ed. H.M. Herner, 28–31. Berkeley: University of California Center for Wildland Resources Report 29.

Millar, C.I., N.L. Stephenson, and S.L. Stephens. 2007. Climate change and forests of the future: Managing in the face of uncertainty. *Ecological Applications* 17: 2145–2151.

Moritz, M.A. 1997. Analyzing extreme disturbance events: Fire in Los Padres National Forest. *Ecological Applications* 7: 1252–1262.

Moritz, M.A. 1999. Controls on disturbance regime dynamics: Fire in Los Padres National Forest. Ph.D. dissertation, University of California, Santa Barbara.

Moritz, M.A. 2003. Spatio-temporal analysis of controls of shrubland fire regimes: Age dependency and fire hazard. *Ecology* 84: 351–361.

Moritz, M.A., and S.L. Stephens. 2008. Fire and sustainability: Considerations for California's altered future climate. *Climatic Change* 87(Suppl 1): S265–S271.

Moritz, M.A., J.E. Keeley, E.A. Johnson, and A.A. Schaffner. 2004. Testing a basic assumption of shrubland fire management: How important is fuel age? *Frontiers in Ecology and the Environment* 2: 67–72.

Moritz, M.A., M.E. Morais, L.A. Summerell, J.M. Carlson, and J. Doyle. 2005. Wildfires, complexity, and highly optimized tolerance. *Proceedings of the National Academy of Sciences of the United States of America* 102: 17912–17917.

Moritz, M.A., T.J. Moody, L.J. Miles, M.M. Smith, and P. de Valpine. 2009. The fire frequency analysis branch of the pyrostatistics tree: Sampling decisions and censoring in fire interval data. *Environmental and Ecological Statistics* 16: 271–289.

Nash, J.C. 1990. *Compact numerical methods for computers: Linear algebra and function mini-misation.* New York: IPO Publishing.

Newman, M.E.J. 2005. Power laws, Pareto distributions and Zipf's law. *Contemporary Physics* 46: 323–351.

Nonaka, E., and T.A. Spies. 2005. Historical range of variability in landscape structure: A simulation study in Oregon, USA. *Ecological Applications* 15: 1727–1746.

Odion, D.C., E.J. Frost, J.R. Strittholt, H. Jiang, D.A. DellaSala, and M.A. Moritz. 2004. Patterns of fire severity and forest conditions in the western Klamath Mountains, northwestern California, U.S.A. *Conservation Biology* 18: 927–936.

Odion, D.C., M.A. Moritz, and D.A. DellaSala. 2009. Alternative community states maintained by fire in the Klamath Mountains, USA. *Journal of Ecology* 98: 96–105.

Parisien, M.A., and M.A. Moritz. 2009. Environmental controls on the distribution of wildfire at multiple spatial scales. *Ecological Monographs* 79: 127–154.

Pausas, J.G., R.A. Bradstock, and D.A. Keith. 2004. Plant functional traits in relation to fire in crown-fire ecosystems. *Ecology* 85: 1085–1100.

Pickett, S.T.A., and P.S. White. 1985. *The ecology of natural disturbance and patch dynamics.* New York: Academic.

Perera, A.H., and L.J. Buse. 2004. Emulating natural disturbance in forest management: An overview. In *Emulating natural forest landscape disturbances: Concepts and applications*, eds. A.H. Perera, L.J. Buse, and M.G. Weber, 3–7. New York: Columbia University Press.

Peterson, G.D. 2002. Contagious disturbance, ecological memory, and the emergence of landscape pattern. *Ecosystems* 5: 329–338.

Peterson, G.D., C.R. Allen, and C.S. Holling. 1998. Ecological resilience, biodiversity, and scale. *Ecosystems* 1: 6–18.

Polakow, D.A., and T.T. Dunne. 1999. Modelling fire-return interval T: Stochasticity and censoring in the two-parameter Weibull model. *Ecological Modelling* 121: 79–102.

Polakow, D.A., and T.T. Dunne. 2001. Numerical recipes for disaster: Changing hazard and the stand-origin-map. *Forest Ecology and Management* 147: 183–196.

Reed, W.J., and K.S. McKelvey. 2002. Power-law behavior and parametric models for the size-distribution of forest fires. *Ecological Modelling* 150: 239–254.

Ricotta, C., G. Avena, and M. Marchetti. 1999. The flaming sandpile: Self-organized criticality and wildfires. *Ecological Modelling* 119: 73–77.

Robert, C., J.M. Carlson, and J. Doyle. 2001. Highly optimized tolerance in epidemic models incorporating local optimization and growth. *Physical Review E* 63: 056122.

Romme, W. 1980. Fire history terminology: report of the ad hoc committee. In *Proceedings of the fire history workshop.* General Technical Report RM-GTR-81, eds. M.A. Stokes and J.H. Dieterich, 135–137. Fort Collins: U.S. Forest Service.

Romme, W.H., E.H. Everham, L.E. Frelich, M.A. Moritz, and R.E. Sparks. 1998. Are large infrequent disturbances qualitatively different from small frequent disturbances? *Ecosystems* 1: 524–534.

Schmidt, K.M., J.P. Menakis, C. Hardy, D.L. Bunnell, and W.J. Hann. 2002. *Development of coarse-scale spatial data for wildland fire and fuel management.* General Technical Report RMRS-GTR-87. Fort Collins: U.S. Forest Service.

Schoennagel, T., T.T. Veblen, W.H. Romme, J.S. Sibold, and E.R. Cook. 2005. ENSO and PDO variability affect drought-induced fire occurrence in Rocky Mountain subalpine forests. *Ecological Applications* 15: 2000–2014.

Schoennagel, T., E.A. Smithwick, and M.G. Turner. 2008. Landscape heterogeneity following large fires: Insights from Yellowstone National Park, USA. *International Journal of Wildland Fire* 17: 742–753.

Solow, A.R. 2005. Power laws without complexity. *Ecology Letters* 8: 361–363.

Song, W.G., F. Weicheng, B.H. Wang, and J.J. Zhou. 2001. Self-organized criticality of forest fire in China. *Ecological Modelling* 145: 61–68.

Strauss, D., L. Bednar, and R. Mees. 1989. Do one percent of forest fires cause ninety-nine percent of the damage? *Forest Science* 35: 319–328.

Swetnam, T.W., and J.L. Betancourt. 1998. Meso-scale disturbance and ecological response to decadal climatic variability in the American Southwest. *Journal of Climate* 11: 3128–3147.

Swetnam, T.W., C.D. Allen, and J.L. Betancourt. 1999. Applied historical ecology: Using the past to manage the future. *Ecological Applications* 9: 1189–1206.

Turcotte, D.L. 1999. Self-organized criticality. *Reports on Progress in Physics* 62: 1377–1429.

Turcotte, D.L., and B.D. Malamud. 2004. Landslides, forest fires, and earthquakes: Examples of self-organized critical behaviour. *Acta Physica* 340: 580–589.

Turner, M.G. 1989. Landscape ecology: The effects of pattern on process. *Annual Review of Ecology and Systematics* 20: 171–197.

Turner, M.G., and W.H. Romme. 1994. Landscape dynamics in crown fire ecosystems. *Landscape Ecology* 9: 59–77.

White, E., B. Enquist, and J.L. Green. 2008. On estimating the exponent of power-law frequency distributions. *Ecology* 89: 905–912.

White, P.S., and A. Jentsch. 2001. The Search for Generality in Studies of Disturbance and Ecosystem Dynamics. *Progress in Botany* 62: 399–450.

Wu, J., and O.L. Loucks. 1995. From balance of nature to hierarchical patch dynamics: A paradigm shift in ecology. *The Quarterly Review of Biology* 70: 439–466.

Yee, T.W. 2006. Constrained additive ordination. *Ecology* 87: 203–213.

Yee, T.W. 2008. The VGAM package. *R News* 8: 28–39.

Zedler, P.H., C.R. Gautier, and G.S. McMaster. 1983. Vegetation change in response to extreme events: The effect of a short interval between fires in California chaparral and coastal scrub. *Ecology* 64: 809–818.

Zhou, T., and J.M. Carlson. 2000. Dynamics and changing environments in highly optimized tolerance. *Physical Review E* 62: 3197–3204.

Zhou, T., J.M. Carlson, and J. Doyle. 2002. Mutation, specialization, and hypersensitivity in highly optimized tolerance. *Proceedings of the National Academy of Sciences of the United States of America* 99: 2049–2054.

Part II
Climate Context

Chapter 4
Climate and Spatial Patterns of Wildfire in North America

Ze'ev Gedalof

4.1 Introduction

Climate interacts with wildfire at a range of spatial and temporal scales. In this chapter I describe a conceptual model that describes how climate (a top-down control) interacts with processes of vegetation development and topography (bottom-up controls) to give rise to characteristic disturbance regimes and observed patterns of wildfire throughout North America. At the shortest timescales (synoptic to seasonal), climate influences fine fuel moisture, ignition frequency, and rates of wildfire spread. At intermediate timescales (annual to interannual), climate affects the relative abundance and continuity of fine fuels, as well as the abundance and moisture content of coarser fuels. At longer timescales (decadal to centennial) climate determines the assemblage of species that can survive at a particular location. Interactions between these species' characteristics and the influence of climatic processes on wildfire activity give rise to the characteristic disturbance regime and vegetation structure at a given location. Large-scale modes of climatic variability such as the El Niño – Southern Oscillation and the Pacific Decadal Oscillation affect patterns in wildfire by influencing the relative frequencies of shorter scale processes. Because the importance of these processes varies depending on topographic position and the ecology of the dominant vegetation the effects of these modes varies both between and within regions. Global climatic change is effectively a centennial to millennial scale process, and so its effects can be understood as resulting from interactions between the observed patterns of higher frequency processes, as well as processes of vegetation change whose temporal evolution exceeds the length of the observational record. Statistical models of future fire that are based on historical fire climate relations and regionally downscaled climate forecasts suggest that in most regions of North America wildfire will increase in frequency over the next several decades. Predictions beyond this interval are probably

Z. Gedalof (✉)
Department of Geography, University of Guelph, Guelph, ON N1G 2W1, Canada
e-mail: zgedalof@uoguelph.ca

D. McKenzie et al. (eds.), *The Landscape Ecology of Fire*, Ecological Studies 213,
DOI 10.1007/978-94-007-0301-8_4, © Springer Science+Business Media B.V. 2011

unreliable as vegetation structure and composition will be changing rapidly response to changing climatic conditions and fire regimes.

Spatial variability in the structure and composition of vegetation occurs as a legacy of interacting processes that are biological, geological, geomorphologica climatic, and anthropogenic in origin. Of the processes that shape ecosystems none is more dramatic or more important (at least in temperate regions) than fire The behavior of individual fires is largely determined by the nature of the fuels, weather, and topography that characterize the site of ignition (Johnson 1992; Agee 1993). Of these factors, weather is the most variable over time (Bessie and Johnson 1995), and is the most poorly understood (e.g., Gedalof et al. 2005). Because the vast majority of area currently burned by wildfire is caused by relatively few fires that burn under extreme weather conditions (Strauss et al. 1989; Gedalof et al. 2005), it is important to understand the causes of variability in extreme fire weather.

The effects of fire weather on fire behavior do not appear to be consistent across space (Jones and Mann 2004). Rather, weather interacts with other factors to give rise to the specific fire regime of a given location. These factors can be generally characterized as being either top-down or bottom-up (Chaps. 1 and 3). Top-down controls include those that originate outside the ecosystem. Of these climate is the most important, although anthropogenic influences are locally important. Bottom-up controls include those that originate inside the ecosystem, such as topography and vegetation dynamics.

Assessments of the relative contributions of top-down and bottom-up controls on wildfire are complicated by many interacting factors, including:

- The climatic history of the earth has not been static at any scale of variability, and will continue to change over the coming decades to centuries (Karl 1985; Meehl et al. 2005)
- Intensive land use by people, including forestry and road building, and grazing by sheep and cattle changed forest conditions in many regions (Madany and West 1983; Belsky and Blumenthal 1997; Heyerdahl et al. 2001)
- Many landscapes may be a legacy of fire use by indigenous peoples, although the pre-settlement fire regime is not well known in most cases (Brown and Hebda 2002; Keeley 2002; Williams 2002; Gedalof et al. 2006)
- Records of fire history are generally short, often lack detailed location information, and are not easily reconciled (Westerling et al. 2003; Gedalof et al. 2005)
- The effects of fire suppression on area burned are uncertain, and controversial in many forest types (Keeley et al. 1999; Johnson et al. 2001; Ward et al. 2001; Bridge et al. 2005).

Despite these challenges, emerging data sets and analytical methods have allowed important insights into the processes that give rise to spatial patterning in severe wildfire years at regional scales (10^3–10^6 km^2), and a coherent conceptual model is emerging. The purpose of this chapter is to summarize recent developments in understanding the role of top-down controls, and in particular climate, on variability in area burned by wildfire. Specifically, I summarize the mechanisms by which top-down controls give rise to widespread severe wildfire years, describe several important patterns of climatic variability and assess their role in giving rise to regional patterns of wildfire, and discuss how vegetation cover and other bottom-up controls modulate the response of a given region to climatic variability to give rise to landscape-scale responses (10^0–10^4 km^2) to these top-down controls. This context is used to understand how climatic change may affect fire frequency over the next several decades to centuries.

4.2   Mechanisms of Top-down Control

Variability in the Earth's climate system represents the most important source of variability in the fire regime of most regions (e.g. Stahle et al. 2000). Properties of the climate system that can affect wildfire include temperature, precipitation, wind speed, relative humidity, and lightning activity. These properties fluctuate in space and time across many orders of magnitude, ranging, for example, from a sunfleck that might dry a few square meters for a minute or two, to a mega-drought that might persist throughout a given region for decades or more (Schroeder 1969; Strauss et al. 1989; Johnson and Wowchuk 1993). The effects of these fluctuations are similarly variable, depending on their characteristic scale and properties of the ecosystem they are incident upon. In the following sections I summarize the main mechanisms by which climatic variability can affect fire, focusing on how the scale of the climatic process involved influences the impact on the fire regime.

4.2.1   Ignition Events

Lightning is the most important natural cause of wildfire ignitions throughout North America (Morris 1934; Rorig and Ferguson 1999; Malamud et al. 2005). Lightning is caused by convection within clouds acting to separate positive and negative charges (Uman 2001; Burrows et al. 2002; van Wagtendonk and Cayan 2008). The convection that gives rise to these charge differentials is most commonly associated with unstable air masses associated with differential surface heating, or diurnal variability in surface temperatures (Uman 2001). Most lightning associated with electrical storms is contained within the cloud (i.e., occurs as intracloud lightning), and is not associated with fire ignitions, but a small proportion of lightning occurs as cloud-to-ground lightning. Surprisingly, most cloud-to-ground lightning strikes are so brief that while they may cause considerable damage to trees they rarely generate sufficient heat to ignite fuels (Latham and Williams 2001). However, about

30% of these strikes are associated with sustained current flows that do generate sufficient heat to potentially ignite fires. Of particular importance are the approximately 10% of strikes that are positive in polarity, which much more commonly sustain the currents needed to ignite fires (Latham and Williams 2001). The causes of positive-polarity lightning are still unknown, but there are distinct regional patterns that may be an important cause of variability in fire frequency. For example, positive-polarity strikes occur most commonly in North America over northern Minnesota and adjacent parts of Ontario and Manitoba, Canada (Lyons et al. 1998). Podur et al. (2003) found evidence for regional increases in lightning ignitions in this region of Canada, which they attributed to localized dry weather and lightning storm occurrence, but increased frequency of positive-polarity lightning could also contribute to this region's anomalously frequent ignitions.

The factors that cause convection, and consequently the frequency of lightning strikes, vary diurnally, by time of year, and between years. At large spatial scales, lightning occurs more frequently in continental than maritime regions, and more frequently at intermediate elevations than at higher elevations (which in turn occurs more frequently than at low elevations). Across North America, the greatest density of lightning strikes occurs in central Florida, and decreases toward the northwest. Relatively few lightning strikes occur west of the Western Cordillera in either the United States or Canada (Huffines and Orville 1999; Burrows et al. 2002). Superimposed on this large-scale pattern there are important regional differences in the frequency of lightning strikes. Topographic variability and land-water temperature differences influence patterns of atmospheric convection, resulting in subregional patterns of lightning variability. For example, in Colorado lightning strikes occur most frequently just east of the Continental Divide (Lopez and Holle 1986). Similarly, in Canada, a regional increase in lightning strikes is found in the foothills region, east of the Rocky Mountains (Burrows et al. 2002).

Surprisingly, however, most researchers have found a poor correspondence between the frequency of lightning strikes and the frequency of ignition events, suggesting that lightning is a necessary but not sufficient condition for wildfire to occur (Morris 1934; Nash and Johnson 1996; Rorig and Ferguson 1999; Latham and Williams 2001). The factors that cause spatial variability in lightning frequency include atmospheric humidity, topography, and surficial properties. These same factors are associated with changes in vegetation type that in turn influence the flammability and continuity of fuels. Consequently, patterns of lightning frequency alone are poor predictors of patterns in wildfire occurrence.

Part of this discrepancy is because for ignition to occur, lightning needs to strike a fuel bed that is sufficiently dry to maintain combustion, and sufficiently continuous for fire to spread. The percentage of successful ignitions per cloud-to-ground lightning strike, termed lightning ignition efficiency, typically ranges between 1 and 4% (Meisner 1993; Latham and Williams 2001). Because the processes that generate lightning require moisture, and are usually associated with precipitation, successful ignitions occur most efficiently under fairly specific conditions (Nash and Johnson 1996). In particular, ignitions occur when fuels are particularly dry due to antecedent weather conditions (see below), and when lightning strikes are not accompanied by precipitation. This "dry" lightning occurs most frequently when the lower atmosphere is particularly unstable, resulting in intense convection and often dry conditions in the lower atmosphere that cause precipitation to evaporate before it reaches the ground (Rorig and Ferguson 1999). These same conditions are also associated with gusty winds that contribute to rapid fire spread. Although dry lightning is probably the most effective cause of ignitions, it is not the only type of lightning that ignites wildfires. Ignitions can also occur in cases where the fuel bed is exceptionally dry, and the precipitation associated with the thunder storm is not sufficient to inhibit burning, during small fast-moving storms that deliver little precipitation to any single location, or when the lightning strikes outside the main plume of the storm (Rorig and Ferguson 2002).

Ignition efficiency differs between various land cover types. Meisner (1993) examined lightning strike and ignition frequency in southern Idaho as functions of the dominant vegetation type. He found that ignition efficiencies ranged from 0.3% (for agricultural crops) to 10% (for logging slash). Mature forests ranged from about 2–4%. Latham and Williams (2001) reached similar conclusions for a more extensive region, and indeed found that some areas of exceptionally high strike density had actually experienced no fires over the duration of their analysis. In California, desert regions experience the most lightning per unit area, but ignitions are very rare due to the discontinuous nature of the fuel bed (van Wagtendonk and Cayan 2008). Krawchuk et al. (2006) found that conifer forests were more likely to burn than nearby deciduous forests. Ignition efficiency also differs between locations within the same basic vegetation type. For example, Díaz-Avalos et al. (2001) found that in the Blue Mountains in Oregon ignition efficiency was higher at lower elevations despite the lower frequency of lightning strikes, and peaked within the central portion of the range—although they were unable to explain the reason for this spatial pattern.

4.2.2 Fire Spread

It is generally recognized that the great majority of area burned by wildfire is caused by relatively few fires that occur under extreme weather conditions (Schroeder 1969; Strauss et al. 1989; Johnson and Wowchuk 1993). For example, one commonly repeated statistic suggests that 99% of the area burned is caused by 1% of the fires. Although the actual figure is probably closer to 90% (Strauss et al. 1989), the importance of relatively few fires causing the bulk of the variability in area burned remains the same. These fires are usually associated with high temperatures, exceptionally low relative humidity, and strong winds (Schroeder 1969; Flannigan and Harrington 1988; Crimmins 2006).

The relationship between fire spread and short-term variations in meteorological variables is reasonably well understood. Early work by Fons (1946) was built on, in particular, by Rothermel (1972, 1983) to develop empirical models of fire spread based on fuel characteristics, slope, and wind speed. Several fire spread simulators based on these mathematical models are now used operationally and in the development of management plans (Finney 1998, 1999; Hargrove et al. 2000; Andrews 2007; Tymstra et al. 2007). These models explain how fire spreads across a given landscape in response to critical fire weather, but do not offer insights into how fire is synchronized across landscapes to give rise to characteristic years of exceptionally high or low fire activity. Schroeder (1969) undertook the first systematic effort to identify meso-scale patterns of atmospheric pressure associated with extreme fire hazard. The patterns most strongly associated with extreme fire hazard are characterized by anomalous high surface pressure. These systems, commonly called blocking ridges, divert moisture away from the region (Wiedenmann et al. 2002). Along their margins (or during their passage) strong pressure gradients contribute to strong winds that cause rapid spread. When blocking ridges are particularly intense the passage of cyclonic storms may produce strong wind and lightning, but little precipitation (Rorig and Ferguson 1999).

A second common set of patterns was associated with air masses that cross mountains (Schroeder 1969). Moisture is lost from these systems due to orographic precipitation along the windward slopes. As the (now) dry air descends the lee slopes it warms by compression, and relative humidity decreases further. Along the eastern slopes of the Rocky Mountains these winds are called Chinooks. Fire danger is greatest when Chinooks are associated with ridges west of the Rocky Mountains that enhance drying due to subsidence, and contribute to strong pressure gradients and the resulting surface winds. An analogous but more severe fire-weather pattern occurs when winds are easterly, i.e. from the continental interior to the coast. In these cases, the air mass is typically dry to start with, and is exceptionally dry when it reaches the coast. These winds are generally termed foëhn winds, but often have local names such as Diablo, sundowner, or Santa Ana winds. In southern California and northern Baja California Santa Ana winds are associated with some of the most extreme wildfires (Keeley et al. 1999; Keeley and Fotheringham 2002) many of which spread into the urban wildland interface resulting in losses of structures and human lives (Keeley et al. 2004).

The synoptic circulation patterns that Schroeder (1969) identified have since been validated using more extensive data and objective analytical techniques (e.g., Skinner et al. 1999, 2002; Gedalof et al. 2005; Crimmins 2006), and confirmed in a large number of case studies (e.g. Countryman et al. 1969; Sando and Haines 1972; Finklin 1973; Street and Alexander 1980). Little work has been done, however, to explicitly link this variability in severe fire *weather* to large-scale ocean–atmosphere interactions. Such a linkage is implicit in analyses that identify climatic patterns at timescales longer than about 10 days (Flannigan and Harrington 1988; Johnson and Wowchuk 1993; Skinner et al. 1999, 2002; Gedalof et al. 2005; Trouet et al. 2006), but none of these analyses discriminated the factors that contribute to fire hazard (such as fuels production and fuels drying) or ignition efficiency from those that contribute to rapid spread. There is evidence that the frequency of extreme events differs depending on the state of large-scale modes of variability, suggesting that such an analysis might prove fruitful. For example, Thompson and Wallace (2001) found that strong winter winds in coastal Washington and Oregon occur approximately three times more often during the positive phase of the Northern Hemisphere Annular Mode than during the negative phase. Similarly, blocking ridges in the Pacific Region occur more commonly during the cold (La Niña) phase of the El Niño Southern Oscillation than during the warm (El Niño) phase (Wiedenmann et al. 2002).

4.2.3 Fuel Moisture

Most land cover types are not so flammable that the above processes alone can explain the regional synchrony of severe fire years. Depending on the dominant vegetation structure, some period of antecedent drought is needed to dry the fuel bed so that fire will spread rapidly (Johnson and Wowchuk 1993; Bessie and Johnson 1995; Meyn et al. 2007; Littell et al. 2009). The relative importance of antecedent drought varies, depending on both mean regional climate and the structure and composition of the fuel bed. In particular, the relative abundance of fine vs. coarse fuels, and the continuity of the fuel bed, determine the importance of antecedent drought in preconditioning stands to burn (Schoennagel et al. 2004, 2005; Gedalof et al. 2005). Because the relative abundance and arrangement of fine fuels differs between land cover types, their drying rate and capacity to carry fire will also differ (Westerling et al. 2003; Gedalof et al. 2005).

Two basic factors regulate the moisture of fuels. First, plant functional type determines the phenology of vegetation (which determines whether foliage and shoots are metabolically active, dormant, or dead) as well as its structure and physiology (which determines its capacity to maintain high moisture levels in either plant tissue or in dead organic matter). Second, antecedent weather determines the moisture available to vegetation, as well as the rate of evaporative and transpirative losses. At their extremes, these processes support the ideas of ignition-limited ecosystems (those with abundant fuels, but that do not burn due to the infrequency of ignition events or the high moisture of fuels) vs. fuel limited ecosystems (those that experience frequent potential ignitions, but that often do not have sufficiently abundant or continuous fuels to allow fire to spread). Meyn et al. (2007) provide a useful conceptual framework for synthesizing these processes, and also identify a third type of ecosystem that does not fit neatly into this dichotomy. They characterize these ecosystems as "biomass poor, rarely dry," and they are both fuels and ignition limited. Examples include subalpine forests, temperate savannas, many wetlands, and some types of chaparral.

High temperatures, low relative humidity, and strong winds in the days to months preceding a potential ignition dry living and dead fuels, and can cause vegetation to senesce. The relative importance of antecedent drying varies by ecosystem type, with some ecosystems requiring much longer periods of time to dry sufficiently to carry fire than others (Westerling et al. 2003). For example, Gedalof et al. (2005) analyzed the relative importance of drought in the months preceding extreme wildfire years in the Pacific Northwest, USA, and found that coastal temperate rainforest experienced large area burned only during years of exceptional drought persisting throughout the winter and spring preceding the fire season. In contrast, dry forest types such as those found in eastern Washington and Oregon experienced extreme wildfire years even in the absence of persistent drought. These differences in sensitivity to antecedent climate can be explained in part by the relative abundances of fine fuels. Many wet forests are characterized by abundant standing and down woody debris that retains moisture effectively (Franklin et al. 1981). These large fuel classes require prolonged dry periods before they become flammable, and they also buffer surface vegetation against prolonged soil-moisture deficits by providing a reservoir of moisture. Closed-canopy forests further buffer surface fuels from drying by reducing insolation, temperature, and windspeeds at the surface, and by helping to maintain high relative humidity (Chen et al. 1999).

In ecosystems with higher relative abundances of fine fuels, such as grasslands, savannas, and chaparral, shorter periods of dry weather are sufficient to precondition ecosystems to burn (Westerling and Swetnam 2003; Gedalof et al. 2005). This difference can be explained in part by the faster drying rate of fine fuels, but is also enhanced by the tendency for these ecosystems to have lower canopy cover, and thus greater evaporation and transpiration from the surface. Many ecosystems dominated by fine fuels also have greater proportions of annual vs. perennial vegetation – meaning that there is more dead fuel at the surface (Knapp 1995). These fuels dry more readily than living vegetation, because they do not maintain internal moisture by using groundwater or resisting transpiration through adaptive measures.

Xeric ecosystems dominated by fine fuels respond to shorter-term variations in fire weather, and are generally more sensitive to the availability of a continuous fine fuel bed (see below), but they are also responsive to seasonal patterns of moisture availability. For example, in southwestern ponderosa pine forests regionally synchronous fire years are strongly associated with drier than average spring conditions (Swetnam and Betancourt 1990). In addition to synchronizing fire activity, large-scale drought may also increase the severity of fire in landscapes more commonly associated with low-severity fire – leading to a complex mosaic of snags, and patches of living trees with heterogeneous age structures (Agee 1998; Baker et al. 2007).

4.2.4 Fuels Production

At seasonal and longer timescales, climatic variability can affect the wildfire regime by modifying the abundance and continuity of fuels, and the relative abundance of fine vs. coarse fuels. As with fuel moisture, the relative importance of antecedent climate in fuels production varies depending on the dominant vegetation present, and the climate of the region (Westerling et al. 2003; Gedalof et al. 2005; Littell et al. 2009). Because the rate of fire spread and intensity at the flaming front is determined primarily by fine fuels, it is largely this fuel component that limits the ignition and spread of fire in most ecosystems (Rothermel 1972; Bessie and Johnson 1995). Most closed-canopy forests have abundant fine fuels, and do not increase in fire hazard with increased production beyond the point at which closed-canopy conditions are achieved, which typically occurs in the first two or three decades of development (Bessie and Johnson 1995; Schimmel and Granström 1997; Keeley et al. 1999; Johnson et al. 2001; Schoennagel et al. 2004).

In ecosystems that are characteristically dry enough that fuels are patchy, climatic conditions conducive to the growth of vegetation may increase the abundance and continuity of fine fuels, increasing the potential for fire during subsequent seasons (Swetnam and Betancourt 1998; Westerling et al. 2003; Collins et al. 2006). Open ponderosa pine forests that are characterized by short fire return intervals often show positive correlations to precipitation in the year(s) preceding regionally synchronous fire years (e.g. Swetnam and Betancourt 1998; Brown and Shepperd 2001; Kitzberger et al. 2007). This relationship is not constant throughout the species' range, however. For example, Sherriff and Veblen (2008) found that antecedent moisture increased fire occurrence in ponderosa pine forests in northern Colorado only at low elevations; at higher elevations this relationship was unimportant. Brown and Shepperd (2001) found that it occurred only in the southernmost portion of their study region in Colorado and Wyoming. In the northern portions of their study region they found that fire was associated only with drought during the year of fire. They also found that stand-replacing fires occurred frequently throughout the study region, even in the pre-suppression era. In the U.S. Southwest, relationships to antecedent moisture are more common due to the generally shorter fire-return interval and warmer mean climatic conditions, which limit fuel accumulation and production respectively (Swetnam and Betancourt 1998; Stephens and Collins 2004), but even there it is generally restricted to ponderosa pine forests and the relationship is not found for mixed conifer forests in the same region (Swetnam and Baisan 1996).

Antecedent moisture plays a particularly strong role in producing fuels and synchronizing fire in ecosystems dominated by annual grasses and herbs (Cable 1975; Knapp 1995; Brooks and Matchett 2006). Indeed, this relationship is sufficiently important in grass dominated ecosystems that in the Great Plains region at long time periods (decades to centuries), fires are more commonly associated with prolonged wet periods than dry ones (Brown et al. 2005) – although conditions are likely dry while fires actually burn. The importance of grasses in producing a continuous fine fuel bed that will carry fire has changed the fire regime of many arid and semi-arid ecosystems where exotic grasses have invaded. For example, in the Intermountain West the introduction of annual grasses, especially cheatgrass (*Bromus tectorum*) has increased the size and frequency of wildfires (Knapp 1995). This change to the fire regime has altered vegetation dynamics, as the affected communities have not evolved with frequent fire (Knapp 1998). The result has been reduced biodiversity, and economic losses associated with lost pasture and suppression efforts.

In North American deserts introduced grasses have caused fires to occur in regions that would historically have experienced little or no fire at all (Brooks and Pyke 2001; Brooks and Matchett 2006). These fires are disrupting regeneration of desert vegetation that is not adapted to fire, and in some regions have converted desert scrub to grassland. In these regions the relationship between fire and climate has changed: whereas historically fire would have occurred very rarely, and only following multiyear to decadal pluvials, it is now occurring following short wet periods lasting perhaps a single season (Brooks and Pyke 2001).

4.3 Patterns of Top-down Control

A distinctive feature of the Earth's climate system is that it varies over time and space in characteristic patterns or "modes" of variability. Some of these modes, such as diurnal or seasonal temperature fluctuations, are readily observable and can be easily explained as a result of the Earth's rotation and revolution around the sun. Other modes are less readily observable, affect different regions of the Earth uniquely, and overlap in time and space (Namias and Cayan 1981). These modes of variability influence patterns of atmospheric pressure, temperature, and precipitation over spatial scales that exceed 10^6 km^2, and over timescales of months to decades or longer (Wallace 2000). They influence fire frequency mainly through their influence on rates of fuel production and drying, but may also influence the frequency of ignitions and the statistics of extreme winds. Their spatial imprint on the wildfire record is a result of interactions between the spatial expression of the mode of climatic variability and the response of individual ecosystems to variability in climate. There are a dizzying array of these modes documented in the literature, although many of them are probably related to each other (Dommenget and Latif 2002), or statistical artifacts rather than separate physical processes (Enfield 1989; McPhaden et al. 2006). Nevertheless, a few of these modes are emerging as fundamental. In this section I review four of the more important modes that impact North America, and summarize their influence on wildfire. I use the term *teleconnection* to explain how climatic variability in one region affects the climate of more distant locations (Wallace and Gutzler 1981).

4.3.1 The El Niño Southern Oscillation

The El Niño Southern Oscillation (ENSO) is the most important source of global climatic variability at interannual timescales. ENSO events result from feedback between the tropical oceans and atmosphere (Wyrtki 1975). During non-ENSO years, the trade winds blow from east to west, and surface waters are pushed away from South America towards Indonesia. These waters warm as they are heated by the sun, and the height of the sea surface increases as water accumulates along the western margin of the Pacific Ocean. On average, the surface height is about 0.5 m higher along the Indonesian coast than along the South American coast (Enfield 1989). This warm water pool heats the air above it, causing it to rise—helping to maintain the east-to-west flow of the trade winds, and bringing the high rainfall typical of Indonesia. Episodically, the trade winds weaken, and the pool of warm water "sloshes" eastwards towards South America. This process further weakens the trade winds, accelerating the eastward movement of the warm water. This system is coupled, in that either the ocean or the atmosphere can initiate the event, and the feedback between them will cause it to strengthen. These events are known as "warm ENSO events" due to the anomalous heating of the Pacific Ocean east of the International Date Line. They are also often simply called El Niño events, although strictly this term applies to only the oceanic component of the system. ENSO events typically initiate in September, are most strongly expressed from December to April, and then decay from September through to March of the following year (Namias 1976; Ropelewski and Halpert 1986; Yarnal and Diaz 1986; Hamilton 1988; Kiladis and Diaz 1989; Sardeshmukh 1990; Diaz and Kiladis 1992).

At the peak of warm events, the weakening of the tropical Pacific trade winds and the redistribution of heat along the equator disrupts the global climate system, redistributing energy and moisture (Trenberth et al. 1998). Although there is considerable variability in the effects of warm ENSO events, the average response during the boreal winter (December to February) includes anomalous dry conditions in the western Pacific, including Indonesia, southeast Asia, and Australia, and warm wet conditions in the central and eastern tropical Pacific (Ropelewski and Halpert 1986; Trenberth et al. 1998). In North America, winter conditions are typically warmer throughout southern Canada and the northern United States. Precipitation effects are more variable, but the Pacific Northwest, USA, is typically drier than normal, while Alaska and the southwestern United States are wetter than normal (Trenberth et al. 1998).

Cool ENSO (or La Niña) events are largely opposite to warm events. They are associated with enhanced trade winds, a larger temperature and height gradient between the western and eastern Pacific Ocean, and approximately the opposite teleconnections. For example, regions of the Earth that are droughty during warm events often are exceptionally wet during cool events (McCabe and Dettinger 2002).

The effects of ENSO events on wildfire vary regionally and sub-regionally, depending on the sign and magnitude of the individual event's effect on climate (Fig. 4.1), and properties of the local vegetation. As the most important mode of global climatic variability, its effect on wildfire spans the globe, with significant effects on every continent except Antarctica (Nkemdirim and Budikova 1996; Kitzberger et al. 2001). Because the ENSO teleconnection to North America is strongest during the boreal winter, in many regions the strongest climatic impact is on total winter snow accumulation (Cayan 1996; Moore 1996). Winter snow accumulation can affect wildfire behavior through two contrasting mechanisms. First, in regions where snow persists into the summer, such as at high elevation, higher than normal snow accumulation will shorten the length of the fire season and help to maintain high moisture levels. These processes collectively decrease the likelihood of fires occurring (Westerling et al. 2006; Heyerdahl et al. 2008). Second, in

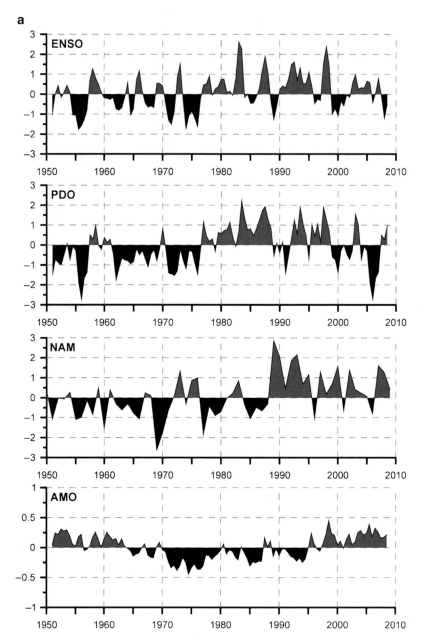

Fig. 4.1 (**a**) Time evolution of the four dominant modes of variability affecting the climate of North America: the El Niño Southern Oscillation (*ENSO*), the Pacific Decadal Oscillation (*PDO*), the Northern Hemisphere Annular Mode (*NAM*), and the Atlantic Multidecadal Oscillation. (**b**) Spatial regressions of each index onto the winter and summer Palner Drought Severity Index (*PDSI*) records for North America. The maps show the typical effect of a one-standard deviation increase in the associated index on regional *PDSI*

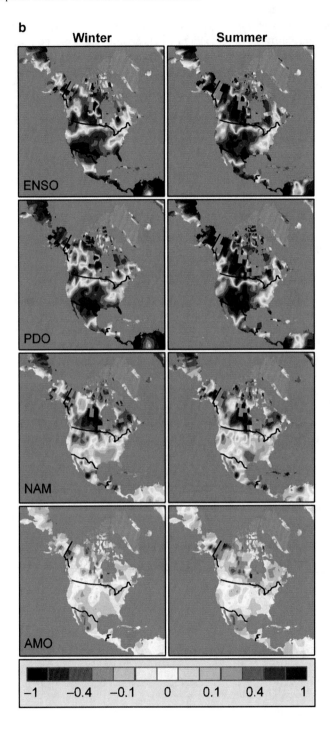

particularly dry regions higher winter precipitation can promote the growth of grasses and shrubs, which increases the abundance and continuity of fine fuels (Westerling et al. 2003; Brown et al. 2005; Gedalof et al. 2005).

The best documented ENSO effects on wildfire occur in the U.S. Southwest, where warm (El Niño) events are associated with increased precipitation and reductions in area burned; cool (La Niña) events exhibit the opposite relationship (Swetnam and Betancourt 1990). Often the most widespread fire years occur when cool (dry) events follow warm (wet) events, and fuel production associated with the warm event is dried and burns during the subsequent cool event (Swetnam and Betancourt 1998). Similar effects occur in the U.S. Southeast, where cool events were found to be associated with decreased rainfall, increased lightning strikes, and consequently more and larger fires (Beckage et al. 2003; Dixon et al. 2008). Because the fire season in the Southeast occurs during winter months ENSO effects are particularly strong, explaining up to 50% of the variability in area burned (Brenner 1991). ENSO teleconnections to the Pacific Northwest are approximately opposite to those in the U.S. Southwest and Southeast, and warm events are associated with drier than normal conditions (Kiladis and Diaz 1989). Because the strongest impact is on winter conditions, though, the effect on fire frequency at most forest types in the Pacific Northwest is small. For example, Norman and Taylor (2003), Gedalof et al. (2005), and Heyerdahl et al. (2008) all found no significant associations between wildfire occurrence and ENSO (but see Hessl et al. 2004), although they did find that it interacted with other processes (see below) significantly. Surprisingly few analyses have been undertaken on the effect of ENSO variability on wildfire activity in the boreal forest. Macias Fauria and Johnson (2006, 2008) found an association between ENSO-like conditions and fire weather for the boreal forest of North America. They did not explicitly separate ENSO conditions from Pacific Decadal Oscillation conditions (see below), however, and they focused on fire weather rather than regionally synchronous fire events. They found that warm ENSO-like conditions were associated with *reduced* wildfire hazard west of the Canadian Rocky Mountains, and *increased* wildfire hazard over the western prairies.

4.3.2   The Pacific Decadal Oscillation

The Pacific Decadal Oscillation (PDO) is an ENSO-like mode of variability that is most strongly expressed in the North Pacific Ocean (Mantua et al. 1997). The precise mechanisms that force it are still unclear, but it does appear to be distinct from ENSO (Zhang et al. 1997; Barlow et al. 2001; Gedalof et al. 2002). The temporal variability in the PDO is characterized by intervals of anomalously warm or cool water in the central North Pacific Ocean that persist for 20–30 years, punctuated by abrupt shifts between phases (Mantua et al. 1997; Minobe 1999; Gedalof and Smith 2001). Superimposed on this low-frequency pattern there is considerable year-to-year variability. The effect on the climate of North America is strongest over the coastal Pacific Northwest, and during winter (Fig. 4.1b). The warm phase of the PDO is associated with slightly elevated winter temperatures, and reduced precipitation, especially west of the Cascade Range. The PDO also affects other regions of North America, most notably the Canadian prairies, where the warm phase is associated with increased drought (Shabbar and Skinner 2004), and the U.S. Southwest, where it is associated with increased precipitation.

The effects of the PDO on wildfire activity are generally weak but significant throughout the Pacific Northwest. Gedalof et al. (2005) found no significant difference in the area burned by wildfire in the Pacific Northwest between the warm and cool phases of the PDO, although they did find a significant correlation between the PDO index and area burned. They also found that seven of the ten largest-fire years followed winters when the PDO index was positive, whereas eight of the ten smallest-fire years occurred following winters when the PDO index was negative. These results together suggest that the PDO exerts important controls on wildfire at annual to interannual scales, if not at interdecadal scales. Similarly, Hessl et al. (2004), examining fire activity in central and eastern Washington, found that the six most regionally coherent fires since 1650 all occurred when the PDO index was positive.

Few analyses of PDO-fire interactions for regions outside the Pacific Northwest have been undertaken to date, although the PDO has been examined for its modulating effect on ENSO. Brown (2006) found an inverse correspondence between wildfire activity in ponderosa pine forests in the Black Hills, South Dakota, and a tree-ring reconstruction of the PDO (cf. Biondi et al. 2001). Similarly, in Mississippi the cool phase of the PDO is associated with increased wildfire activity (Dixon et al. 2008). However there are two good reasons to believe that it may be an important determinant of fire activity in at least the arid Southwest and possibly other grass-dominated ecosystems. First, the influence of the PDO is primarily on winter precipitation, which has more of an effect on fuels production than on fuel moisture during the fire season. Second, several analyses of grass-fire interactions have noted a correspondence between wildfire and several wet years followed by one dry one. The persistent nature of the PDO should affect the frequency of these types of events.

The PDO has been found to interact with other modes of climatic variability to influence the fire regime. Gershunov and Barnett (1998) examined climatic conditions for various combinations of ENSO and PDO phases, and found that the two modes enhance each other when they are in the same phase (e.g., an El Niño event during the warm phase of the PDO), and they offset each other when they are in opposite phases (e.g., a La Niña event during the warm phase of the PDO). A number of fire-climate studies have found these interactions to be an important source of variability in regionally synchronous fire years. Corresponding warm phases are associated with increased wildfire activity in pine forests in northeastern California (Norman and Taylor 2003), in subalpine forests in Yellowstone and Jasper National Parks, in the central United States and southern Canadian Rocky Mountains, respectively (Schoennagel et al. 2005), and in Douglas-fir and ponderosa pine forests in the Pacific Northwest (Heyerdahl et al. 2008). Corresponding cool phases are associated with increased wildfire activity in subalpine forests in Rocky Mountain National Park, in the southern U.S. Rocky Mountains (Schoennagel et al. 2005), in ponderosa pine forests in northern Colorado (Sherriff and Veblen 2008) and South Dakota (Brown 2006; Kitzberger et al. 2007). An analysis of fire weather in the Canadian boreal forest suggests that west of the Rocky Mountains, increased fire hazard is associated with the cool phases of ENSO and the PDO; the opposite relationship was found for regions east of the Rocky Mountains (Macias Fauria and Johnson 2006).

4.3.3 The Northern Hemisphere Annual Mode

The Atlantic Ocean exerts a smaller influence on the climate of North America than the Pacific Ocean does, due to the prevailing westerly circulation in mid latitudes. Nevertheless, several related modes of climatic variability may exert important controls that are relevant to patterns of wildfire. The Northern Hemisphere Annular Mode (NAM) is a pattern of variability in atmospheric pressure that is characterized by out-of-phase differences between the polar and subpolar sectors (Thompson and Wallace 1998, 2000). Although some disagreement exists as to which is the fundamental process (Ambaum et al. 2001), the NAM is very closely related to the North Atlantic Oscillation (NAO) and the Arctic Oscillation (AO), and they are assumed to be the same process for the discussion here.

In North America the positive phase of the NAM is associated with anomalous warm temperatures in the eastern half of the continent, south of the Great Lakes (Thompson and Wallace 2001; Visbeck et al. 2001) (Fig. 4.1b). Precipitation relationships are weak, but slightly drier than normal conditions occur through much of central North America. A wide range of winter climatic extremes have been found to be associated with the NAM, including short-term events (rather than mean conditions), and events in western North America (Thompson and Wallace 2001). The summer climate of North America has not been analyzed in this fashion, but it seems likely that similar relationships would exist for variables that could affect fire such as temperature, lightning, strong winds, and precipitation.

The role of the NAM in forcing wildfire has not been widely studied. However Macias Fauria and Johnson (2006; see also Le Goff et al. 2007) found that the positive phase of the NAM was associated with almost 70% of large fires in eastern Canada; the negative phase was associated with increases in fire in Alaska and the Northwest Territories. Dixon et al. (2008) found complex relationships between variability in the NAM and area burned in Mississippi: Significant negative correlations were found between total area burned in March and April (the height of the fire season) and the February state of the NAM. Curiously, area burned in October, November, and December was positively correlated to the NAM during September. They attribute these opposing relationships to differences in the seasonal expression of the NAM. In the late winter, the negative phase of the NAM is associated with drier than normal conditions, which would reduce fuel moisture. In the summer, the positive phase is associated with increased convection, dry lightning, and strong winds, which would increase the frequency of ignitions and the rate of spread. Given that these processes have distinct effects that are modulated by vegetation structure and composition it is possible that they have distinct spatial expressions within this region.

4.3.4 The Atlantic Multidecadal Oscillation

The Atlantic Multidecadal Oscillation (AMO) is a slowly changing pattern of variability in North Atlantic surface temperatures. It was first identified by Schlesinger and Ramankutty (1994), and later named by Kerr (2000) It is characterized by alternate warming and cooling in the Atlantic Ocean north of the equator, with a period of approximately 65–80 years (Delworth and Mann 2000; Enfield et al. 2001). The warm phase of the AMO is associated with decreased rainfall throughout most of central North America, but increased rainfall in Florida and in some regions of the Pacific Northwest (Enfield et al. 2001).

Because of its slowly changing nature the AMO does not exert a strong control on year-to-year variability in wildfire (Fig. 4.1b). However it may influence patterns of vegetation structure and composition, fuel production and accumulation, and the frequency of ignitions, which in turn feed back to modulate the fire regime at longer timescales (Sibold and Veblen 2006; Schoennagel et al. 2007). The AMO has also been found to interact with other modes of variability to influence regional synchrony of wildfires. Throughout most of the U.S. Western Interior, years when the positive phase of the AMO corresponds to negative phases of ENSO and the PDO are associated with the most regionally synchronous wildfire (Kitzberger et al. 2007). This same pattern has been found in subalpine forests (Schoennagel et al. 2007) and ponderosa pine forests (Sherriff and Veblen 2008) in Colorado. South of central Colorado this association shifts, and wildfires are most commonly associated with combined negative phases of the AMO, ENSO, and the PDO (Kitzberger et al. 2007). In the Pacific Northwest fires are associated with the negative phase of the AMO combined with the positive phases of ENSO and the PDO (Kitzberger et al. 2007). The AMO has not been found to influence fire in the boreal forest (Macias Fauria and Johnson 2006; Le Goff et al. 2007). Relationships in the U.S. Southeast are inconsistent, but weakly negative (Guyette et al. 2006; Dixon et al. 2008).

4.4 Fire in the Future

The Earth's climate is changing in response to the actions of people (Solomon et al. 2007), and fire regimes will change in response to changing climates. The effects of climate change will differ regionally due to variability in the magnitude and seasonality of climatic changes, as well as differences in how vegetation and fire respond to climate. One approach to predicting how fire will change in response to climate change is to model the current relationship between fire and climate, and then use future climate projections to assess change. Nearly all of these efforts predict a substantial increase in wildfire activity over the next century, including analyses of the Canadian boreal forest (Stocks et al. 1998; Gillett et al. 2004) and the western United States (McKenzie et al. 2004). Exceptions to this general pattern are rare, but include regions where fine-fuel abundance and continuity are more important than flammability, such as in the deserts of eastern California (Westerling and Bryant 2008).

A second approach to forecasting fire activity in the future involves using downscaled climate projections to assess fire hazard based on operational guidelines or process based models (e.g. Torn and Fried 1992). These models have the advantage of reducing the complexities introduced by land use change, fire management practices, and vegetation change and focusing simply on the climate-associated fire hazard compared to today (Brown et al. 2004). They also have the advantage of forecasting fire hazard indexes in use by managers, providing a recognized "currency" for planning purposes. Flannigan and Van Wagner (1991) examined seasonal fire severity for the Canadian boreal forest under a range of climate projections and determined that annual area burned would increase by 46% with a doubling in carbon dioxide (CO_2) (see also Flannigan et al. 2000, 2001). Brown et al. (2004) evaluated fire hazard in the western United States, and determined that the number of days of severe fire weather will increase throughout their study region by up to 2 weeks per year by 2089. These effects are strongest in the northern Rocky Mountains, the Great Basin, and the Southwest. Although all of these models predict an overall increase in wildfire there are considerable regional differences in the magnitude and even the sign of the change, depending on the projection used (Flannigan et al. 2001).

Most analyses of the effects of climatic change on wildfire have focused on temperature and precipitation as the driving variables, but other approaches are possible. For example Miller and Schlegel (2006) modeled the occurrence of Santa Ana winds under a range of future climate scenarios, and concluded that Santa Ana occurrence "may significantly increase the extent of California coastal areas burned by wildfires, loss of life, and property." Price and Rind (1994) modeled thunderstorm activity and concluded that lightning activity in the United States will increase by 26% and annual area burned would increase by 78%.

These efforts may provide insights into fire activity for the next several decades, but they assume that vegetation structure and composition are static. However, as vegetation responds to more frequent fire and changing climate, there may be rapid changes to ecosystem structure and composition (Bachelet et al. 2001b). Consequently these forecasts are probably unreliable beyond a few decades. There have been a number of efforts to model the interaction between climatic change, wildfire, and vegetation (e.g. Neilson and Drapek 1998; Bachelet et al. 2001a, 2003; Thonicke et al. 2001). To date, these efforts have produced variable results, depending on assumptions about the role of atmospheric carbon, nitrogen limitations, and disturbance (Running 2008). Neilson and Drapek (1998) provide one perspective on why this exercise is both critically important and particularly challenging. In a comparison of short-term CO_2 effects on vegetation distribution vs. long-term combined temperature and CO_2 effects, they found that in many locations the trajectory of ecosystems reversed direction over time. For example, near-term expansion of grasslands into arid lands and westward expansion of eastern temperate mixed forest reverse in the long-term, and substantial area is ultimately lost. Fire and other disturbances are the most likely mechanism of vegetation dieback.

These predictions seem dire, but there is a growing body of evidence that fire is already increasing in severity or frequency. For example, Westerling et al. (2006) found that annual area burned by wildfire in the western United States has increased by a factor of more than 6.5 relative to 1970. They attributed this change to earlier snowmelt and longer fire seasons. Similarly, Kurz and Apps (1999; see also Kurz et al. 2008) concluded that fire and insect disturbance have caused the Canadian boreal forest to become a source of carbon since 1979, in contrast to the preceding 60 years. There is also a growing body of evidence to suggest that climatically induced forest dieback is underway at many locations in western North America: Breshears et al. (2005) found evidence for drought-induced dieback of two-needle pinyon pine (*Pinus edulis*) in the southwestern United States. van Mantgem et al. (2009) documented increased mortality among a wide range of species and age classes throughout the western United States and Canada that they attributed to climate-induced water deficits. Logan and Powell (2001) found that recent warming has allowed the mountain pine beetle (*Dendroctonus ponderosae*) to expand its range to higher elevations and thereby attack whitebark pine (*Pinus albicaulis*), leaving behind "ghost forests." These changes alter forest habitat quality, but also the abundance and moisture of fuels, and the likely development of forest ecosystems over the coming decades.

4.5 Summary and Conclusions

Climatic processes act as top-down controls on regional patterns of fire ignition, rate of spread, fuel moisture, and fuel abundance and continuity. Lightning is the most important natural cause of fire ignition. Lightning frequency varies at continental, regional, and local scales, with areas of convergence and convection experiencing the highest frequency of lightning strikes. In mountainous regions the greatest frequency of lightning strikes often occurs at intermediate elevations. Lightning frequency alone is a poor predictor of the number of ignitions or total area burned, partly because lightning is often accompanied by precipitation, but also because not all vegetation types are equally flammable and ignition efficiency varies between land cover types. Following ignition, rates of spread are most rapid when strong winds, low humidity and high temperatures coincide. Regionally synchronous conditions conducive to rapid rates of spread are associated with several specific synoptic circulation types. In particular, persistent blocking ridges often contribute to the development of strong pressure gradients and intense cyclonic activity that contributes to rapid fire spread. Another important set of circulation patterns is associated with air masses that cross mountains. These masses lose moisture as they are pushed up the windward side of mountains, and warm and dry rapidly as they descend the leeward side. The most severe of such patterns is probably the Santa Ana winds, which are associated with extreme wildfire hazard in southern California.

Patterns of ignition and spread depend on slower varying patterns in fuel moisture and fuel abundance. Fuel moisture is a function of climate over the days to years preceding ignition. In regions such as the coastal temperate rainforest where there are abundant coarse fuels, a seasonally wet climate, high soil water-storage capacity, and dense canopy cover, extended periods of antecedent drying are a necessary precondition to wildfires. At the other extreme, arid and semiarid ecosystems frequently have conditions conducive to ignition and spread, but require anomalously wet conditions over the preceding seasons in order to produce the continuous fine fuels required for fire spread. In between these extremes these processes interact, depending on such factors as long-term changes in mean climate, which determine dominant vegetation types; slope, aspect, and soil properties, which influence soil moisture and microclimate, contributing to variability in vegetation structure and composition; and recent disturbance history, which affects the abundance of fuel and the developmental stage of vegetation.

These processes operate at different scales, and interact to give rise to regionally synchronous wildfire years that differ depending on properties of the affected ecosystems. The oft-cited dichotomy of "fuel vs. climate" fails to incorporate the full range of possible relationships between top-down and bottom-up processes in regulating the fire regime. For example, ponderosa pine forests in the U.S. Southwest are both fuel- and ignition-limited, and respond to interannual variability in fuels production but to sub-seasonal variability in drought. Several important patterns of climatic variability influence the processes that control wildfire across a range of temporal and spatial scales. Globally, ENSO is the most important such pattern of variability, but over North America it interacts with the PDO, the NAM, and the AMO to produce regionally synchronous variability in patterns of wildfire. In the coming decades, fire is likely to be an important agent of ecosystem change, as climatic change and exotic species increase the frequency and magnitude of wildfire nearly everywhere in North America.

References

Agee, J.K. 1993. *Fire ecology of Pacific Northwest forests*. Washington: Island Press.
Agee, J.K. 1998. The landscape ecology of western forest fire regimes. *Northwest Science* 72: 24–34.
Ambaum, M.H.P., B.J. Hoskins, and D.B. Stephenson. 2001. Arctic oscillation or north atlantic oscillation? *Journal of Climate* 14: 3495–3507.

Andrews, P.L. 2007. BehavePlus fire modeling system: Past, present, and future. In *Proceedings of seventh symposium on fire and forest meteorological society*, Bar Harbor. http://ams.confex.com/ams/pdfpapers/126669.pdf. Accessed 4 Mar 2010.

Bachelet, D., J.M. Lenihan, C. Daly, R.P. Neilson, D.S. Ojima, and W.J. Parton. 2001a. *MC1: A dynamic vegetation model for estimating the distribution of vegetation and associated ecosystem fluxes of carbon, nutrients, and water.* General Technical Report PNW-GTR-508. Portland: U.S. Forest Service.

Bachelet, D., R.P. Neilson, J.M. Lenihan, and R.J. Drapek. 2001b. Climate change effects on vegetation distribution and carbon budget in the United States. *Ecosystems* 4: 164–185.

Bachelet, D., R.P. Neilson, T. Hickler, R.J. Drapek, J.M. Lenihan, M.T. Sykes, B. Smith, S. Sitch, and K. Thonicke. 2003. Simulating past and future dynamics of natural ecosystems in the United States. *Global Biogeochemical Cycles* 17: 14-1. doi:10.1029/2001GB001508.

Baker, W.L., T.T. Veblen, and R.L. Sherriff. 2007. Fire, fuels and restoration of ponderosa pine–Douglas fir forests in the Rocky Mountains, USA. *Journal of Biogeography* 34: 251–269.

Barlow, M., S. Nigam, and E.H. Berbery. 2001. ENSO, Pacific decadal variability, and U.S. summertime precipitation, drought, and stream flow. *Journal of Climate* 14: 2105–2128.

Beckage, B., W.J. Platt, M.G. Slocum, and B. Panko. 2003. Influence of the El Niño southern oscillation on fire regimes in the Florida Everglades. *Ecology* 84: 3124–3130.

Belsky, A.J., and D.M. Blumenthal. 1997. Effects of livestock grazing on stand dynamics and soils in upland forests of the interior West. *Conservation Biology* 11: 315–327.

Bessie, W.C., and E.A. Johnson. 1995. The relative importance of fuels and weather on fire behavior in subalpine forests. *Ecology* 76: 747–762.

Biondi, F., A. Gershunov, and D.R. Cayan. 2001. North Pacific decadal climate variability since 1661. *Journal of Climate* 14: 5–10.

Brenner, J. 1991. Southern oscillation anomalies and their relationship to wildfire activity in Florida. *International Journal of Wildland Fire* 1: 3–78.

Breshears, D.D., N.S. Cobb, P.M. Richd, K.P. Price, C.D. Allen, R.G. Balice, W.H. Romme, J.H. Kastens, M.L. Floyd, J. Belnap, J.J. Anderson, O.B. Myers and C.W. Meyer. 2005. Regional vegetation die-off in response to global-change-type drought. *Proceedings of the National Academy of Sciences* 102: 15144–15148.

Bridge, S.R., K. Miyanishi, and E.A. Johnson. 2005. A critical evaluation of fire suppression effects in the boreal forest of Ontario. *Forest Science* 51: 41–50.

Brooks, M.L., and J.R. Matchett. 2006. Spatial and temporal patterns of wildfires in the Mojave Desert, 1980–2004. *Journal of Arid Environments* 67: 148–164.

Brooks, M.L., and D.A. Pyke. 2001. Invasive plants and fire in the deserts of North America. In *Proceedings of the invasive species workshop: The role of fire in the control and spread of invasive species. Fire conference 2000: The first national congress on fire ecology, prevention, and management,* vol. Miscellaneous Publication No. 11, eds. K.E.M. Galley and T.P. Wilson, 1–14. Tallahassee: Tall Timbers Research Station.

Brown, P.M. 2006. Climate effects on fire regimes and tree recruitment in Black Hills ponderosa pine forests. *Ecology* 87: 2500–2510.

Brown, K.J., and R.J. Hebda. 2002. Ancient fires on southern Vancouver Island, British Columbia, Canada: A change in causal mechanisms at about 2,000 ybp. *Environmental Archaeology* 7: 1–12.

Brown, P.M., and W.D. Shepperd. 2001. Fire history and fire climatology along a 5° gradient in latitude in Colorado and Wyoming, USA. *Paleobotanist* 50: 133–140.

Brown, T.J., B.L. Hall, and A.L. Westerling. 2004. The impact of twenty-first century climate change on wildland fire danger in the western United States: An applications perspective. *Climatic Change* 62: 365–388.

Brown, K.J., J.S. Clark, E.C. Grimm, J.J. Donovan, and P.G. Mueller. 2005. Fire cycles in North American interior grasslands and their relation to prairie droughts. *Proceedings of the National Academy of Sciences of the United States of America* 102: 8865–8870.

Burrows, W.R., P. King, P.J. Lewis, B. Kochtubajda, B. Snyder, and V. Turcotte. 2002. Lightning occurrence patterns over Canada and adjacent United States from lightning detection network observations. *Atmosphere-Ocean* 40: 59–81.

Cable, D.R. 1975. Influence of precipitation on perennial grass production in the semidesert southwest. *Ecology* 56: 981–986.

Cayan, D.R. 1996. Interannual climate variability and snowpack in the western United States. *Journal of Climate* 9: 928–948.

Chen, J., S.C. Saunders, T.R. Crow, R.J. Naiman, K.D. Brosofske, G.D. Mroz, B.L. Brookshire, and J.F. Franklin. 1999. Microclimate in forest ecosystem and landscape ecology. *Bioscience* 49: 288–297.

Collins, B.M., P.N. Omi, and P.L. Chapman. 2006. Regional relationships between climate and wildfire-burned area in the interior West, USA. *Canadian Journal of Forest Research* 36: 699–709.

Countryman, C.M., M.H. McCutchan, and B.C. Ryan. 1969. *Fire weather and fire behavior at the 1968 Canyon fire.* Research Paper PSW-55. Berkeley: U.S. Forest Service.

Crimmins, M.A. 2006. Synoptic climatology of extreme fire weather conditions across the southwest United States. *International Journal of Climatology* 26: 1001–1016.

Delworth, T.L., and M.E. Mann. 2000. Observed and simulated variability in the Northern Hemisphere. *Climate Dynamics* 16: 661–676.

Diaz, H.F., and G.N. Kiladis. 1992. Atmospheric teleconnections associated with the extreme phases of the Southern Oscillation. In *El Niño: Historical and paleoclimatic aspects of the Southern Oscillation*, eds. H.F. Diaz and V. Markgraf, 7–28. Cambridge: Cambridge University Press.

Díaz-Avalos, C., D.L. Peterson, E. Alvarado, S.A. Ferguson, and J.E. Besag. 2001. Space-time modelling of lightning-caused ignitions in the Blue Mountains, Oregon. *Canadian Journal of Forest Research* 31: 1579–1593.

Dixon, P.G., G.B. Goodrich, and W.H. Cooke. 2008. Using teleconnections to predict wildfires in Mississippi. *Monthly Weather Review* 136: 2804–2811.

Dommenget, D., and M. Latif. 2002. A cautionary note on the interpretation of EOFs. *Journal of Climate* 15: 216–225.

Enfield, D.B. 1989. El Niño, past and present. *Reviews of Geophysics* 27: 159–187.

Enfield, D.B., A.M. Mestas-Nuñez, and P.J. Trimble. 2001. The Atlantic multidecadal oscillation and its relation to rainfall and river flows in the continental U.S. *Geophysical Research Letters* 28: 2077–2080.

Finklin, A.I. 1973. *Meteorological factors in the Sundance Fire run.* General Technical Report INT-6. Ogden: U.S. Forest Service.

Finney, M.A. 1998. *FARSITE: Fire area simulator–Model development and evaluation.* Research Paper RMRS-RP_004. Ogden: U.S. Forest Service.

Finney, M.A. 1999. Mechanistic modelling of landscape fire patterns. In *Spatial modeling of forest landscape change: Approaches and applications*, eds. D.J. Mladenoff and W.L. Baker, 186–209. Cambridge: Cambridge University Press.

Flannigan, M.D., and J.B. Harrington. 1988. A study of the relation of meteorological variables to monthly provincial area burned by wildfire in Canada (1953–1980). *Journal of Applied Meteorology* 27: 441–452.

Flannigan, M.D., and C.E. van Wagner. 1991. Climate change and wildfire in Canada. *Canadian Journal of Forest Research* 21: 66–72.

Flannigan, M.D., B.J. Stocks, and B.M. Wotton. 2000. Climate change and forest fires. *The Science of the Total Environment* 262: 221–229.

Flannigan, M., I. Campbell, M. Wotton, C. Carcaillet, P. Richard, and Y. Bergeron. 2001. Future fire in Canada's boreal forest: Paleoecology results and general circulation model–Regional climate model simulations. *Canadian Journal of Forest Research* 31: 854–864.

Fons, W.T. 1946. Analysis of fire spread in light forest fuels. *Journal of Agricultural Research* 72: 93–121.

Franklin, J.F., K. Cromack, W. Denison, A. McKee, C. Maser, J. Sedell, F. Swanson, and G. Juday. 1981. *Ecological characteristics of old-growth Douglas-fir forests.* General Technical Report PNW-118. Portland: U.S. Forest Service.

Gedalof, Z., and D.J. Smith. 2001. Interdecadal climate variability and regime-scale shifts in Pacific North America. *Geophysical Research Letters* 28: 1515–1518.

Gedalof, Z., N.J. Mantua, and D.L. Peterson. 2002. A multi-century perspective of variability in the Pacific Decadal Oscillation: new insights from tree rings and coral. *Geophysical Research Letters* 29: 4. doi:10.1029/2002GL015824.

Gedalof, Z., D.L. Peterson, and N.J. Mantua. 2005. Atmospheric, climatic and ecological controls on extreme wildfire years in the northwestern United States. *Ecological Applications* 15: 154–174.

Gedalof, Z., M.G. Pellatt, and D.J. Smith. 2006. From prairie to forest: Three centuries of environmental change at Rocky Point, Vancouver Island, BC. *Northwest Science* 80: 34–46.

Gershunov, A., and T.P. Barnett. 1998. Interdecadal modulation of ENSO teleconnections. *Bulletin of the American Meteorological Society* 79: 2715–2725.

Gillett, N.P., F.W. Zwiers, A.J. Weaver, and M.D. Flannigan. 2004. Detecting the effect of climate change on Canadian forest fires. *Geophysical Research Letters* 31: 4. doi:10.1029/2004GL020876.

Guyette, R.P., M.A. Spetich, and M.C. Stambaugh. 2006. Historic fire regime dynamics and forcing factors in the Boston Mountains, Arkansas, USA. *Forest Ecology and Management* 234: 293–304.

Hamilton, K. 1988. A detailed examination of the extratropical response to tropical El Niño / Southern Oscillation events. *Journal of Climatology* 8: 67–86.

Hargrove, W.W., R.H. Gardner, M.G. Turner, W.H. Romme, and D.G. Despain. 2000. Simulating fire patterns in heterogeneous landscapes. *Ecological Modelling* 135: 253–263.

Hessl, A., D. McKenzie, and R. Schellhaas. 2004. Drought and Pacific decadal oscillation linked to fire occurrence in the inland Pacific Northwest. *Ecological Applications* 14: 425–442.

Heyerdahl, E.K., L.B. Brubaker, and J.K. Agee. 2001. Spatial controls of historical fire regimes: A multiscale example from the interior West, USA. *Ecology* 82: 660–678.

Heyerdahl, E.K., D. McKenzie, L.D. Daniels, A.E. Hessl, J.S. Littell, and N.J. Mantua. 2008. Climate drivers of regionally synchronous fires in the inland northwest (1651–1900). *International Journal of Wildland Fire* 17: 40–49.

Huffines, G.R., and R.E. Orville. 1999. Lightning ground flash density and thunderstorm duration in the continental United States: 1989–96. *Journal of Applied Meteorology* 38: 1013–1019.

Johnson, E.A. 1992. *Fire and vegetation dynamics: Studies from the North American boreal forest.* Cambridge: Cambridge University Press.

Johnson, E.A., and D.R. Wowchuk. 1993. Wildfires in the southern Canadian Rocky Mountains and their relationship to mid-tropospheric anomalies. *Canadian Journal of Forest Research* 23: 1213–1222.

Johnson, E.A., K. Miyanishi, and S.R.J. Bridge. 2001. Wildfire regime in the boreal forest and the idea of suppression and fuel buildup. *Conservation Biology* 15: 1554–1557.

Jones, P.D., and M.E. Mann. 2004. Climate over past millennia. *Reviews of Geophysics* 42: 42. doi:10.1029/2003RG000143.

Karl, T.R. 1985. Perspective on climate change in North America during the twentieth century. *Physical Geography* 6: 207–229.

Keeley, J.E. 2002. Native American impacts on fire regimes of the California coastal ranges. *Journal of Biogeography* 29: 303–320.

Keeley, J.E., and C.J. Fotheringham. 2002. Historic fire regime in southern California shrublands. *Conservation Biology* 15: 1536–1548.

Keeley, J.E., C.J. Fotheringham, and M. Morais. 1999. Reexamining fire suppression impacts on brushland fire regimes. *Science* 284: 1829–1832.

Keeley, J.E., C.J. Fotheringham, and M.A. Moritz. 2004. Lessons from the October 2003 wildfires in southern California. *Journal of Forestry* 102: 26–31.

Kerr, R.A. 2000. A North Atlantic climate pacemaker for the centuries. *Science* 288: 1984–1985.

Kiladis, G.N., and H.R. Diaz. 1989. Global climatic anomalies associated with extremes in the Southern Oscillation. *Journal of Climate* 2: 1069–1090.

Kitzberger, T., T.W. Swetnam, and T.T. Veblen. 2001. Inter-hemispheric synchrony of forest fires and the El Niño-southern oscillation. *Global Ecology and Biogeography* 10: 315–326.

Kitzberger, T., P.M. Brown, E.K. Heyerdahl, T.W. Swetnam, and T.T. Veblen. 2007. Contingent Pacific-Atlantic Ocean influence on multicentury wildfire synchrony over western North America. *Proceedings of the National Academy of Sciences* 104: 543–548.

Knapp, P.A. 1995. Intermountain West lightning-caused fires: Climatic predictors of area burned. *Journal of Range Management* 48(8): 5–91.

Knapp, P.A. 1998. Spatio-temporal patterns of large grassland fires in the intermountain West, U.S.A. *Global Ecology and Biogeography* 7: 259–272.

Krawchuk, M.A., S.G. Cumming, M.D. Flannigan, and R.W. Wein. 2006. Biotic and abiotic regulation of lightning fire initiation in the mixedwood boreal forest. *Ecology* 87: 458–468.

Kurz, W.A., and M.J. Apps. 1999. A 70-year retrospective analysis of carbon fluxes in the Canadian forest sector. *Ecological Applications* 9: 526–547.

Kurz, W.A., G. Stinson, G.J. Rampley, C.C. Dymond, and E.T. Neilson. 2008. Risk of natural disturbances makes future contribution of Canada's forests to the global carbon cycle highly uncertain. *Proceedings of the National Academy of Sciences of the United States of America* 105: 1551–1555.

Latham, D., and E. Williams. 2001. Lightning and forest fires. In *Forest fires: Behaviour and ecological effects*, eds. E.A. Johnson and K. Miyanishi, 375–418. San Diego: Academic.

Le Goff, H., M.D. Flannigan, Y. Bergeron, and M.P. Girardin. 2007. Historical fire regime shifts related to climate teleconnections in the Waswanipi area, central Quebec, Canada. *International Journal of Wildland Fire* 16: 607–618.

Littell, J.S., D. McKenzie, D.L. Peterson and A.L. Westerling. 2009. Climate and wildfire area burned in western U.S. ecoprovinces, 1916–2003. *Ecological Applications* 19: 1003–1021.

Logan, J.A., and J.A. Powell. 2001. Ghost forests, global warming, and the mountain pine beetle (Coleoptera: Scolytidae). *American Entomologist* 47: 160–173.

Lopez, R.E., and R.L. Holle. 1986. Diurnal and spatial variability of lightning activity in northeastern Colorado and central Florida during the summer. *Monthly Weather Review* 114: 1288–1312.

Lyons, W.A., M. Uliasz, and T.E. Nelson. 1998. Large peak current cloud-to-ground lightning flashes during the summer months in the contiguous United States. *Monthly Weather Review* 126: 2217–2233.

Macias Fauria, M., and E.A. Johnson. 2006. Large-scale climatic patterns control large lightning fire occurrence in Canada and Alaska forest regions. *Journal of Geophysical Research* 111: 17. doi:10.1029/2006JG000181.

Macias Fauria, M., and E.A. Johnson. 2008. Climate and wildfires in the North American boreal forest. *Philosophical Transactions of the Royal Society of London. Series B: Biological Sciences* 363: 2317–2329.

Madany, M.H., and N.E. West. 1983. Livestock grazing-fire regime interactions within montane forests of Zion National Park, Utah. *Ecology* 64: 661–667.

Malamud, B.D., J.D.A. Millington, and G.L.W. Perry. 2005. Characterizing wildfire regimes in the United States. *Proceedings of the National Academy of Sciences of the United States of America* 102: 4694–4699.

Mantua, N.J., S.R. Hare, Y. Zhang, J.M. Wallace, and R.C. Francis. 1997. A Pacific interdecadal climate oscillation with impacts on salmon production. *Bulletin of the American Meteorological Society* 78: 1069–1079.

McCabe, G.J., and M.D. Dettinger. 2002. Primary modes and predictability of year-to-year snowpack variations in the western United States from teleconnections with Pacific Ocean climate. *Journal of Hydrometeorology* 3: 13–25.

McKenzie, D., Z. Gedalof, D.L. Peterson, and P. Mote. 2004. Climatic change, wildfire, and conservation. *Conservation Biology* 18: 890–902.

McPhaden, M.J., S.E. Zebiak, and M.H. Glantz. 2006. ENSO as an integrating concept in Earth science. *Science* 314: 1740–1745.

Meehl, G.A., W.M. Washington, W.D. Collins, J.M. Arblaster, A. Hu, L.E. Buja, W.G. Strand, and H. Teng. 2005. How much more global warming and sea level rise? *Science* 307: 1769–1773.

Meisner, B.N. 1993. *Correlation of National Fire Danger Rating System indices and weather data with fire reports.* Final Report, Interagency Agreement No. R500A20021. Riverside: U.S. Forest Service.

Meyn, A., P.S. White, C. Buhk, and A. Jentsch. 2007. Environmental drivers of large, infrequent wildfires: The emerging conceptual model. *Progress in Physical Geography* 31: 287–312.

Miller, N.L., and N.J. Schlegel. 2006. Climate change projected fire weather sensitivity: California Santa Ana wind occurrence. *Geophysical Research Letters* 33: 5. doi:10.1029/2006GL025808.

Minobe, S. 1999. Resonance in bidecadal and pentadecadal climate oscillations over the North Pacific: Role in climatic regime shifts. *Geophysical Research Letters* 26: 855–858.

Moore, R.D. 1996. Snowpack and runoff responses to climatic variability, southern Coast Mountains, British Columbia. *Northwest Science* 70: 321–333.

Morris, W.G. 1934. Lightning storms and fires on the national forests of Oregon and Washington. *Monthly Weather Review* 62: 370–375.

Namias, J. 1976. Some statistical and synoptic characteristics associated with El Niño. *Journal of Physical Oceanography* 6: 130–138.

Namias, J., and D.R. Cayan. 1981. Large-scale air-sea interactions and short-scale climatic fluctuations. *Science* 214: 869–876.

Nash, C.H., and E.A. Johnson. 1996. Synoptic climatology of lightning-caused forest fires in subalpine and boreal forests. *Canadian Journal of Forest Research* 26: 1859–1874.

Neilson, R.P., and R.J. Drapek. 1998. Potentially complex biosphere responses to transient global warming. *Global Change Biology* 4: 505–521.

Nkemdirim, L.C., and D. Budikova. 1996. The El Niño southern oscillation has a truly global impact: A preliminary report on the ENSO project of the commission on climatology. *International Geographical Union Bulletin* 46: 27–37.

Norman, S.P., and A.H. Taylor. 2003. Tropical and north Pacific teleconnections influence fire regimes in pine-dominated forests of north-eastern California, USA. *Journal of Biogeography* 30: 1081–1092.

Podur, J., D.L. Martell, and F. Csillag. 2003. Spatial patterns of lightning-caused forest fires in Ontario, 1976–1998. *Ecological Modelling* 164: 1–20.

Price, C., and D. Rind. 1994. The impact of a $2 \times CO_2$ climate on lightning-caused fires. *Journal of Climate* 7: 1484–1494.

Ropelewski, C.F., and M.S. Halpert. 1986. North American precipitation and temperature patterns associated with the El Niño southern oscillation (ENSO). *Monthly Weather Review* 114: 2352–2362.

Rorig, M.L., and S.A. Ferguson. 1999. Characteristics of lightning and wildland fire ignition in the Pacific Northwest. *Journal of Applied Meteorology* 38: 1565–1575.

Rorig, M.L., and S.A. Ferguson. 2002. The 2000 fire season: Lightning-caused fires. *Journal of Applied Meteorology* 41: 786–791.

Rothermel, R.C. 1972. *A mathematical model for fire spread predictions in wildland fuels.* Research Paper INT-115. Ogden: U.S. Forest Service.

Rothermel, R.C. 1983. *How to predict the spread and intensity of forest and range fires.* General Technical Report INT-143. Ogden: U.S. Forest Service.

Running, S.W. 2008. Ecosystem disturbance, carbon, and climate. *Science* 321: 652–653.

Sando, R.W., and D.A. Haines. 1972. *Fire weather and behavior of the Little Sioux fire.* Research Paper NC-76. St. Paul: U.S. Forest Service.

Sardeshmukh, P.D. 1990. Factors affecting the extratropical anomalies associated with individual ENSO events. In *International TOGA scientific conference proceedings*, 85. WMO/TD-No. 379. Geneva: World Meteorological Organization.

Schimmel, J., and A. Granström. 1997. Fuel succession and fire behavior in the Swedish boreal forest. *Canadian Journal of Forest Research* 27: 1207–1216.

Schlesinger, M.E., and N. Ramankutty. 1994. An oscillation in the global climate system of period 65–70 years. *Nature* 367: 723–726.

Schoennagel, T., T.T. Veblen, and W.H. Romme. 2004. The interaction of fire, fuels, and climate across Rocky Mountain forests. *Bioscience* 54: 661–676.

Schoennagel, T., T.T. Veblen, W.H. Romme, J.S. Sibold, and E.R. Cook. 2005. ENSO and PDO variability affect drought-induced fire occurrence in Rocky Mountain subalpine forests. *Ecological Applications* 15: 2000–2014.

Schoennagel, T., T.T. Veblen, D. Kulakowski, and A. Holz. 2007. Multidecadal climate variability and climate interactions affect subalpine fire occurrence, western Colorado (USA). *Ecology* 88: 2891–2901.

Schroeder, M.J. 1969. *Critical fire weather patterns in the conterminous United States.* Technical Report WB 8. Silver Spring: Environmental Science Services Administration.

Shabbar, A., and W. Skinner. 2004. Summer drought patterns in Canada and the relationship to global sea surface temperatures. *Journal of Climate* 17: 2866–2880.

Sherriff, R.L., and T.T. Veblen. 2008. Variability in fire–climate relationships in ponderosa pine forests in the Colorado Front Range. *International Journal of Wildland Fire* 17: 50–59.

Sibold, J.S., and T.T. Veblen. 2006. Relationships of subalpine forest fires in the Colorado Front Range with interannual and multidecadal-scale climatic variation. *Journal of Biogeography* 33: 833–842.

Skinner, W.R., B.J. Stocks, D.L. Martell, B. Bonsal, and A. Shabbar. 1999. The association between circulation anomalies in the mid-troposphere and area burned by wildfire in Canada. *Theoretical and Applied Climatology* 63: 89–105.

Skinner, W.R., M.D. Flannigan, B.J. Stocks, D.L. Martell, B.M. Wotton, J.B. Todd, J.A. Mason, K.A. Logan, and E.M. Bosch. 2002. A 500 hPa synoptic wildland fire climatology for large Canadian forest fires, 1959–1996. *Theoretical and Applied Climatology* 71: 157–169.

Solomon, S., M.M. Qin, M. Chen, K.B. Marquis, M. Averyt, and H.L. Miller. 2007. *Climate change 2007, the physical science basis: Contribution of working group I to the fourth assessment report of the intergovernmental panel on climate change.* Cambridge: Cambridge University Press.

Stahle, D.W., E.R. Cook, M.K. Cleaveland, M.D. Therrell, D.M. Meko, H.D. Grissino-Mayer, E. Watson, and B.H. Luckman. 2000. Tree-ring data document 16th century megadrought over North America. *Eos Transactions of the American Geophysical Union* 81: 124–125.

Stephens, S.L., and B.M. Collins. 2004. Fire regimes of mixed conifer forests in the north-central Sierra Nevada at multiple spatial scales. *Northwest Science* 78: 12–23.

Stocks, B.J., M.A. Fosberg, T.J. Lynham, L. Mearns, B.M. Wotton, Q. Yang, J.-Z. Jin, K. Lawrence, G.R. Hartley, J.A. Mason, and D.W. McKenney. 1998. Climate change and forest fire potential in Russian and Canadian boreal forests. *Climatic Change* 38: 1–13.

Strauss, D., L. Bednar, and R. Mees. 1989. Do one percent of forest fires cause ninety-nine percent of the damage. *Forest Science* 35: 319–328.

Street, R.B., and M.E. Alexander. 1980. *Synoptic weather associated with five major forest fires in Pukaskwa National Park.* Reg. Int. Rep. SSD-80-2. Toronto: Environment Canada, Atmospheric Environment Service.

Swetnam, T.W., and C.H. Baisan. 1996. Historical fire regime patterns in the southwestern United States since AD 1700. In *Fire effects in southwest forests: Proceedings of the second La Mesa Fire symposium*, tech. ed. C.D. Allen, 11–32. General Technical Report RM-286. Fort Collins: U.S. Forest Service.

Swetnam, T.W., and J.L. Betancourt. 1990. Fire-southern oscillation relations in the southwestern United States. *Science* 249: 1017–1020.

Swetnam, T.W., and J.L. Betancourt. 1998. Mesoscale disturbance and ecological response to decadal climatic variability in the American Southwest. *Journal of Climate* 11: 3128–3147.

Thompson, D.W.J., and J.M. Wallace. 1998. The Arctic oscillation signature in the wintertime geopotential height and temperature fields. *Geophysical Research Letters* 25: 1297–1300.

Thompson, D.W.J., and J.M. Wallace. 2000. Annular Modes in the extratropical circulation. Part I: Month-to-month variability. *Journal of Climate* 13: 1000–1016.

Thompson, D.W.J., and J.M. Wallace. 2001. Regional climate impacts of the Northern Hemisphere Annular Mode. *Science* 293: 85–89.

Thonicke, K., S. Venevsky, S. Sitch, and W. Cramer. 2001. The role of fire disturbance for global vegetation dynamics: Coupling fire into a dynamic global vegetation model. *Global Ecology and Biogeography* 10: 661–677.

Torn, M.S., and J.S. Fried. 1992. Predicting the impacts of global warming on wildland fire. *Climatic Change* 21: 257–274.

Trenberth, K.E., G.W. Branstator, D. Karoly, A. Kumar, N.-C. Lau, and C. Ropelewski. 1998. Progress during TOGA in understanding and modeling global teleconnections associated with tropical sea surface temperatures. *Journal of Geophysical Research* 103: 14291–14324.

Trouet, V., A.H. Taylor, A.M. Carleton, and C.N. Skinner. 2006. Fire-climate interactions in forests of the American Pacific coast. *Geophysical Research Letters* 33: 5. doi:10.1029/2006GL027502.

Tymstra, C., M.D. Flannigan, O.B. Armitage, and K. Logan. 2007. Impact of climate change on area burned in Alberta's boreal forest. *International Journal of Wildland Fire* 16: 153–160.

Uman, M.A. 2001. *The lightning discharge*. Mineola: Dover Publications.

van Mantgem, P.J., N.L. Stephenson, J.C. Byrne, L.D. Daniels, J.F. Franklin, P.Z. Fulé, M.E. Harmon, A.J. Larson, J.M. Smith, A.H. Taylor, and T.T. Veblen. 2009. Widespread increase of tree mortality rates in the western United States. *Science* 323: 521–524.

van Wagtendonk, J.W., and D.R. Cayan. 2008. Temporal and spatial distribution of lightning strikes in California in relation to large-scale weather patterns. *Fire Ecology* 4: 34–56.

Visbeck, M.H., J.W. Hurrell, L. Polvani, and H.M. Cullen. 2001. The north Atlantic oscillation: Past, present, and future. *Proceedings of the National Academy of Sciences of the United States of America* 98: 12876–12877.

Wallace, J.M. 2000. North Atlantic oscillation/annular mode: Two paradigms – One phenomenon. *Quarterly Journal Royal Meteorological Society* 126: 791–805.

Wallace, J.M., and D.S. Gutzler. 1981. Teleconnections in the geopotential height field during the Northern Hemisphere winter. *Monthly Weather Review* 109: 784–812.

Ward, P.C., A.G. Tithecott, and B.M. Wotton. 2001. Reply: A re-examination of the effects of fire suppression in the boreal forest. *Canadian Journal of Forest Research* 31: 1467–1480.

Westerling, A.L., and B.P. Bryant. 2008. Climate change and wildfire in California. *Climatic Change* 87(Suppl 1): S231–S249.

Westerling, A.L., and T.W. Swetnam. 2003. Interannual to decadal drought and wildfire in the Western US. *Eos Transactions of the American Geophysical Union* 84: 545–560.

Westerling, A.L., A. Gershunov, T.J. Brown, D.R. Cayan, and M.D. Dettinger. 2003. Climate and wildfire in the western United States. *Bulletin of the American Meteorological Society* 84: 595–604.

Westerling, A.L., H.G. Hidalgo, D.R. Cayan, and T.W. Swetnam. 2006. Warming and earlier spring increase western U.S. forest wildfire activity. *Science* 313: 940–943.

Wiedenmann, J.M., A.R. Lupo, I.I. Mokhov, and E.A. Tikhonova. 2002. The climatology of blocking anticyclones for the northern and southern hemispheres: Block intensity as a diagnostic. *Journal of Climate* 15: 3459–3473.

Williams, G.W. 2002. Aboriginal use of fire: Are there any "natural" plant communities? In *Wilderness and political ecology: Aboriginal influences and the original state of nature*, ed. C. Kay, 179–214. Salt Lake City: University of Utah Press.

Wyrtki, K. 1975. El Niño - The dynamic response of the equatorial Pacific Ocean to atmospheric forcing. *Journal of Physical Oceanography* 5: 572–584.

Yarnal, B., and H.F. Diaz. 1986. Relationships between extremes of the southern oscillation and the winter climate of the Anglo-American Pacific coast. *Journal of Climatology* 6: 197–219.

Zhang, Y., J.M. Wallace, and D. Battisti. 1997. ENSO-like interdecadal variability: 1900–1993. *Journal of Climate* 10: 1004–1020.

Chapter 5
Climatic Water Balance and Regional Fire Years in the Pacific Northwest, USA: Linking Regional Climate and Fire at Landscape Scales

Jeremy S. Littell and Richard B. Gwozdz

5.1 Introduction

Fire and water are linked across multiple spatial and temporal scales. Climate provides a top-down control on fire regimes (Gedalof et al. 2005; Littell et al. 2009a, b; Chap. 4), via seasonal-to-multidecadal patterns of temperature and precipitation and their interaction. At fine scales, fuel structure and composition interact with micro-meteorology to affect fire intensity and fire spread (Rothermel 1972). At all scales, water relations provide the physical basis for understanding the variability in fire activity and the landscape patterns it produces.

The paleoecological record (both tree-ring and sediment charcoal fire histories) and the modern record document strong associations between fire and climate (e.g., Clark 1990; Swetnam and Betancourt 1990; McKenzie et al. 2004 and references therein; Littell et al. 2009a). Due to decreasing fuel moisture, warmer drier conditions should be associated with increased fire activity. Indeed, interannual relationships between climate (drought indices, precipitation, temperature) and drought are frequently implicated in the number of fires and the area burned by fires in the western United States. Fire histories from tree-ring data indicate, however, that the relationship between climate and fire varies considerably with the type of forest in question. For example, regional composite fire histories from the U.S. Pacific Northwest and northern Rocky Mountains suggest that drought and warmer temperatures in spring and summer of the fire season are associated with regionally synchronous fire in forested ecosystems and that antecedent conditions were comparatively unimportant (Heyerdahl et al. 2008a,b). In contrast, for open woodland in the southwestern United States, drought is still implicated in the year of fire, but antecedent increases in moisture availability are also associated with synchronous fire years (Swetnam and Betancourt 1998; Grissino-Mayer and Swetnam 2000; Brown et al. 2008).

J.S. Littell (✉)
CSES Climate Impacts Group, University of Washington, Seattle, WA 98195_5672, USA
e-mail: jlittell@uw.edu

D. McKenzie et al. (eds.), *The Landscape Ecology of Fire*, Ecological Studies 213,   117
DOI 10.1007/978-94-007-0301-8_5, © Springer Science+Business Media B.V. 2011

The twentieth-century record also shows that area burned by fire is strongly related to climate, although there is considerable sub-regional variation associated with vegetation type (Westerling et al. 2003; Littell et al. 2009a). In forested eco-systems of the Sierra Nevada, Cascade Range, and northern Rocky Mountains, USA, warm dry summers or growing seasons are strongly associated with high area burned, reflecting regionally synchronous fire activity. In drier forests and shrub-lands of the southern Rocky Mountains and desert Southwest, the strongest cli-mate predictors of area burned were wetter and sometimes cooler years preceding the fire season.

The primary mechanisms that relate climate variation to fire occurrence and area are (1) the availability and continuity of fuels and (2) the fuel-independent conditions that influence fire spread, such as short-term weather and fire suppres-sion. Climatic influences on fire fall largely into the first category by influencing the rate of fuel production and the moisture content of live and dead fuels. In eco-systems where area burned and regional fire synchrony are positively related to temperature and drought, the controlling mechanism would appear to be the dry-ing of existing fuels below some threshold fuel moisture that substantially increases flammability (Romme and Despain 1989; Johnson and Wowchuk 1993; Nash and Johnson 1996; Littell et al. 2009a). These ecosystems are typically pro-ductive enough that they are not fuel limited—fuel buildup is important, particu-larly on longer time scales of decades or even centuries. The limiting factor, however, appears to be fuel condition (Rollins et al. 2002; Littell et al. 2009a) and suggests a lack of energy required to dry fuels sufficiently for combustion. Boreal and cool temperate forests are examples of ecosystems in which fire is likely energy-limited. The spatial arrangement and continuity of fuels in these systems also does not vary much from year to year, at least on average. In contrast, fuels in water-limited systems are frequently dry enough to carry fire, but there is typi-cally less fuel and both canopy and surface fuels may be patchier. Increased fire activity in these systems is associated with wetter cooler conditions in the year or years prior to fire. The climatic correlations appear to suggest that fuel production and fuel continuity are facilitated by climate conditions that favor vegetation, and if so, the spatial continuity of surface fuels would vary on the time scale of years (Rollins et al. 2002; Littell et al. 2009a). Across the western United States, there likely exists a gradient such that vegetation types fall in between these two extremes of climate influence on fuel availability through fuel moisture (energy-limited) and on fuel availability through fuel production (water-limited). For example, Milne et al. (2002) demonstrated that there are continuous scale-invariant relation-ships between vegetation pattern and the biophysical gradient between energy- and water-limited vegetation.

If there is a multi-scale relationship that relates climate, fire occurrence, and area burned consistently via fuels, it must consider both temperature and precipitation and the relative role of each in affecting the likelihood of fire occurrence and spread. In this chapter, we are interested in these relationships on time scales from months to years. We focus on the propensity for antecedent climate conditions to precondition landscapes such that large areas can burn rather than on the weather conditions that cause fire fronts and fire behavior to produce large fires over hours to weeks.

Fire histories and 20th century studies relate fire occurrence and area burned to temperature and precipitation, and frequently to indices of ocean-atmosphere circulation such as the El Niño Southern Oscillation (ENSO) or the Pacific Decadal Oscillation (PDO). However, these approaches are unsatisfying in their physical approach to climate as plants and fuels "sense" it. During the late 20th century, unusually warm springs and longer summer dry seasons were associated with increased numbers of fires westwide (Westerling et al. 2006), and the connection between these seasonal effects appears to be in water deficits during the fire season (Littell et al. 2009a, b). Water-balance deficit (DEF) is defined as the difference between potential evapotranspiration (PET, driven by temperature, solar radiation, wind, etc.) and actual evapotranspiration (AET, driven by water availability). When PET exceeds AET, water-balance deficit is positive (vegetation is water-limited). DEF in particular is a useful predictor of coarse vegetation properties such as the distribution of biomes or vegetation types (Stephenson 1990; Neilson 1995) and a component of many hydrologic and biogeochemical models. This biophysical grounding for the relationship between fire and climate is potentially more climatically appropriate than indices representing modes of ocean-atmosphere interactions for the analysis of fire-climate relationships and their consequences at multiple scales. Persistent ocean-atmosphere variation (such as PDO and ENSO), while an important part of the climate system, varies through time and regionally in the degree to which it controls climate variables that directly affect the probability of fire ignition, spread, and ultimately area and severity. Large-scale circulation patterns and their influences on regional climate are increasingly uncoupled from local climatic variation across landscapes; water balance variables provide the potential to integrate across multiple scales.

We propose that water-balance variables should capture the climatic mechanisms that limit and facilitate fire, likely via their effects on fuels (Littell et al. 2009a). The fine fuels that carry fires are dynamic in terms of their moisture status, and fuel availability fluctuates with weather, but as resistance to high frequency (hourly or daily) fluctuations in water balance increases with fuel size, monthly and seasonal water balance should more closely approximate the moisture status of fuels. For live plant tissues, as soil water is depleted, plants have less water to draw on to meet the demands of transpiration. Tissue water should be reduced as droughts progress and DEF increases. PET may also provide a good indication of moisture content of small-diameter dead fuels during the fire season because its aggregated value over the fire season should track the frequency of days when small diameter fuels can burn even though PET varies greatly at daily and sub-daily time steps. In contrast, AET, and therefore DEF, cannot be mechanistically related to dead fuel moisture, because its calculation considers water loss from the soil via transpiration. However, DEF may provide an indication of dead fuel moisture because it incorporates both the evaporative demand associated with PET and the supply of precipitation that is one factor controlling AET. Seasonally aggregated AET or DEF may also indicate how early in the year small-diameter surface fuels become sensitive to atmospheric variability. For example, years with early snow-melt are more likely to have increased DEF and in these years local meteorological control of surface fuel availability will also occur earlier. Finally, in ecosystems with predominately fine and medium fuels, the production of fuels is partially controlled by favorable climate for plant growth in the year(s) preceding fire, and lower DEF (either less PET or increased AET or both) may lead to more abundant fuels that subsequently become available.

We suggest that the same ecohydrological principles (e.g., Milne et al. 2002) should apply to vegetation, water, and fire at multiple scales. At finer scales, local controls on water balance (slope, aspect, terrain shading, and soil properties) are superimposed on regional controls (climate, orographic, and elevation patterns), but the overall effect should remain; where fuel is not limited, zones of greater DEF are more likely to burn. This was clearly demonstrated in modeled data generated by Miller and Urban (1999). By showing that similar theoretical constructs govern the variation in hydrology, vegetation, and fire, we may also facilitate the incorporation of fire into ecohydrological models that operate at landscape scales. This chapter thus provides a mechanistic basis for applying climatically driven water-balance fire controls to the landscape ecology of fire. In this chapter, we explore the utility of water-balance components as predictors of the area burned by fire within coarse vegetation types of the Pacific Northwest and the northern Rocky Mountains, with the eventual goal of a scalable approach to climatic facilitation and limitation of area burned, one that is applicable not only to ecoregions but also to watersheds or other landscapes in which water relations can be linked to the contagious properties of fire.

5.2 Methods: Identifying Relationships Between Water Balance and Area Burned

We extend the work of Littell et al. (2009a), which established temperature and precipitation relationships for area burned in ecoprovinces in the western U.S., in two ways. First, we investigate PET, AET, and DEF deficit as potential climate predictors instead of temperature and precipitation. Second, we partition the climatic control of area burned by ecosystem vegetation more finely by using ecosections (Bailey 1995) in the Pacific Northwest and northern Rocky Mountains (Table 5.1). Ecosections are reasonably homogeneous geographic areas that have similar biophysical and orographic properties (Bailey 1995), and also have coarsely similar vegetation across groups of ecosections within an ecoprovince. We focus on the climatic limitation and facilitation of fire in each ecosection within the U.S. Columbia River Basin (CRB), including coastal areas that drain to the Pacific Ocean (Fig. 5.1, Table 5.2). These ecosections range in vegetation type from cool, moist, temperate maritime forests along the Washington and Oregon coasts to rainshadow desert in the interior Columbia Basin, with intermediate montane forest, subalpine forest, and shrub-dominated ecosystems.

Table 5.1 Bailey (1994) Ecosections used in this study

Ecoprovince	Ecosection	Ecosection code
Cascade Mixed Forest—Coniferous Forest—Alpine Meadow	Eastern cascades	M242C
	Oregon And Washington Coast Ranges	M242A
	Western Cascades	M242B
Great Plains—Palouse Dry Steppe	Palouse Prairie	331A
Intermountain Semidesert	Columbia Basin	342I
	High Lava Plains	342H
	Northwestern Basin And Range	342B
	Owyhee Uplands	342C
	Snake River Basalts	342D
Middle Rocky Mountain Steppe—Coniferous Forest—Alpine Meadow	Beaverhead Mountains	M332E
	Bitterroot Valley	M332B
	Blue Mountains	M332G
	Challis Volcanics	M332F
	Idaho Batholith	M332A
Northern Rocky Mountain Forest—Steppe—Coniferous Forest—Alpine Meadow	Bitterroot Mountains	M333D
	Flathead Valley	M333B
	Northern Rockies	M333C
	Okanogan Highlands	M333A
Pacific Lowland Mixed Forest	Willamette Valley and Puget Trough	242A

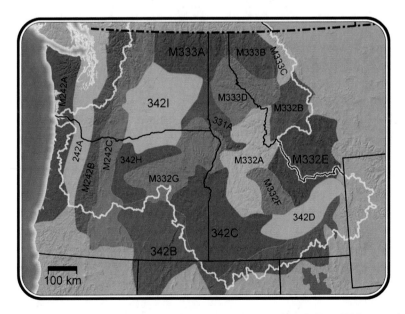

Fig. 5.1   U.S. Columbia River Basin (*outlined in yellow*) and associated Bailey (1995) ecosections. Ecosection names and climate variables are summarized in Table 5.1

Table 5.2 1980–2006 mean climate by ecosection code[1] (driving data and output from VIC model)

Ecosection code	Ann T_max (°C)	Ann T_min (°C)	Ann PPT (mm)	Oct–Mar PPT (mm)	JJA[2] PPT (mm)	%PPT JJA (mm)	Ann PET (mm)	PET JJA (mm)	DEF JJA (mm)	ANN PPT–JJA PET (mm)
M242C	13.0	−1.1	885	685	69	0.10	672	415	200	471
M242A	14.7	4.5	2339	1845	141	0.08	669	355	0	1983
M242B	12.8	0.6	2139	1629	157	0.10	667	374	26	1764
331A	14.8	1.4	588	354	92	0.26	647	393	157	195
342I	16.5	2.9	290	197	36	0.18	665	410	340	−119
342H	15.6	0.5	356	224	51	0.23	543	354	249	2
342B	14.8	−1.0	347	201	55	0.27	506	345	215	2
342C	16.3	0.3	333	207	46	0.22	527	358	217	−25
342D	14.4	−1.4	315	170	62	0.37	591	396	272	−81
M332E	10.9	−4.5	475	220	123	0.56	548	376	182	99
M332B	11.0	−3.5	636	349	132	0.38	604	386	185	250
M332G	14.0	−1.0	587	379	80	0.21	622	403	187	183
M332F	10.6	−5.3	597	349	109	0.31	530	371	124	226
M332A	11.5	−3.9	901	572	129	0.23	624	401	67	501
M333D	12.8	−1.5	1001	655	138	0.21	698	421	98	580
M333B	11.6	−2.8	759	450	143	0.32	665	407	182	352
M333C	9.3	−4.5	1214	747	208	0.28	558	360	61	853
M333A	13.2	−0.1	628	393	105	0.27	669	419	228	209
242A	15.8	4.4	1445	1089	115	0.11	703	396	57	1049

[1] See Table 5.1 for code definitions

[2] JJA = June, July, August; Ann = Oct–Sep hydrologic year

We collected records of the annual area burned for management units in the CRB study region from online interagency fire datasets (FAMWEB, National Fire and Aviation Management 2007) for federal lands managed by the U.S. Forest Service, National Park Service (NPS), Bureau of Land Management (BLM), and Bureau of Indian Affairs (BIA). Area-burned data were available from 1970, but records between 1970 and 1980 were inconsistent within or between agencies (indicated by years recorded as zero when fires were known to have occurred in the management units), so we restricted our analyses to the period 1980–2006, which was the last complete year available at the time of analysis. We removed duplicate fire records (cross-agency listings of fire events) and obvious errors (e.g., fires reported in the wrong state for a known management unit), and compiled a database of the area burned in each year in each agency unit (National Forest ranger districts, NPS parks, BLM districts, and BIA reservations). We then aggregated the individual unit time series into the ecosections that contain them. This yielded a 27-year time series for each of the 19 CRB ecosections.

We developed climate variables for each ecosection (Table 5.2) from gridded observed data and output from the Variability Infiltration Capacity (VIC) hydrologic model (Liang et al. 1994; Hamlet and Lettenmaier 2005; Elsner et al. 2010). The VIC driving data are gridded (1/16-degree grid, Elsner et al. 2010) observed climate data derived from National Climatic Data Center Cooperative Observer network daily station data as the primary sources for precipitation and temperature values. These data are adjusted by a method described by Hamlet and Lettenmaier (2005), which corrects for temporal inhomogeneities in the raw gridded data using a set of temporally consistent and quality-controlled index stations from the U.S. Historical Climatology Network. This approach eliminates spurious trends in the gridded historical data from inclusion of stations with records that are shorter than the length of the gridded data set. The gridded estimates are then adjusted for orographic effects using the PRISM climatology for 1971–2001 (Daly et al. 1994, 2002) following methods outlined in Maurer et al. (2002). The VIC model then uses these gridded outputs and additional parameter files for topography, vegetation, soil conditions, and other factors to calculate hydrologic variables such as snow-water equivalent, runoff, evapotranspiration, and streamflow. In this chapter, we use ecosection-averaged VIC monthly or seasonal temperature (T), precipitation (PPT), and several derived variables, including potential evapotranspiration (PET), actual evapotranspiration (AET), and water balance deficit (DEF, or PET-AET). VIC calculates PET with a Penman-Monteith equation with canopy resistance set to zero. AET is calculated with the same equation, but with canopy resistance as a function of minimum canopy resistance, soil moisture stress, and leaf area index (Liang et al. 1994). In a few models, we also used April 1 snow water equivalent (SWE). We expected statistical models of the interactions between temperature and precipitation should be less effective than the physically meaningful integration represented by water-balance deficit, and that both approaches are more proximate to fire than ENSO or PDO, which cause variation in regional climate.

5.2.1 Data Analysis

Using the complete array of monthly and annual predictor variables in regression would have had an unacceptably high probability of generating spurious relationships with only 27 years of data. Yet multivariate data reduction techniques, such as principal components analysis, would have confounded our ability to examine specific climatic variables, leaving only the aggregate ability to explain variance in the area-burned time series. To minimize the probability of spurious relationships and still identify the best climate predictor variables, we relied on an exploratory analysis using Pearson product-moment correlations between monthly T, PPT, PET, AET, and DEF variables and log-transformed ecosection area-burned time series. This helped determine if seasonal aggregations of the variables might better explain the variance in area burned than monthly means that would be correlated and introduce collinearity into multiple regression models. When several months had comparable sign and magnitude and significant correlations with area burned, we grouped them into seasonal variables. We also used these correlations as first-pass estimates of the best explanatory variables in multiple regression models of area burned as a function of climate.

We used seasonal or monthly climate variables as predictors in multiple linear regression models of area burned for each ecosection. We iteratively entered the most highly correlated seasonal or monthly variables (and in many cases, their interactions) into predictive models and used Akaike's Information Criterion (AIC, Akaike 1974) to compare nested models. When the AIC could not be reduced further by addition of significantly correlated variables, we considered the model final. We retained predictors only if $p(t) \leq 0.05$, unless subsequent interactions required the retention of predictors with $p(t) \geq 0.05$ for estimation of main effects.

To better understand the spatial arrangement of climatically driven deficit and its potential role in area burned, we used longer-term data (Littell et al. 2009a) on area burned to define years to include in composite maps of climatic variables for the 10% highest and 10% lowest fire years. Specifically, we calculated the total annual area burned in Washington, Oregon, Idaho, and Montana from 1916–2006 and used these values to rank years from lowest to highest. We then used VIC estimates of PET and AET to develop gridded maps of water balance deficit for the composite mean of the low and high fire years.

5.3 Results

The mean annual and percentage area burned by fire varied by orders of magnitude across the 19 ecosections (Table 5.3). Areas of large human population or relatively high agricultural or other human management had low median and mean areas burned (Puget Trough and Willamette Valley, Flathead Valley, Columbia Basin sections),

Table 5.3 Annual area burned statistics for 1980–2006 by ecosection code[1]

Ecosection code	Area (10⁶ ha)	Mean (ha)	Standard deviation (ha)	Median (ha)	Mean (%)	Standard deviation (%)	Median (%)
M242C	5.8	26909	35555	7630	0.5	0.6	0.1
M242A	4.1	302	486	127	0.0	0.0	0.0
M242B	4.0	938	2032	198	0.0	0.1	0.0
331A	1.7	1605	3463	348	0.1	0.2	0.0
342I	5.5	177	229	74	0.0	0.0	0.0
342H	2.1	6789	8080	3887	0.3	0.4	0.2
342B	11.5	27319	26211	17763	0.2	0.2	0.2
342C	7.4	71331	63089	60014	1.0	0.9	0.8
342D	2.7	41645	55962	15800	1.5	2.1	0.6
M332E	5.1	8474	33847	740	0.2	0.7	0.0
M332B	2.0	8989	25928	317	0.4	1.3	0.0
M332G	4.5	20954	27802	2459	0.5	0.6	0.1
M332F	1.4	3751	7392	616	0.3	0.5	0.0
M332A	4.3	35350	69318	10713	0.8	1.6	0.2
M333D	3.3	3926	11149	200	0.1	0.3	0.0
M333B	2.1	4000	11121	70	0.2	0.5	0.0
M333C	1.1	8405	26920	217	0.8	2.4	0.0
M333A	3.4	18949	23593	11045	0.6	0.7	0.3
242A	3.9	12	28	2	0.0	0.0	0.0

[1] See Table 5.1 for code definitions

whereas Intermountain basin and range and drier forest vegetation types had high median and mean areas burned (Owyhee Uplands, Northwestern Basin and Range, Snake River Basalts, and Okanogan Highlands). The standard deviation was roughly proportional to the mean for most of the drier ecosections, but increased to two, three, or four times the mean in the wetter forested ecosections.

Interannual area burned varied substantially within ecosections and five (M242C, 342C, 342D, M332G, M332A) of the 19 ecosections contributed 68% of the annual area burned in regional fire years (Table 5.3). The regional time series was not indicative of a significantly large trend because the interannual variability in area burned was so large. However, there was a positive increase of approximately 10,300 ha year^{-1} regionally averaged over the study period, although 2007 and 2008 data would probably decrease the observed trend (Fig. 5.2).

Correlation analyses indicated that water balance variables (PET and AET) were marginally more frequently correlated with area burned than temperature and precipitation (significant $r = \pm 0.323$ for n = 27, df = 25, t ≥ 1.708, $\alpha = 0.05$) (Fig. 5.3). Summer (JJA) precipitation (negative) and temperature (positive) were correlated with annual area burned in most ecosections, and summer (JJA) correlations with PET were consistently positive whereas late-summer (JAS) correlations with AET

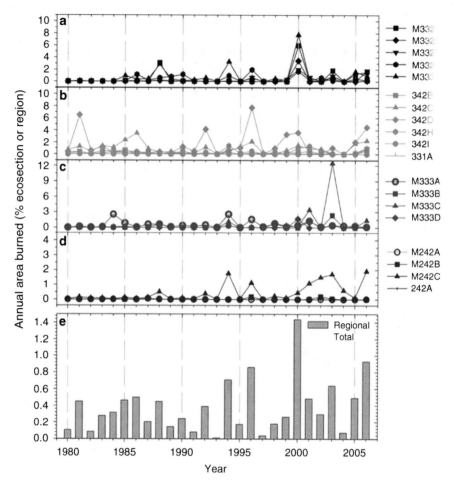

Fig. 5.2 Time series of percentage area burned for ecosections (**a–d**) in this study and total for the Columbia River Basin (**e**). See Table 5.1 for ecosection codes

were consistently negative. In some ecosections, fire was associated negatively with year prior growing season maximum temperature and positively with year prior growing season precipitation and AET. Warm Januarys were positively correlated with area burned as well (Fig. 5.3).

We developed regression models for 18 of the 19 ecosections; the Puget Trough/ Willamette Valley had no statistically acceptable climate predictors. Regressions explained 25–78% of the area burned in each ecosection (Table 5.4), with an average of 54%.

The first predictor in eight of 19 models was summer (months JA, JJA, or JJAS) deficit or PET, whereas for the Northern Rocky Mountains and Okanogan Highlands

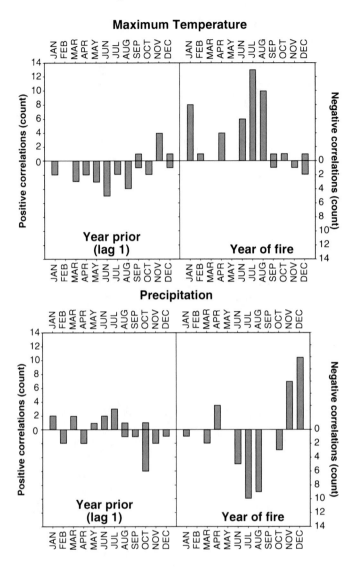

Fig. 5.3 Counts of significant ($\alpha = 0.05$) correlations between area burned and monthly climate variables in year prior to and year of fire. Counts pooled across 19 ecosections

ecosections JA maximum temperature (Tmax) was the first predictor. All ecosections in the Intermountain Semidesert ecoprovince had antecedent climate conditions that were more important than year-of-fire climate, with subsequent model terms associated with warmer fire seasons ("L1" variables—Table 5.4). Many variables in these models for arid and sparsely vegetated ecosections were consistent

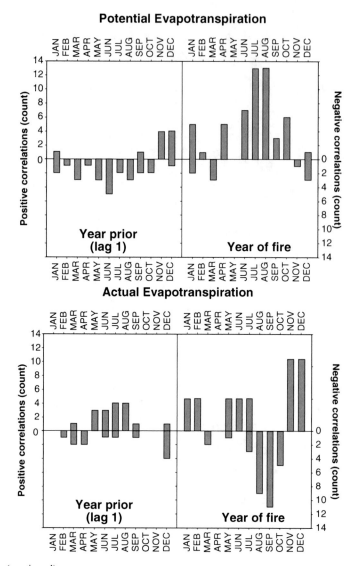

Fig. 5.3 (continued)

with vegetation production (e.g., less PET in prior autumn associated with increased fire activity in the Owyhee Uplands ecosection or lag 1 April SWE in the High Lava Plains ecosection), but many were also consistent with vegetation drying (e.g., a negative relationship with JA PPT in the Snake River Basalt section).

PET was a prominent predictor variable in several of the forest ecosection models, so we investigated the relationship between summer PET and area burned in the

Table 5.4 Regression models for Columbia River Basin ecosections. "L1" refers to a variable that represents previous-year conditions. "±" refer to signs of coefficients. Colons between variables indicate an interaction term. Consecutive months use first letter abbreviations (e.g., JJA = June, July, August total (PET, AET, DEF, PPT) or average (T); single months use three letter month abbreviations

Ecosection code	Best model	R^2
M242C	JJAS.DEF + L1.MAM.DEF	0.527
M242A	Oct,DEF	0.253
M242B	-Dec.DEF + JA.Tmax	0.405
331A	JJAS.DEF + L1.MAM.DEF + JJAS.DEF:L1.MAM.DEF	0.584
M332E	JJA.PET	0.591
M332B	JJAS.PET	0.627
M332G	JJAS.PET	0.497
M332F	JJA.PPT	0.552
M332A	JJA.PET + AS.AET	0.629
342I	L1.Dec.DEF + Jan.PET	0.511
342H	L1.JJ.PET + L1.Apr.SWE + L1.JJ.PET:L1.Apr.SWE	0.468
342B	L1.JFM.PET + L1.SO.PET + L1.JJA.AET + JJ.PET + -L1.JFM. PET:L1.SO.PET	0.734
342C	-L1.SO.PET + JJ.PET	0.305
342D	JF.Tmax + -JA.PPT + -L1.MJ.PET	0.781
M333D	JA.PET	0.480
M333B	AS.DEF	0.584
M333C	JA.Tmax	0.612
M333A	JA.Tmax + L1.JJAS.DEF + L1.Mar.PPT	0.598
242A	No model	

[1] See Table 5.1 for code definitions

forested ecosections. JJAS PET explained greater than 33% of the variance in 8 of 12 forested ecosections (Fig. 5.4), and these were generally interior mountain forests. The relationship between JJA deficit and area burned across all ecosections (i.e., including the mostly nonforested ecoprovince 342, Intermountain semi-desert) suggests either a threshold or unimodal relationship–more arid ecosystems appear to have a poor relationship with JJA deficit, but so do the most maritime ecosystems. The relationship between log-transformed area burned and a deficit gradient appears nonlinear (Fig. 5.5).

The 10% high fire years occurred before 1932 or after 1995 and include 1917, 1919, 1926, 1929, 1931, 1996, 200, 2003, and 2006. The 10% low fire years were primarily in the middle of the twentieth century and include 1937, 1950, 1953, 1956, 1964, 1965, 1975, 1978 and 1993. Figure 5.6 shows the composite mean June–August deficit for the high and low fire years. Averaged over the entire study area, low fire years have JJA deficit only about 32 mm (ecosection means range from about −10 to −52 mm) less than normal, while high fire years have JJA deficit about 36 mm (ecosection means range from about +8 to +72 mm) greater than normal. These averages do not adequately describe the spatial variability in the anomalies, however, with significant variability within sections (Figs. 5.6 and 5.7).

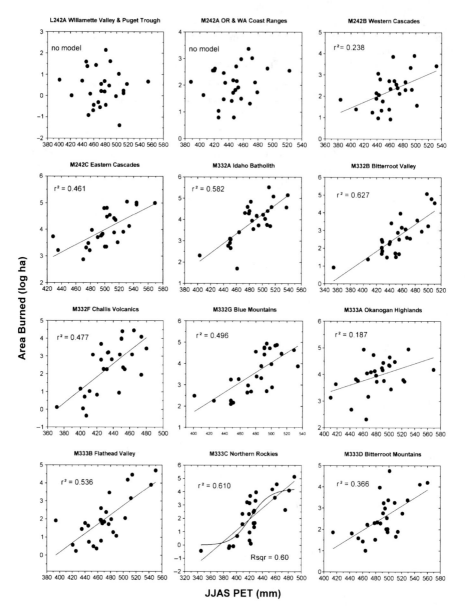

Fig. 5.4 Simple linear regressions of log(area burned) on seasonal (JJAS) PET for forested or partially forested. Fit in M333C (*bottom center*) is similar for linear and non-linear regressions. 242A and M242A did not yield significant JJAS PET regressions

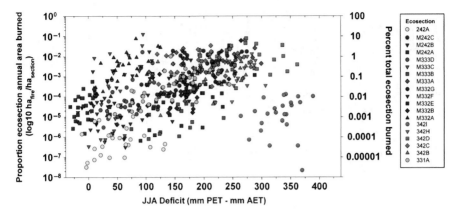

Fig. 5.5 Proportion ecosection area burned (logarithmically scaled) plotted against summer (JJA) deficit for Columbia River Basin ecosections

5.4 Discussion

The area burned by fire in ecosections of the Pacific Northwest and the northern Rocky Mountains is significantly related to climate in all but the most maritime (242A) ecosections. More than half the variability in area burned can be explained by just a few climate variables. This implies that despite fire suppression, climate was still a strong determinant of area burned in the late 20th century. The correlations (Fig. 5.3) and regression models (Table 5.4) point to similar water-deficit constraints to those identified in Littell et al. (2009a) and those implied by summer precipitation and temperature relationships in the models used by McKenzie et al. (2004) to project future area burned at the state level. The relationship between climate and area burned varies with ecosystem vegetation, and a gradient of sensitivities exists between (1) those fire regimes that are driven primarily by water deficit during the fire season (characterized by high temperature, low precipitation, high potential evapotranspiration and low actual evapotranspiration) and (2) those fire regimes that are driven by a combination of climate conditions, some facilitating vegetation growth (such as higher snow water equivalent) and others drying vegetation (such as high temperature or deficit). In other words, the biophysical mechanisms relating climate and fire are not necessarily either (1) or (2), but may fall between them according to vegetation sensitivity to changes in the primary climatically limiting and facilitating factors.

The observed regional patterns of area burned reflect a continuous gradient in continentality of precipitation and temperature (Table 5.2, Figs. 5.5 and 5.7). As conditions become more continental, area burned increases with increasing water-balance deficit until vegetation becomes sufficiently water-limited that fuel connectivity is decreased (e.g., transition from forest to grassland or shrubland) and area burned begins to decline with further increases in deficit. In contrast, we observed that area burned increased with increasing prior winter precipitation in water-limited vegetation. It is therefore possible that increasing temperature and thus PET could decrease area burned in the future in water-limited systems if no increase in winter

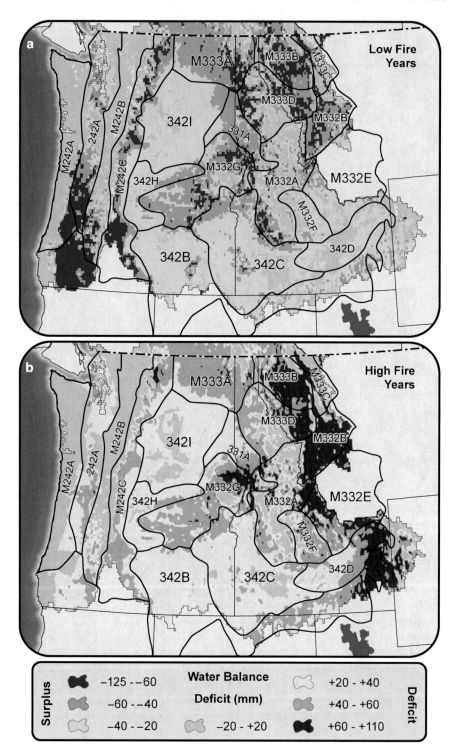

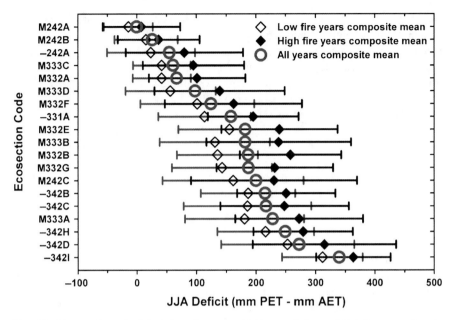

Fig. 5.7 Mean and standard deviation of summer (JJA) water balance deficit for ecosections in the Columbia River Basin. Solid black diamonds and bars (*right*) indicate statistics calculated across VIC cells for the high fire years composite, while open diamonds and grey bars (*left*) indicate statistics for the low fire years composite. The light gray circles indicate means across all years and all cells

precipitation occurred (Littell et al. 2009b). These patterns of fire activity and their climatic correlates parallel vegetation responses along water-balance gradients described by Stephenson (1990) and Neilson (1995), and have implications for modeling biome and finer scale vegetation (e.g., Neilson 1995).

Potential evapotranspiration, actual evapotranspiration, and deficit were usually better predictors in regression models of area burned than temperature or precipitation and their interactions. These variables better capture the loss of water from vegetation and fuels, as well as the storage and release of water from soil and snow. As noted above, the importance of temperature and precipitation on fuel condition and amount is in their control of moisture deficit. Different combinations of temperature and precipitation may produce equal deficit, so it makes sense that these variables have less predictive power. Alternatively, water-balance variables provide a more direct indication of moisture deficit and thus a more direct measure of landscape preconditioning with respect to large fires.

Fig. 5.6 Composite water balance deficit across all ecosections for the 10% (*n*=9) low fire years (*top*) and 10% (*n*=9) high fire years (*bottom*), 1916–2006. Darker shading corresponds to lower water balance deficit (surplus, more water availability) in the *top panel* and to higher water balance deficit (less water availability) in the *bottom panel*

Temperature and precipitation were still strongly correlated with area burned, but using them to develop predictive models of fire at multiple scales is problematic. For example, PET is directly modified by topography because solar radiation is either reduced or increased depending on landscape positions, so spatial variability can be translated directly into distance in climate space represented by PET, whereas estimating the topographic influence on temperature and precipitation independently is more difficult. Deficit can represent either high water demand or low ability to meet that demand or both, but the consequences for fire area burned are clear. Higher deficits are associated with increasing area burned – to a point, particularly in forest ecosystems. Once a certain level of average deficit is reached, however, forests are no longer the dominant vegetation type and the relationship between area burned and climate shifts to vegetation facilitation rather than vegetation drying. In this study, this transition appears to occur somewhere around 200–220 mm, or between the Eastern Cascades and the Okanogan Highlands (Fig. 5.5).

These relationships point to mechanisms that can be empirically tested and for which data exist to conduct modeling experiments. The question remains whether landscapes have different fuel availability or fuel continuity in years when deficit is higher regionally. That deficit (which is implicitly related to soil, fuel, and foliar moisture) was a good predictor of area burned strongly suggests that landscape variability in fuel availability limits fires in some years. Furthermore, it suggests that during extreme fire years climate causes a substantial change in the spatial pattern of water balance and fuel availability that produces a large increase in area burned. If landscape spatial pattern of fuels does not covary with deficit, however, then we must look to weather during fire events rather than hydrologic controls over fuel availability. But the role of climate in preconditioning larger areas to burn is quite clear.

5.4.1   *Linking Water Balance and Fire at Finer Scales*

Our models suggest that hydro-climate exerts strong controls on area burned in some ecoregions. But are spatial and temporal patterns of water balance useful for modeling fire at finer scales? Results of other modeling experiments suggest that they are. For example, Miller and Urban (1999) used a fine-scale forest-dynamics model that incorporates the effects of water balance on fuel load and moisture to model fire across a 2,000-m elevational gradient. Patterns of water balance had a reduced impact on fire where fuel was limited, but a stronger effect at elevations and slope facets with unlimited fuels.

The physical controls that shape patterns of water balance, flammability, and fuel buildup at fine scales ($10–10^2$ m) suggest that patterns of water balance may be related to the size of an individual fire event or its burn pattern. Climate and topography interact to produce landscape variability in water balance and fuel availability. Temperature-induced elevation gradients are positively related to snow storage, snowmelt, and AET, while negatively related to snowmelt, PET, and DEF.

Superimposed on elevation gradients are patterns of PET caused by the effect of slope angle and aspect on heat load. In water-limited environments, deficits and fuel flammability are likely to be less extreme at high elevations and on north-facing (northern hemisphere) slopes. In moderate climatic years, these areas of greater moisture may constitute a barrier to fire or at least a zone of slower spread. In years of extreme drought, fine-scale patterns of water balance are likely to be quite different. In those years, precipitation may be so reduced that energy received at high elevations and northern slopes may be sufficient to transfer most of the water stored in plants, fuels, and soils to the atmosphere. Alternatively, during abnormally warm years PET increases at all locations, which increases the loss of water stored in snow, soil, plants, and dead fuel. In either case, topographic "refugia" for moisture become smaller or less numerous and water deficits of higher magnitude occur on a larger proportion of the landscape. Drought therefore increases the average deficit for the entire ecoregion, while increasing either the number or size of patches of burnable fuel. The increase in the number of burnable patches means that a randomly placed ignition is more likely to ignite. It also means that a spreading fire is less likely to encounter barriers or impediments to fire spread caused by high fuel moisture contents. During extreme drought, corridors of fire spread become more numerous, larger, and more connected. This concept is consistent with our statistical models of fire extent for energy-limited ecoregions, but its operational scale is much finer.

The ability of water-balance spatial patterns to create barriers to or corridors for fire spread is difficult to infer from modern, historic, and paleo-era fire records. Patterns of fire spread are confounded by wind, slope, fuel type and load in addition to the preconditioned hydro-climatic state of the landscape. Some studies have shown that topographic locations associated with lower deficits (high elevations, north slopes) experience fire less frequently (Camp et al. 1997; Beaty and Taylor 2001), and that vegetation type often covaries with low-deficit topography. Barriers caused by fuel moisture are therefore difficult to discern from barriers due to fuel type. Assessing the effect of water-balance pattern on fire events may require alternative research approaches that complement the more traditional assessments that use geographic patterns of fire scars.

Hydrologic models provide one way to examine the effect of climate on fine-scale water balance patterns, and in turn, on area burned. The distributed hydrology soil and vegetation model (DHSVM—Wigmosta et al. 1994) links water and energy cycles at fine spatial scales (30–150 m). By applying the model to an entire ecoregion, annual metrics of water balance pattern could be derived. One such metric would be the value of deficit at the 40.275 percentile of a cumulative probability distribution function of all pixels from a map of fire season water deficit. This percentile is the value of critical percolation threshold (pc) of a square 2D matrix with eight neighbors. The pc indicates the proportion of the pixels that must be burnable to ensure propagation across the full length of the landscape matrix. If we assume that deficits are a sign of how well a pixel acts as a barrier or corridor to fire spread, higher deficits at the 40.275 percentile indicate more connectivity with respect to fire spread. The time series of pc scores could then be compared to the annual time series of area burned. The strength of the correlation between these two series would provide a more explicit indication of how of water-balance patterns effect connectivity of the landscape with respect to fire.

Landscape fire succession models (LFSMs) may provide another way to examine the effect of water balance on fire activity at scales much finer than ecosections. Several researchers have developed and used LFSM experiments to explore the effect of climatic variation on vegetation and fire (Keane et al. 2004). Incorporation of realistic fire spread in such models is hampered by the large data requirements and long computation times implicit in simulation of large landscape for long time periods (e.g. Keane et al. 1999). As an alternative, some modelers have turned to water-balance calculations as a surrogate. For example, the LandClim model (Schumacher et al. 2006) uses water balance variables to calculate fuel availability at very fine scales (25 m). Let us assume, following our work here, that local water deficits are indicative of local fuel availability during the fire season. High deficit (DEF) increases the probability of fire ignition and spread, whereas low deficit impedes fire spread by isolating ignited fires or increasing the distance and travel time between burnable fuels. Instead of calibrating fire regime parameters to reproduce observed fire size distributions, as is current practice in landscape fire models, one could let simulated fire size and severity emerge from a stochastic implementation of the deficit/fire mechanism. Ensemble simulations, under an equilibrium climate (e.g., representing our 1980–2006 database) should reproduce the aggregate statistical properties of fire sizes. This would illustrate the sensitivity of fire size to the spatial variability of deficit. The sensitivity of simulations for different ecoregions should provide an indication as to whether fire there is limited by fuel moisture or fuel availability (Littell et al. 2009a). By applying this same approach to landscapes before and after a century of vegetation and fuels succession, we may be able to assess whether a landscape has shifted from energy to moisture limitation after an extended period of climate change.

5.4.2   Implications for Future Landscapes and Modeling

Understanding the relationships between climate, fire, and fuels at multiple scales would aid the development of fire models used to study climate change impacts and develop policy. At coarse scales, the relation of area burned to climate can provide state and regional agency managers with general indications of how and to what degree they may need to change the management of land, water, and air resources. At fine scales, models incorporating hydro-climatic controls on fire occurence, behavior, spread, and severity will enable study of the successional response of forests to climate, and the cascading effects from that response (e.g., streamflow, wildlife habitat, timber production, etc.). These models could be used as gaming tools to help management increase the resilience of landscapes to expected changes in fire regime (Chap. 3).

Our results show the sensitivity of area burned to variability in ecohydrological variation. Shifts in the climate of the twenty-first century are associated with increasing area burned in many of the ecosections analyzed here, and the effects on vegetation due to changes in fire size and potentially severity would likely be profound (Littell et al. 2009b). The same changes in climate also increase the potential for increasing spatial percolation of low fuel moisture, which in turn can exceed thresholds of connectivity on landscapes. That water balance variables are closely related to fire, vegetation, and fuel processes suggests some interesting possibilities for integrated landscape models that use a gradient approach to water balance. Incorporating water relations directly or semi-directly into landscape simulation models may obviate the need to specify fire regime parameters, as many of the current generation of models do (Keane et al. 2004), and instead allow them to become emergent properties of landscapes under future climate.

References

Akaike, H. 1974. A new look at the statistical model identification. *IEEE Transactions on Automation and Control* 19: 716–723.

Bailey, R.G. 1994. *Ecoregions and subregions of the United States* (map). Revised. Washington: U.S. Geological Survey.

Bailey, R.G. 1995. *Description of the ecoregions of the United States*. Washington: U.S. Forest Service, Miscellaneous Publication 1391 (revised).

Beaty, R.M., and A.H. Taylor. 2001. Spatial and temporal variation of fire regimes in a mixed conifer forest landscape, Southern Cascades, California, USA. *Journal of Biogeography* 28: 955–966.

Brown, P.M., E.K. Heyerdahl, S.G. Kitchen, and M.H. Weber. 2008. Climate effects on historical fires (1630–1900) in Utah. *International Journal of Wildland Fire* 17: 28–39.

Camp, A., C. Oliver, P. Hessburg, and R. Everett. 1997. Predicting late-successional fire refugia pre-dating European settlement in the Wenatchee Mountains. *Forest Ecology and Management* 95: 63–77.

Clark, J.S. 1990. Fire and climate change during the last 750 yr in northwestern Minnesota. *Ecological Monographs* 60: 135–159.

Daly, C., R.P. Neilson, and D.L. Phillips. 1994. A statistical-topographic model for mapping climatological precipitation over mountainous terrain. *Journal of Applied Meteorology* 33: 140–158.

Daly, C., W.P. Gibson, G. Taylor, G.L. Johnson, and P. Pasteris. 2002. A knowledge-based approach to the statistical mapping of climate. *Climate Research* 22: 99–113.

Elsner, M.M., L. Cuo, N. Voisin, J. Deem, A.F. Hamlet, J.A. Vano, K.E.B. Mickelson, S.Y. Lee, and D.P. Lettenmaier. 2010. Implications of 21st century climate change for the hydrology of Washington State. *Climatic Change* 102: 225–260, doi: 10.1007/s10584-010-9855-0.

Gedalof, Z., D.L. Peterson, and N.J. Mantua. 2005. Atmospheric, climatic and ecological controls on extreme wildfire years in the northwestern United States. *Ecological Applications* 15: 154–174.

Grissino-Mayer, H.D., and T.W. Swetnam. 2000. Century-scale climate forcing of fire regimes in the American Southwest. *The Holocene* 10: 207–214.

Hamlet, A.F., and D.P. Lettenmaier. 2005. Production of temporally consistent gridded precipitation and temperature fields for the continental U.S. *Journal of Hydrometeorology* 6: 330–336.

Heyerdahl, E.K., D. McKenzie, L.D. Daniels, A.E. Hessl, J.S. Littell, and N.J. Mantua. 2008a. Climate drivers of regionally synchronous fires in the inland Northwest (1651–1900). *International Journal of Wildland Fire* 17: 40–49.

Heyerdahl, E.K., P. Morgan, and J.P. Riser II. 2008b. Multi-season climate synchronized historical fires in dry forests (1650–1900), Northern Rockies, USA. *Ecology* 89: 705–716.

Johnson, E.A., and D.R. Wowchuk. 1993. Wildfires in the southern Canadian Rocky Mountains and their relationships to mid-tropospheric anomalies. *Canadian Journal of Forest Research.* 23: 1213–1222.

Keane, R.E., P. Morgan, and J.D. White. 1999. Temporal patterns of ecosystem processes on simulated landscapes in Glacier National Park, Montana, USA. *Landscape Ecology* 14: 311–329.

Keane, R.E., G.J. Cary, I.D. Davies, M.D. Flannigan, R.H. Gardner, S. Lavorel, J.M. Lenihan, C. Li, and T.S. Rupp. 2004. A classification of landscape fire succession models: Spatial simulations of fire and vegetation dynamics. *Ecological Modelling* 179: 3–27.

Liang, X., D.P. Lettenmaier, E.F. Wood, and S.J. Burges. 1994. A simple hydrologically based model of land-surface water and energy fluxes for general circulation models. *Journal of Geophysical Research-Atmospheres* 99: 14415–14428.

Littell, J.S., D. McKenzie, D.L. Peterson, and A.L. Westerling. 2009a. Climate and wildfire area burned in western U.S. ecoprovinces, 1916–2003. *Ecological Applications* 19: 1003–1021.

Littell, J.S., Oneil, E.E., McKenzie, D., Hicke,J.A., Lutz, J.A., Norheim,R.A., and M.M. Elsner. 2009b. Implications of climate change for the forests of Washington State. In *The Washington climate change impacts assessment: Evaluating Washington's future in a changing climate*, chapter 6. Manuscript in press.

Maurer, E.P., A.W. Wood, J.C. Adam, D.P. Lettenmaier, and B. Nijssen. 2002. A long-term hydrologically based dataset of land surface fluxes and states for the conterminous United States. *Journal of Climate* 15: 3237–3251.

McKenzie, D., Z. Gedalof, D.L. Peterson, and P.M. Mote. 2004. Climatic change, wildfire, and conservation. *Conservation Biology* 18: 890–902.

Miller, C., and D.L. Urban. 1999. A model of surface fire, climate, and forest pattern in the Sierra Nevada, California. *Ecological Modelling* 114: 113–135.

Milne, B.T., V.K. Gupta, and C. Restrepo. 2002. A scale-invariant coupling of plants, water, energy, and terrain. *EcoScience* 9: 191–199.

Nash, C.H., and E.A. Johnson. 1996. Synoptic climatology of lightning-caused forest fires in subalpine and boreal forests. *Canadian Journal of Forest Research* 26: 1859–1874.

National Fire and Aviation Management. 2007. Online. http://fam.nwcg.gov/fam-web/weatherfirecd/fire_files.htm. Accessed 20 Nov 2007.

Neilson, R.P. 1995. A model for predicting continental-scale vegetation distribution and water balance. *Ecological Applications* 5: 362–385.

Rollins, M.G., P. Morgan, and T.W. Swetnam. 2002. Landscape-scale controls over 20th century fire occurrence in two large Rocky Mountain (USA) wilderness areas. *Landscape Ecology* 17: 539–557.

Romme, W.H., and D.G. Despain. 1989. Historical perspective on the Yellowstone fires of 1988. *BioScience* 39: 695–699.

Rothermel, R.C. 1972. *A mathematical model for predicting fire spread in wildland fuels. Research Paper INT-115* Ogden: U.S. Forest Service.

Schumacher, S., B. Reineking, J. Sibold, and H. Bugmann. 2006. Modeling the impact of climate and vegetation on fire regimes in mountain landscapes. *Landscape Ecology* 21: 539–554.

Stephenson, N.L. 1990. Climatic control of vegetation distribution: the role of the water balance. *American Naturalist* 135: 649–670.

Swetnam, T.W., and J. Betancourt. 1990. Fire-Southern Oscillation relationships in the southwestern United States. *Science* 249: 1017–1020.

Swetnam, T.W., and J. Betancourt. 1998. Mesoscale disturbance and ecological response to decadal climate variability in the American Southwest. *Journal of Climate* 11: 3128–3147.

Westerling, A.L., A. Gershunov, T.J. Brown, D.R. Cayan, and M.D. Dettinger. 2003. Climate and wildfire in the western United States. *Bulletin of the American Meteorological Society* 84: 595–604.

Westerling, A.L., H.G. Hidalgo, D.R. Cayan, and T.W. Swetnam. 2006. Warming and earlier spring increase western U.S. forest wildfire activity. *Science* 313: 940–943.

Wigmosta, M.S., L.W. Vail, and D.P. Lettenmaier. 1994. A distributed hydrology-vegetation model for complex terrain. *Water Resources Research* 30: 1665–1679.

Chapter 6
Pyrogeography and Biogeochemical Resilience

Erica A.H. Smithwick

6.1 Introduction

The response of biogeochemical fluxes to perturbation has been a major focus of ecosystem studies for decades (Bormann and Likens 1979a; Vitousek and Melillo 1979; West et al. 1981). Perturbation is a fundamental component of conceptualizations of system resilience (Holling 1973; Carpenter et al. 2001; Folke et al. 2004), complex adaptive cycles (Norberg and Cumming 2008), self-organization (Rietkerk et al. 2004a), cross-scale interactions (Allen and Holling 2002; Peters et al. 2004; Allen 2007), and abrupt shifts in ecosystem behavior (Groffman et al. 2006; Sonderegger et al. 2009). To date, however, linkages between disturbance biogeochemistry and ecosystem resilience are lacking even though biogeochemical cycles underlie ecosystem function and energy flux (Giblin et al. 1991; Hedin et al. 1995; Schlesinger 1997) and are susceptible to disruption following disturbance (Vitousek and Reiners 1975; Woodmansee and Wallach 1981; Campbell et al. 2009).

I review how biogeochemical cycles are influenced by fire, introduce the concept of *biogeochemical resilience*, and discuss how this perspective can be useful for refining global concepts of pyrogeography (Krawchuk et al. 2009) at landscape scales relevant to fire management. I rely on the Greater Yellowstone Ecosystem as a case study for examining fire biogeochemistry in the context of a stand-replacing fire regime to ask whether this system is resilient to shifts in climate and fire frequency. Finally, I conclude with recommendations for including the concept of biogeochemical resilience into the study and management of fire across complex landscapes.

E.A.H. Smithwick (✉)
Department of Geography & Intercollege Graduate Program in Ecology,
The Pennsylvania State University, 318 Walker Building,
University Park, PA 16802-5011, USA
e-mail: smithwick@psu.edu

D. McKenzie et al. (eds.), *The Landscape Ecology of Fire*, Ecological Studies 213,   143
DOI 10.1007/978-94-007-0301-8_6, © Springer Science+Business Media B.V. 2011

6.2 Fire Biogeochemistry

Biogeochemistry refers to the flow of nutrients or elements through an ecosystem. It focuses on agents of elemental transformations (e.g., organisms, including microbiota), elemental pool sizes, system input and output flows, and ecological drivers that modify fluxes and nutrient dynamics. By connecting the biological, geological, and chemical components of ecosystems, it addresses the accumulation and distribution of matter in the form of vegetation and the subsequent cycling of this matter through rocks, soil, and water. The agents of biogeochemical fluxes operate at multiple scales, both temporally and spatially, from global to microbial. For example, microbes modify fluxes over spatial grains of millimeters or less, and temporal grains of seconds to days (Groffman et al. 2009). In contrast, many slow agents of biogeochemical flux operate at spatial scales of meters to tens of meters and temporal grains of years to decades. For example, old-growth trees in the Pacific Northwest serve as storage for carbon stocks for hundreds of years in the absence of disturbance (Smithwick et al. 2002). At the extreme, volcanic eruptions influence nutrient cycles on centennial to millennial scales (Huebert et al. 1999; Mather et al. 2004; Oppenheimer et al. 2005).

Fire modifies nutrient availability (Wan et al. 2001; Certini 2005), microbial community composition (Hart et al. 2005b; Mabuhay et al. 2006; Hamman et al. 2007), carbon storage (Kurz and Apps 1999; Bond-Lamberty et al. 2007; Dore et al. 2008), and productivity (Turner et al. 2004; Kurz et al. 2008). Until recently, however, understanding of post-fire biogeochemistry was based on prescribed fire or slash burns, with relatively little information from natural or severe fires (Smithwick et al. 2005). From these and other studies, we know that a fire event directly changes the distribution and mass of nutrient elements in ecosystems through pyrolysis (thermal decomposition of organic matter by fire), volatilization (gaseous loss through combustion), and ash deposition. Specifically, nutrient cycling is modified by shifts in the abiotic template, e.g., shifts in temperature (Pietikainen et al. 2000; Choromanska and DeLuca 2002), moisture (O'Neill et al. 2006), pH, and charcoal deposition (DeLuca et al. 2006). Modifications of biotic substrate quality and quantity, microbial pool size, and stoichiometry can also affect post-fire nutrient cycling (Hart et al. 2005b; Grady and Hart 2006). Nutrients are also modified by fire through translocation, leaching, plant uptake, shifts in plant community composition including N-fixers, and mineral soil interactions dependent on soil exchange capacity and base saturation (Woodmansee and Wallach 1981; Certini 2005; Hart et al. 2005a; Smithwick et al. 2005).

Wildfire can also modify hydrologic fluxes by modifying soil structure (e.g., repellecy, hydrophobicity), infiltration rates (including canopy interception), overland flow, and post-fire peak flow rates (DeBano 2000; Moody et al. 2009; de Blas et al. 2010). These hydrologic changes can interact with biogeochemistry by modifying leaching, lateral nutrient flows, and erosion rates (Minshall et al. 1997; Minshall et al. 2001). The hydrologic and geomorphic changes to soils following wildfire are complex and often indirect (Shakesby and Doerr 2006). For example, riparian biogeochemistry is affected by fire through shifts in vegetation patterns between upper and lower slope positions that shift the spatial location of hotspots of nutrient accumulation and loss (Jacobs et al. 2007).

The effects of fire on biogeochemical fluxes and pools depends on the initial state of the system prior to the fire, the characteristics of the fire event itself (e.g., intensity), and the post-fire ecosystem response, including changes to soils and vegetation (Fig. 6.1). Shifts in physical, chemical, and biotic factors before, during, and after the fire event interact to affect post-fire biogeochemistry. For example, recognizing pre-fire spatial variation in hydrology (Littell and Gwozdz, Chap. 5), soils, topography, climate, fuels, and land use can inform understanding of fire patterns (Parshall and Foster 2002; Cary et al. 2006; Poage et al. 2009) and subsequent shifts in biogeochemistry. Post-fire recovery is dependent on existing vegetation adaptations to fire (Bond et al. 2004; Buhk and Hensen 2006). These post-fire productivity rates affect biogeochemical fluxes by affecting nutrient uptake and loss (Vitousek and Reiners 1975). Less well recognized are the direct physio-chemical changes in soil structure and properties during the fire event and the immediate (<1 year) post-fire period (Andreu et al. 1996; Neary et al. 1999). Biogeochemical patterns will be determined by dynamics and patterns during each phase, the consequences of which are likely to have complex feedbacks on ecosystem function (and potentially, future fire events through modification of fuel structure and abundance).

Reflecting the multi-scale nature of biogeochemical fluxes, post-fire biogeochemistry has been studied at local (Binkley et al. 1992; Leduc and Rothstein 2007),

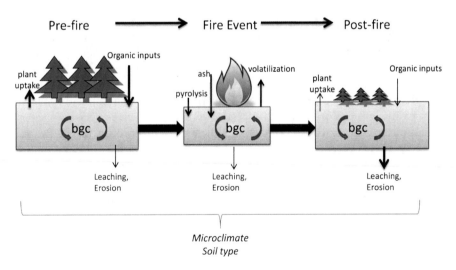

Fig. 6.1 Biogeochemical (bgc) flows before, during, and after a stand-replacing fire event. The magnitude of biogeochemical flows (thickness of arrows) differs depending on vegetation uptake and other organic inputs, soil microbial activity (mineralization/immobilization), fire intensity, soil type, and microclimatic conditions

regional (Boerner 1982; Bond-Lamberty et al. 2006), and global (Schultz et al. 2008; Randerson et al. 2009) scales. At local scales, fire influences biogeochemistry by removing vegetation, depositing ash, elevating soil temperatures, modifying microclimatic conditions, and potentially shifting microbial community composition (Raison 1979). At regional scales, fire modifies biogeochemical patterns by shifting the age-class mosaic and influencing landscape carbon flux (Kurz and Apps 1999; Euskirchen et al. 2002). At global scales, fire modifies global biogeochemistry by contributing emissions to the atmosphere (Auclair and Carter 1993; Lehsten et al. 2009), modifying surface albedo (Randerson et al. 2006), and influencing the distribution of vegetation types (Bachelet et al. 2001; Bowman 2005).

In practice, studies tend to be focused at a single scale (e.g., microbial or tree) and rarely are multiple scales compared. Single-scale studies provide little synthetic information that can be used to forecast future dynamics or extrapolate to broader extents because drivers are complex and interact across scales (Allen 2007; Falk et al. 2007). For example, understanding the response of soil biogeochemical processes to the perturbation of fire is difficult because an important proximate cause of transformations, the microbial community, is influenced by soil conditions that reflect centuries to millennia of vegetation-soil feedbacks. As such, the influence of a single fire event on belowground dynamics reflects the event itself as well as historical legacies of past fire events and dynamics (Fraterrigo et al. 2006).

Further hindering landscape perspectives on fire biogeochemistry are the limited number of studies that link biogeochemical responses to explicit variations in fire regime characteristics such as intensity, frequency, and seasonality. This limitation is exacerbated by the paucity of studies of fire biogeochemistry across the full complement of fire-prone or fire-dependent ecosystems. Fire regimes vary geographically with spatial patterns in topography, climate, and historical land management (Foster et al. 1998; Schoennagel et al. 2004; Mermoz et al. 2005). Where fire intervals are long, studies of fire biogeochemistry necessarily use forest chronosequences (Bond-Lamberty et al. 2006; Yermakov and Rothstein 2006; Leduc and Rothstein 2007; Smithwick et al. 2009a), which assume implicitly that fire seasonality and severity are relatively stable across time. Some studies of fire biogeochemistry have used experimental gradients (Monleon et al. 1997; Lynham et al. 1998; Harden et al. 2004; Mills and Fey 2005) where fire behavior and frequency are manipulated, but the relevance of such controlled experiments to the range of natural fire conditions on complex landscapes is unclear. Without a geographically explicit template for understanding the spatial and temporal variation in fire regimes, linkages between fire patterns and biogeochemistry are likely to be poorly characterized at broader scales of space and time and difficult to extrapolate to other places and times.

Climate change is likely to lead to novel disturbance regimes (Moritz and Stephens 2008; Power et al. 2008; Krawchuk et al. 2009) and novel vegetation distributions (Williams and Jackson 2007; Iverson et al. 2008). As a result, biogeochemical fluxes could be modified directly through changes in fire severity and frequency, and indirectly through changes in vegetation type and abundance that may promote or inhibit fire or modify surface energy budgets. The relative balance of shifts in biogeochemistry due to direct (disturbance events) versus indirect (e.g., vegetation shifts in response to disturbance, shifts in albedo) factors is not known. Moreover, it is unclear whether changes in biogeochemistry associated with shifts in disturbance regimes, vegetation, and climate will mediate, constrain, or amplify ecological feedbacks to the climate system. Constraining uncertainties in biogeochemical fluxes, including indirect responses to disturbance, is a key priority for improving forecasts of ecosystem feedbacks to climate change (Rustad 2008; Campbell et al. 2009).

6.3 Pyrogeography

Pyrogeography is the study of the spatial differentiation of the causes and ecological consequences of fire, a concept that has been previously defined at continental and global scales (Bond et al. 2005; Krawchuk et al. 2009). Broadly, pyrogeography reflects the notion that fire regimes are spatially contingent, representing the confluence of abiotic and biotic drivers (including available fuel, weather, and ignition sources) that are spatially patterned and temporally variable. Several recent studies that have explored global fire patterns exemplify the importance of understanding spatial dimensions of fire dynamics for inferring ecological process. Bond et al. (2005) suggested that global patterns in vegetation are driven largely by the presence or absence of fire, especially in southern Africa. Krawchuk et al. (2009) used statistical models relating resource availability and climate to identify fire-prone and non-fire-prone regions of the globe and to forecast future fire patterns under climate change. They determined that future patterns of fire activity can be compensatory, suggesting that multiple spatially explicit drivers are needed to determine future patterns of fire activity. On a finer scale, Smithwick et al. (2009b) showed that ecosystem response to future climate is determined largely by the recovery trajectories initiated by past fire events, which are heterogeneous across broad landscapes (Turner et al. 1994). Thus, spatial patterns of fire dynamics may affect regional to global ecosystem feedbacks by shifting the spatial distribution of future vegetation types, future fire activity, and the magnitude of future carbon sequestration.

At broad scales, fire biogeochemistry is conditioned by global constraints of latitude (i.e., its influence on post-fire albedo or seasonality) and broad-scale drivers of ignition sources. At landscape scales, fire biogeochemistry is determined by spatially complex, proximate, and contingent drivers such as topography, weather, and fuel condition, but few landscape models that include fire are coupled with mechanistic biogeochemical subroutines (Scheller and Mladenoff 2007). Those that do (e.g., FIRE-BGC (Keane et al. 1996)) often sacrifice detail in biogeochemical processes to preserve model stability at broader levels of model organization. As a result, the study of fire biogeochemistry continues to be a research frontier in landscape and regional modeling.

The study of how disturbances modify ecosystem function across landscapes is not new. Indeed, the study of variation in fire patterns and effects has been at the forefront of landscape ecology for decades. For example, large infrequent disturbances are known to imprint legacies on landscape structure and function across broad spatial and temporal scales, indicating the importance of perturbation for understanding landscape equilibrium (Foster and Boose 2001; Turner and Dale 2001). Similarly, disturbances such as fire have long been known to both respond to and create landscape pattern (Romme 1982; Yang et al. 2008). Partially as a result of these lessons, heterogeneity of fire regimes is now at the forefront of new approaches for fire management. In African savannas, for example, there is increasing recognition that regular burning cycles fail to maintain desired ecological responses to fire, and often lead to negative effects such as the loss of indigenous species (Du Toit et al. 2003; Rogers 2003). Landscape perspectives on fire heterogeneity have been critical for advancing the field of fire ecology and provide an important lens for interpreting pyrogeography at global scales.

6.4 Biogeochemical Resilience

In addition to a continued focus on fire biogeochemistry and geographic patterns in fire regimes, unraveling complex interactions among fire patterns and processes and their potential for change requires a new framework that is able to incorporate nonlinearities, thresholds, and resilience of coupled biogeochemical systems (Peters et al. 2004; Falk et al. 2007; Peters et al. 2007). To address this need, here I introduce the concept of *biogeochemical resilience*. Conceptually, biogeochemical resilience focuses research on the biogeochemical drivers, stressors, and feedbacks that are useful for identifying shifts in system states in response to disturbance. Critical to the concept of biogeochemical resilience is recognition that biogeochemical drivers, stressors, and feedbacks are (1) spatially differentiated prior to the disturbance event, (2) sensitive to physio-chemical modifications during the disturbance event, and (3) coupled to vegetation recovery patterns and rates following the disturbance. Identifying and understanding shifts in system behavior resulting from altered disturbance regimes or coupled vegetation dynamics requires attention to the critical biogeochemical processes that underlie these transitions. As such, recognition of biogeochemical resilience is critical for identifying key feedbacks of the terrestrial biosphere to altered climate and disturbance regimes.

System resilience emerges from biogeochemical and ecological dynamics (Fig. 6.2). For example, ecological resilience may include consideration of species composition, trophic interactions, and intra-specific competition, all of which may be influenced and modified by disturbance such as fire (e.g., Zimmermann et al. 2010). Hypothetically, however, shifts in species composition may not necessarily change biogeochemical stocks or fluxes. A system that has crossed an ecological threshold to a new system state, defined by interactions among vegetation species, could still be considered biogeochemically resilient because pool sizes and flows

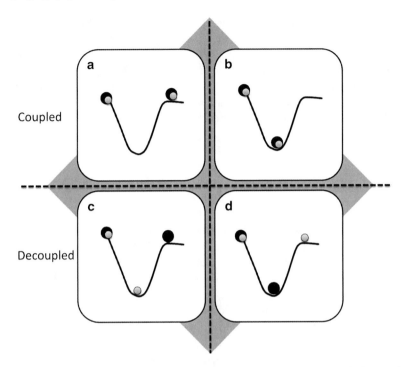

Fig. 6.2 Conceptual 'ball-cusp' (Gunderson 2000) representation of biogeochemical (gray circle) and ecosystem (black circle) system states in response to fire perturbation over short timescales. In all four figures (**a–d**), the biogeochemical system begins as a component of (embedded within) the ecosystem. In figures **a** and **b**, biogeochemical and overall ecosystem dynamics remain coupled following a fire event; in **a**, the system is resilient, retaining pre-fire interactions and feedbacks, whereas **b** represents a system that has shifted to a new system state with novel sets of biogeochemical and ecological feedbacks and interactions. In figures **c** and **d**, fire acts to decouple biogeochemical dynamics from other ecosystem processes. In **c**, biogeochemical function does not limit or determine vegetation recovery (e.g., serotiny governs recovery trajectories independent of soil nutrient availability), whereas in **d**, biogeochemical function recovers quickly to its pre-fire conditions, but vegetation is not resilient and moves to a new system state (e.g., nutrient pools and stocks are similar but species composition and interactions are altered)

are largely conserved. Conversely, fire may modify nutrient cycling over short time-frames without modifying species composition over the long term. If a system is nutrient-limited, post-fire system resilience would be a function of the degree to which biogeochemical and ecological processes are decoupled by the disturbance event.

It is critical to define the characteristics of a biogeochemical system for which resilience might be evaluated (Carpenter et al. 2001). Because biogeochemistry is defined by both pools and flows (Odum 1960, 1968), biogeochemical resilience must incorporate changes in both pool sizes and input and output rates. The balance of pool size and flows in ecosystems was described by Odum (1960) as the ecological analogue of Ohm's Law. In the biogeochemical context of an ecosystem, if inputs (e.g., rate of mineralization) increase, but outputs (rate of decomposition) do not, then pool size should increase. However, it is possible that rates of inputs and outputs could both change by the same magnitude and in the same direction such that pools sizes remain the same. If the size of elemental pools is the object of interest, a lack of change in this variable may indicate buffering in the system to perturbation, and potentially increased resistance. Recognition that both rates and pool sizes must be considered in understanding of biogeochemical fluxes is not new (Davidson et al. 1992) and must similarly be considered when inferring system response to perturbation.

Biogeochemical resilience must consider complex adaptive cycles that include perturbation and recovery explicitly. Holling and Gunderson (2002) introduced the idea of *panarchy*, describing how all systems (not just ecological) move from a phase of overexploitation to one of conservation, followed by release, and then re-organization. Therefore, it is critical that biogeochemical resilience be considered not only in response to the initial perturbation (post-disturbance studies), but also in the context of longer-term ecosystem recovery. Early ecosystem studies on disturbance and successional dynamics recognized this pulse-recovery dynamic (Vitousek and Reiners 1975; Gorham et al. 1979; Vitousek and Melillo 1979) but these early studies were limited by notions of mass balance and equilibria. Pulse-recovery schematics of biogeochemical processes in these studies typically described an immediate post-disturbance flux (e.g., nitrate leaching following clear cut), followed by a recovery period governed by rates of net ecosystem productivity (e.g., nitrate uptake in regrowing vegetation), and culminating in "old growth" conditions in which mass input and output fluxes were balanced. These dynamics are now recognized to be over-simplified and not likely to apply across all ecosystems or disturbance events (e.g., Turner et al. 2007a).

As spatial perspectives of ecosystem dynamics have broadened over the past several decades, the general pulse-recovery schematic has been complemented by new conceptual models to explore disturbance dynamics at landscape scales, e.g., space-time considerations of stability (Baker 1989; Turner et al. 1993), spatially explicit modeling of disturbance and succession across complex biotic and abiotic templates (He and Mladenoff 1999), and ecological memory (Peterson 2002). This has increased appreciation of the heterogeneity and variability of disturbance effects on ecological processes (Fraterrigo and Rusak 2008). Similarly, concepts of resilience, complex adaptive cycles, and panarchy that emerged decades ago (e.g., Holling 1973, 1986) have evolved to include multi-scale drivers (Allen 2007; Falk et al. 2007), cross-scale interactions (Peters et al. 2004), catastrophic shifts (Scheffer et al. 2009), and a greater appreciation of how spatial interactions govern landscape resilience (Nystrom and Folke 2001; Rietkerk et al. 2004; van Nes and Scheffer 2005; Dakos et al. 2009).

Building on the evolution of the pulse-recovery and resilience literature, biogeochemical resilience attends to several key themes, including spatial patterning of biogeochemical flows and pools, their modification by disturbance, and the degree to which perturbations of biogeochemical dynamics influence the resilience of both ecosystems and landscapes. For example, modification of biogeochemical processes by disturbance can modify vegetation trajectories, potentially shifting the system to a new state characterized by a new set of internal dynamics. In contrast, biogeochemical dynamics (e.g., dominance of post-fire landscapes by nitrogen fixers) has the potential to buffer ecosystems from large biogeochemical fluxes following disturbance, maintaining the pre-disturbance system state. Largely unanswered is the degree to which coupled biogeochemical-vegetation systems would need to be perturbed (by fire) to shift into a new system state. The relative influence of these processes must be considered across heterogeneous landscapes, in which both fire disturbance and biogeochemical flows spatially interact.

6.5   Example: The Greater Yellowstone Ecosystem

The Greater Yellowstone Ecosystem (GYE), located in northwestern Wyoming, USA, provides a good case study for exploring the resilience of a nutrient-limited fire-prone ecosystem. Stand-replacing fires characterize the disturbance regime of the GYE, with an average fire return interval between 150 and 300 years (Schoennagel et al. 2003). The most common tree species in the region is lodgepole pine (*Pinus contorta* var. *latifolia*), but subalpine fir (*Abies lasiocarpa* (Hook.) Nutt.), Engelmann spruce (*Picea engelmannii* Parry), whitebark pine (*Pinus albicaulis* Engelm.) and Douglas-fir (*Pseudotsuga menziesii*) are also present. Soils are dominantly derived from rhyolite or tuff (shallow inceptisols). Previous research indicates that nitrogen is a limiting nutrient to the productivity of lodgepole pine forests (Fahey et al. 1985; Prescott et al. 1989), although results are conflicting (Brockley 2003; Romme et al. 2009), suggesting other nutrient limitations. Water is also tightly linked to fire size and distribution in sub-alpine forests of the Rocky Mountains (Westerling et al. 2006).

Numerous studies, not reviewed here, have characterized the heterogeneous ecosystem recovery of lodgepole pine forests following large fires of 1988 (Tinker et al. 1994; Turner et al. 1994, 1997; Litton et al. 2003; Kashian et al. 2004), which burned approximately 35% of the park. Of interest to the concept of biogeochemical resilience are subsequent studies that explored how patterns of nitrogen availability and mineralization were related to heterogeneous patterns of vegetation recovery following stand-replacing fire, and if so, at what scale. As summarized in Turner et al. (2007a), nitrogen availability was influenced by stand-replacing fire, although it is not clear how severe fires in the GYE affect overall biogeochemical dynamics that would include consideration of other nutrients (e.g., phosphorus, boron), carbon, or hydrologic fluxes. Under what conditions is the Yellowstone landscape biogeochemically resilient to large fires? Are these conditions likely for the future, or are thresholds in coupled biogeochemical-vegetation dynamics likely? Would conditions that favor resilience in one biogeochemical variable also favor resilience in other variables?

Recently, Schoennagel et al. (2008) synthesized available studies in the post-fire GYE landscape to explore multi-scale, pattern-process interactions (Fig. 6.3). At the broadest scale, historical patterns in stand-replacement fire set the landscape

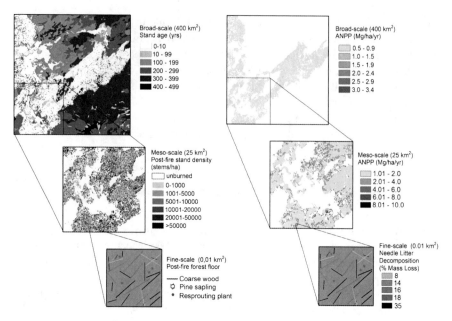

Fig. 6.3 Spatial patterns of variation are roughly parallel in structural and compositional (*left*) vs. functional (*right*) charcteristics of landscapes at three scales (From Schoennagel et al. (2008) with permission)

age-class mosaic (Romme and Despain 1989). At medium scales, the 1988 fires produced heterogeneous patterns in seedling regeneration, with rapid regeneration in some areas (Turner et al. 1994; Turner et al. 1997). At fine scales, structural variation and fire severity in burned stands were heterogeneous due to the combination of dead wood, regenerating vegetation, and open areas. In turn, these patterns produced heterogeneous aboveground net primary production and nitrogen cycling at each scale (Turner et al. 2004; Metzger et al. 2008). However, post-fire stand structure and function appear to converge over time through both stand in-filling and stem exclusion, resulting in relatively similar conditions among mature forests with different fire histories (Kashian et al. 2005a, b) (Fig. 6.4).

Biogeochemical resilience in the GYE is generally supported by these multi-scale studies. First, although stand-replacing fires like the ones in 1988 kill trees, relatively little biomass was consumed by the 1988 fires because the fire was carried through the canopy, leaving large wood and soil, the largest pools of carbon in forested systems, intact (Tinker and Knight 2000; Kashian et al. 2006). Second, lodgepole pine rapidly regenerated following the 1988 fires, largely facilitated by interactions of fire severity and pre-fire serotiny (Turner et al. 2003). Serotinous cones were also observed as early as 15 years following the 1988 fire (Turner et al. 2007b). Third, recent modeling and chronosequence work shows that most nitrogen and carbon storage recovered quickly (Bradford et al. 2008; Smithwick et al. 2009a, b), well within the typical fire return interval for lodgepole-pine forests.

Fig. 6.4 Chronosequence of lodgepole pine forest development in Yellowstone National Park, USA. (**a**) 2 years, (**b**) 15 years, (**c**) 99 years, and (**d**) 297 years following fire

Taken together, low amounts of elemental losses, combined with rapid but variable regeneration of vegetation and recovery of carbon and nitrogen stocks, suggest that the Yellowstone system is biogeochemically resilient to large stand-replacing fires, although the role of other nutrients (e.g., phosphorus) is still largely unexplored.

On the other hand, subalpine Rocky Mountain forests may be increasingly susceptible to large stand-replacing fires under scenarios of climate change (Running 2006; Westerling et al. 2006). Rapid post-fire regeneration observed after the 1988 fires was largely a result of pre-fire levels of serotiny, which is understood to be an evolutionary adaptation to fire (Richardson 1998; Schoennagel et al. 2003; Bond et al. 2004) and which may confer resilience where it currently exists. Over short time frames, however, evolutionary adaptation to fire via serotiny is not a mechanism that is likely to confer resilience for areas that have lower levels of serotiny (e.g., high-elevation forests). Forests may also be increasingly susceptible to multiple disturbance events, e.g., beetles (Bebi et al. 2003; Raffa et al. 2008). Finally, although there are indications of enhanced productivity under certain climate change scenarios (Smithwick et al. 2009b), studies in Canada have shown that increases in productivity would need to be substantial to offset losses from disturbance (Kurz et al. 2008). Equally likely are precipitation deficits that both reduce productivity and increase disturbance frequency and severity (Running 2006; Running 2008). Together, these factors suggest possible sources of ecological vulnerability of the Yellowstone landscape if fire regimes change rapidly under novel climate conditions. A key unknown in these scenarios is the degree to which biogeochemical feedbacks could mediate the response of lodgepole pine forests to these stressors. Would changes in climate favor species or vegetation types (e.g., grasses) with different nutrient-cycling strategies? Would shifts in hydrology, nutrient availability and substrate quality limit vegetation recovery rates and patterns under novel disturbance-climate regimes? Combining field and modeling studies to explore these interactions would lead to new insights about the role of biogeochemistry for understanding resilience of Yellowstone landscape.

6.6 Looking Forward: Biogeochemical Resilience and the Landscape Ecology of Fire

Biogeochemistry is a key frontier for understanding responses of terrestrial ecosystems to climate change. Unfortunately, a lack of understanding of the role of biogeochemistry in disturbance dynamics at regional to global scales currently limits our ability to forecast feedbacks between the terrestrial biosphere and the climate system. Coupling biogeochemistry with the landscape ecology of fire would help to understand, model, and forecast trajectories of future landscape change. The concept of biogeochemical resilience can be useful for identifying the conditions that lead to tipping points (Wall 2007; Kriegler et al. 2009) in response to fire across complex landscapes. Towards this goal, several new research areas are identified below.

6.6.1 Identify the Conditions under Which Interactions of Post-fire Biogeochemistry and Vegetation Shift Systems to Alternate States

Disturbance modifies the distribution and form of aboveground biomass (Chap. 1). Biomass reductions affect ecosystem carbon storage over short temporal scales and modify nutrient cycling over longer temporal scales as this new vegetation is incorporated into litter and soil. However, biogeochemical processes are also modified directly by disturbance through physiochemical processes or shifts in microbial communities that mediate immediate post-fire response (Raison 1979; Neary et al. 1999). These processes are likely to be short-lived (Wan et al. 2001; Smithwick et al. 2005), and over time, biogeochemical dynamics will be determined by the regrowing vegetation (Hart et al. 2005b). However, the effect of fire on microbial community composition, the biotic agents of biogeochemical function, may be critical for interpreting future response to fire over the short term (months to years). Over longer (decadal and centennial) timescales, coupled interactions of biogeochemistry with the recovering vegetation and associated ecological processes (e.g., species competition) will be important. At broad *spatial* scales, biogeochemical cycling may be constrained by fundamental limitations in critical nutrients, determined by variation in substrate. As a result, some post-disturbance ecosystems may be more vulnerable than others to large shifts in disturbance regimes depending on the degree to which they are buffered by their pre-existing chemical template. Future work should prioritize studies that couple biogeochemical and vegetative perspectives on post-fire recovery to identify drivers that facilitate or limit resilience (Naeem 1996).

6.6.2   Compare Models with Empirical Data from Multiple Scales of Space and Time

Projections involving single drivers of ecosystem response to disturbance are more likely wrong than not. Confronting multiple hypotheses in a scaled perspective is necessary. For example, although climate may drive vegetation conditions at global scales, fire processes are determined by forces acting across scales, from local-scale fuel and weather conditions and ignition sources to large-scale ocean-atmosphere interactions (Falk et al. 2007). Consequently, matching the scale of the driver to the response variable is critical. Statistical modeling to capture the behavior of fine-scale phenomena and extrapolate global fire patterns (Krawchuk et al. 2009) shows promise as a methodology to understand spatial and temporal variability in fire. Similar approaches have been addressed with ecological models to capture fine-scale behavior of individual trees to encapsulate ecosystem behavior (e.g., Medvigy et al. 2009). These and similar approaches are needed at landscape scales to determine reciprocal pattern-process interactions of fire regimes.

6.6.3   Use Concepts of Equilibrium to Explore Conditions that Promote Resilience

Input and output budgets have long been used to identify equilibrium pool sizes and observe responses to perturbation (Odum 1969; Bormann and Likens 1979b).

Although we have an increased appreciation for multiple stable states, cross-scale interactions, and threshold dynamics in ecological systems, it is important to recall that ecosystems are ultimately biogeochemical and are governed by inputs, outputs, and pool sizes, each of which is potentially modified by fire. We should use the growing body of empirical knowledge of fire biogeochemistry to refine and explore the conditions under which pools and fluxes of key ecosystem elements promote resilience. Mass-balance approaches that describe ecosystem energetics (Chap. 1) must be combined with dynamic watershed and landscape models that allow for ecological contingencies and spatial complexities (Chaps. 3 and 5).

6.6.4 Establish a General Framework for Biogeochemical Resilience across a Variety of Ecosystems and Disturbance Regimes, and Over a Broader Range of Biogeochemical Fluxes

While it is straightforward for ecologists to confront complexity by doing more single-variable studies, it is critical that a new research framework be developed to test dynamics of biogeochemistry and disturbance across ecosystems and spatiotemporal scales. We must expand our research into under-studied ecosystems and across broad biogeographic gradients, recognizing that these systems have different historical legacies and future trajectories. For example, fire has a long and controversial history as a contributor to the development of eastern forests in the U.S. (Abrams 1992; Foster et al. 2002; Parshall and Foster 2002), but projections of fire-prone southern species moving into the northeast under climate change (Iverson et al. 2008) suggest that the past is only a loose guide for the future (Williams and Jackson 2007). We must recognize that fire regimes are variable across space and time and continue to explore experimental and observational studies of biogeochemistry across the full spectrum of fire-regime characteristics such as seasonality, severity, and frequency across geographically diverse and novel systems.

In conclusion, there is an opportunity in landscape fire science to borrow pyrogeographic perspectives of global fire patterns to explore landscape biogeochemistry. Just as pyrogeography assumes spatial complexity in the causes and consequences of fire, biogeochemical perspectives of fire recognizes that biogeochemical drivers, stressors, and feedbacks are spatially differentiated. Biogeochemical resilience further recognizes that fire modifies pools and fluxes both during and after the fire event through both direct physio-chemical reactions and post-fire vegetative coupling. In order to provide a useful framework for understanding fire on changing landscapes, concepts of resilience must fully appreciate the coupling of biogeochemical and ecological responses to fire over space and time.

References

Abrams, M.D. 1992. Fire and the development of oak forests in eastern North-America: Oak distribution reflects a variety of ecological paths and disturbance conditions. *Bioscience* 42: 346–353.

Allen, C.D. 2007. Interactions across spatial scales among forest dieback, fire, and erosion in northern New Mexico landscapes. *Ecosystems* 10: 797–808.

Allen, C.R., and C.S. Holling. 2002. Cross-scale structure and scale breaks in ecosystems and other complex systems. *Ecosystems* 5: 315–318.

Andreu, V., J.L. Rubio, J. Forteza, and R. Cerni. 1996. Postfire effects on soil properties and nutrient losses. *International Journal of Wildland Fire* 6: 53–58.

Auclair, A.N.D., and T.B. Carter. 1993. Forest wildfires as a recent source of CO_2 at northern latitudes. *Canadian Journal of Forest Research* 23: 1528–1536.

Bachelet, D., R.P. Neilson, J.M. Lenihan, and R.J. Drapek. 2001. Climate change effects on vegetation distribution and carbon budget in the United States. *Ecosystems* 4: 164–185.

Baker, W.L. 1989. Effect of scale and spatial heterogeneity on fire-interval distributions. *Canadian Journal of Forest Research* 19: 700–706.

Bebi, P., D. Kulakowski, and T.T. Veblen. 2003. Interactions between fire and spruce beetles in a subalpine Rocky Mountain forest landscape. *Ecology* 84: 362–371.

Binkley, D., D. Richter, M.B. David, and B.A. Caldwell. 1992. Soil chemistry in a loblolly/longleaf pine forest with interval burning. *Ecological Applications* 2: 157–164.

Boerner, R.E.J. 1982. Fire and nutrient cycling in temperate ecosystems. *Bioscience* 32: 187–192.

Bond, W.J., K.J.M. Dickinson, and A.F. Mark. 2004. What limits the spread of fire-dependent vegetation? Evidence from geographic variation of serotiny in a New Zealand shrub. *Global Ecology and Biogeography* 13: 115–127.

Bond, W.J., F.I. Woodward, and G.F. Midgley. 2005. The global distribution of ecosystems in a world without fire. *The New Phytologist* 165: 525–538.

Bond-Lamberty, B., S.T. Gower, C. Wang, P. Cyr, and H. Veldhuis. 2006. Nitrogen dynamics of a boreal black spruce wildfire chronosequence. *Biogeochemistry* 81: 1–16.

Bond-Lamberty, B., S.D. Peckham, D.E. Ahl, and S.T. Gower. 2007. Fire as the dominant driver of central Canadian boreal forest carbon balance. *Nature* 450: 89–92.

Bormann, F.H., and G.E. Likens. 1979a. Catastrophic disturbance and the steady-state in northern hardwood forests. *American Scientist* 67: 660–669.

Bormann, F.H., and G.E. Likens. 1979b. *Pattern and process in a forested ecosystem*. New York: Springer.

Bowman, D. 2005. Understanding a flammable planet-climate, fire, and global vegetation patterns. *The New Phytologist* 165: 341–345.

Bradford, J.B., R.A. Birdsey, L.A. Joyce, and M.G. Ryan. 2008. Tree age, disturbance history, and carbon stocks and fluxes in subalpine Rocky Mountain forests. *Global Change Biology* 14: 2882–2897.

Brockley, R.P. 2003. Effects of nitrogen and boron fertilization on foliar boron nutrition and growth in two different lodgepole pine ecosystems. *Canadian Journal of Forest Research* 33: 988–996.

Buhk, C., and I. Hensen. 2006. "Fire seeders" during early post-fire succession and their quantitative importance in south-eastern Spain. *Journal of Arid Environments* 66: 193–209.

Campbell, J.L., L.E. Rustad, E.W. Boyer, S.F. Christopher, C.T. Driscoll, I.J. Fernandez, P.M. Groffman, D. Houle, J. Kiekbusch, A.H. Magill, M.J. Mitchell, and S.V. Ollinger. 2009. Consequences of climate change for biogeochemical cycling in forests of northeastern North America. *Canadian Journal of Forest Research* 39: 264–284.

Carpenter, S., B. Walker, J.M. Anderies, and N. Abel. 2001. From metaphor to measurement: Resilience of what to what? *Ecosystems* 4: 765–781.

Cary, G.J., R.E. Keane, R.H. Gardner, S. Lavorel, M.D. Flannigan, I.D. Davies, C. Li, J.M. Lenihan, T.S. Rupp, and F. Mouillot. 2006. Comparison of the sensitivity of landscape-fire-succession models to variation in terrain, fuel pattern, climate and weather. *Landscape Ecology* 21: 121–137.

Certini, G. 2005. Effects of fire on properties of forest soils: A review. *Oecologia* 143: 1–10.

Choromanska, U., and T.H. DeLuca. 2002. Microbial activity and nitrogen mineralization in forest mineral soils following heating: evaluation of post-fire effects. *Soil Biology and Biochemistry* 34: 263–271.

Dakos, V., E. Beninca, E.H. van Nes, C.J.M. Philippart, M. Scheffer, and J. Huisman. 2009. Interannual variability in species composition explained as seasonally entrained chaos. *Proceedings of the Royal Society B-Biological Sciences* 276: 2871–2880.

Davidson, E.A., S.C. Hart, and M.K. Firestone. 1992. Internal cycling of nitrate in soils of a mature coniferous forest. *Ecology* 73: 1148–1156.

de Blas, E., M. Rodriguez-Alleres, and G. Almendros. 2010. Speciation of lipid and humic fractions in soils under pine and eucalyptus forest in northwest Spain and its effect on water repellency. *Geoderma* 155: 242–248.

DeBano, L.F. 2000. The role of fire and soil heating on water repellency in wildland environments: A review. *Journal of Hydrology* 231: 195–206.

DeLuca, T.H., M.D. MacKenzie, M.J. Gundale, and W.E. Holben. 2006. Wildfire-produced charcoal directly influences nitrogen cycling in ponderosa pine forests. *Soil Science Society of America Journal* 70: 448–453.

Dore, S., T.E. Kolb, M. Montes-Helu, B.W. Sullivan, W.D. Winslow, S.C. Hart, J.P. Kaye, G.W. Koch, and B.A. Hungate. 2008. Long-term impact of a stand-replacing fire on ecosystem CO2 exchange of a ponderosa pine forest. *Global Change Biology* 14: 1801–1820.

Du Toit, S.R., K.H. Robers, and H.C. Biggs, eds. 2003. *The Kruger experience: Ecology and management of savanna heterogeneity*. Washington: Island.

Euskirchen, E.S., J. Chen, H. Li, E.J. Gustafson, and T.R. Crow. 2002. Modeling landscape net ecosystem productivity (LandNEP) under alternate management regimes. *Ecological Modelling* 154: 75–91.

Fahey, T., J.B. Yavitt, J.A. Pearson, and D.H. Knight. 1985. The nitrogen cycle in lodgepole pine forests, southeastern Wyoming. *Biogeochemistry* 1: 257–275.

Falk, D.A., C. Miller, D. McKenzie, and A.E. Black. 2007. Cross-scale analysis of fire regimes. *Ecosystems* 10: 809–823.

Folke, C., S. Carpenter, B. Walker, M. Scheffer, T. Elmqvist, L. Gunderson, and C.S. Holling. 2004. Regime shifts, resilience, and biodiversity in ecosystem management. *Annual Review of Ecology, Evolution, and Systematics* 35: 557–581.

Foster, D.R., and E.R. Boose. 2001. Patterns of forest damage resulting from catastrophic wind in central New England, USA. *Journal of Ecology* 80: 79–98.

Foster, D.R., G. Motzkin, and B. Slater. 1998. Land-use history as long-term broad-scale disturbance: Regional forest dynamics in central New England. *Ecosystems* 1: 96–119.

Foster, D.R., S. Clayden, D.A. Orwig, B. Hall, and S. Barry. 2002. Oak, chestnut and fire: climatic and cultural controls of long-term forest dynamics in New England, USA. *Journal of Biogeography* 29: 1359–1379.

Fraterrigo, J.M., and J.A. Rusak. 2008. Disturbance-driven changes in the variability of ecological patterns and processes. *Ecology Letters* 11: 756–770.

Fraterrigo, J.M., T.C. Balser, and M.G. Turner. 2006. Microbial community variation and its relationship with nitrogen mineralization in historically altered forests. *Ecology* 87: 570–579.

Giblin, A.E., K.J. Nadelhoffer, G.R. Shaver, J.A. Laundre, and A.J. McKerrow. 1991. Biogeochemical diversity along a riverside toposequence in Arctic Alaska. *Ecological Monographs* 61: 415–435.

Gorham, E., P.M. Vitousek, and W.A. Reiners. 1979. The regulation of chemical budgets over the course of terrestrial ecosystem succession. *Annual Review of Ecology and Systematics* 10: 53–84.

Grady, K.C., and S.C. Hart. 2006. Influences of thinning, prescribed burning, and wildfire on soil processes and properties in southwestern ponderosa pine forests: A retrospective study. *Forest Ecology and Management* 234: 123–135.

Groffman, P., J. Baron, T. Blett, A. Gold, I. Goodman, L. Gunderson, B. Levinson, M. Palmer, H. Paerl, G. Peterson, N. Poff, D. Rejeski, J. Reynolds, M. Turner, K. Weathers, and J. Wiens. 2006. Ecological thresholds: The key to successful environmental management or an important concept with no practical application? *Ecosystems* 9: 1–13.

Groffman, P.M., K. Butterbach-Bahl, R.W. Fulweiler, A.J. Gold, J.L. Morse, E.K. Stander, C. Tague, C. Tonitto, and P. Vidon. 2009. Challenges to incorporating spatially and temporally explicit phenomena (hotspots and hot moments) in denitrification models. *Biogeochemistry* 93: 49–77.

Gunderson, L.H. 2000. Ecological resilience–In theory and application. *Annual Review of Ecology and Systematics* 31: 425–439.

Hamman, S.T., I.C. Burke, and M.E. Stromberger. 2007. Relationships between microbial community structure and soil environmental conditions in a recently burned system. *Soil Biology and Biochemistry* 39: 1703–1711.

Harden, J.W., Neff, J.C., Sandberg, D.V.,Turetsky, M.R., Ottmar, R.D., Gleixner, G., Fries, T.L., and K.L. Manies. 2004. Chemistry of burning the forest floor during the FROSTFIRE experimental burn, interior Alaska, 1999. *Global Biogeochemical Cycles* 18:GB3014, doi:3010.1029/2003 GB002194.

Hart, S.C., A.T. Classen, and R.J. Wright. 2005a. Long-term interval burning alters fine root and mycorrhizal dynamics in a ponderosa pine forest. *Journal of Applied Ecology* 42: 752–761.

Hart, S.C., T.H. DeLuca, G.S. Newman, M.D. MacKenzie, and S.I. Boyle. 2005b. Post-fire vegetative dynamics as drivers of microbial community structure and function in forest soils. *Forest Ecology and Management* 220: 166–184.

He, H.S., and D.J. Mladenoff. 1999. Spatially explicit and stochastic simulation of forest-landscape fire disturbance and succession. *Ecology* 80: 81–99.

Hedin, L.O., J.J. Armesto, and A.H. Johnson. 1995. Patterns of nutrient loss from unpolluted old-growth temperate forests: Evaluation of biogeochemical theory. *Ecology* 76: 493–509.

Holling, C.S. 1973. Resilience and stability of ecological systems. *Annual Review of Ecology and Systematics* 4: 1–23.

Holling, C.S. 1986. The resilience of terrestrial ecosystems: Local surprise and global change. In *Sustainable development of the biosphere*, eds. W.C. Clark and W.E. Munn, 293–320. London: Cambridge University Press.

Holling, C.S., and L.H. Gunderson. 2002. Resilience and adaptive cycles. In *Panarchy: Understanding transformations in human and natural systems*, eds. L.H. Gunderson and C.S. Holling, 25–62. Washington: Island.

Huebert, B., P. Vitousek, J. Sutton, T. Elias, J. Heath, S. Coeppicus, S. Howell, and B. Blomquist. 1999. Volcano fixes nitrogen into plant-available forms. *Biogeochemistry* 47: 111–118.

Iverson, L.R., A.M. Prasad, S.N. Matthews, and M. Peters. 2008. Estimating potential habitat for 134 eastern US tree species under six climate scenarios. *Forest Ecology and Management* 254: 390–406.

Jacobs, S.M., J.S. Bechtold, H.C. Biggs, N.B. Grimm, S. Lorentz, M.E. McClain, R.J. Naiman, S.S. Perakis, G. Pinay, and M.C. Scholes. 2007. Nutrient vectors and riparian processing: A review with special reference to African semiarid savanna ecosystems. *Ecosystems* 10: 1231–1249.

Kashian, D.M., D.B. Tinker, M.G. Turner, and F.L. Scarpace. 2004. Spatial heterogeneity of lodgepole pine sapling densities following the 1988 fires in Yellowstone National Park, Wyoming, U.S.A. *Canadian Journal of Forest Research* 34: 2263–2276.

Kashian, D.M., M.G. Turner, and W.H. Romme. 2005a. Variability in leaf area and stemwood increment along a 300-year lodgepole pine chronosequence. *Ecosystems* 8: 48–61.

Kashian, D.M., M.G. Turner, W.H. Romme, and C.G. Lorimer. 2005b. Variability and convergence in stand structure with forest development on a fire-dominated landscape. *Ecology* 86: 643–654.

Kashian, D.M., M.G. Ryan, W.H. Romme, D.B. Tinker, and M.G. Turner. 2006. Carbon cycling on landscapes with stand-replacing fire. *Bioscience* 56: 598–606.

Keane, R.E., K.C. Ryan, and S.W. Running. 1996. Simulating effects of fire on northern Rocky Mountain landscapes with the ecological process model FIRE-BGC. *Tree Physiology* 16: 319–331.

Krawchuk, M.A., M.A. Moritz, M.-A. Parisien, J. Van Dorn, and K. Hayhoe. 2009. Global pyrogeography: The current and future distribution of wildfire. *PLoS ONE* 4: e5102.

Kriegler, E., J.W. Hall, H. Held, R. Dawson, and H.J. Schellnhuber. 2009. Imprecise probability assessment of tipping points in the climate system. *Proceedings of the National Academy of Sciences of the United States of America*. doi:10.1073/pnas.0809117106.

Kurz, W.A., and M.J. Apps. 1999. A 70-year retrospective analysis of carbon fluxes in the Canadian forest sector. *Ecological Applications* 9: 526–547.

Kurz, W.A., G. Stinson, and G. Rampley. 2008. Could increased boreal forest ecosystem productivity offset carbon losses from increased disturbances? *Philosophical Transactions of the Royal Society B-Biological Sciences* 363: 2261–2269.

Leduc, S.D., and D.E. Rothstein. 2007. Initial recovery of soil carbon and nitrogen pools and dynamics following disturbance in jack pine forests: A comparison of wildfire and clearcut harvesting. *Soil Biology and Biochemistry* 39: 2865–2876.

Lehsten, V., K. Tansey, H. Balzter, K. Thonicke, A. Spessa, U. Weber, B. Smith, and A. Arneth. 2009. Estimating carbon emissions from African wildfires. *Biogeosciences* 6: 349–360.

Litton, C.M., M.G. Ryan, D.B. Tinker, and D.H. Knight. 2003. Below- and aboveground biomass in young post-fire lodgepole pine forests of contrasting tree density. *Canadian Journal of Forest Research* 33: 351–363.

Lynham, T.J., G.M. Wickware, and J.A. Mason. 1998. Soil chemical changes and plant succession following experimental burning in immature jack pine. *Canadian Journal of Soil Science* 78: 93–104.

Mabuhay, J., Y. Isagi, and N. Nakagoshi. 2006. Wildfire effects on microbial biomass and diversity in pine forests at three topographic positions. *Ecological Research* 21: 54–63.

Mather, T.A., A.G. Allen, B.M. Davison, D.M. Pyle, C. Oppenheimer, and A.J.S. McGonigle. 2004. Nitric acid from volcanoes. *Earth and Planetary Science Letters* 218: 17–30.

Medvigy, D., S.C. Wofsy, J.W. Munger, D.Y. Hollinger, and P.R. Moorcroft. 2009. Mechanistic scaling of ecosystem function and dynamics in space and time: Ecosystem Demography model version 2. *Journal of Geophysical Research* 114: G01002. doi:10.1029/2008JG000812.

Mermoz, M., T. Kitzberger, and T.T. Veblen. 2005. Landscape influences on occurrence and spread of wildfires in Patagonian forests and shrublands. *Ecology* 86: 2705–2715.

Metzger, K.L., E.A.H. Smithwick, D.B. Tinker, W.H. Romme, T.C. Balser, and M.G. Turner. 2008. Influence of coarse wood and pine saplings on nitrogen mineralization and microbial communities in young post-fire *Pinus contorta*. *Forest Ecology and Management* 256: 59–67.

Mills, A.J., and M.V. Fey. 2005. Interactive response of herbivores, soils and vegetation to annual burning in a South African savanna. *Austral Ecology* 30: 435–444.

Minshall, G.W., C.T. Robinson, and D.E. Lawrence. 1997. Postfire responses of lotic ecosystems in Yellowstone National Park, U.S.A. *Canadian Journal of Fisheries and Aquatic Science* 54: 2509–2525.

Minshall, G.W., J.T. Brock, D.A. Andrews, and C.T. Robinson. 2001. Water quality, substratum and biotic responses of five central Idaho (USA) streams during the first year following the Mortar Creek fire. *International Journal of Wildland Fire* 10: 185–199.

Monleon, V.J., K. Cromack Jr., and J.D. Landsberg. 1997. Short- and long-term effects of prescribed underburning on nitrogen availability in ponderosa pine stands in central Oregon. *Canadian Journal of Forest Research* 27: 369–378.

Moody, J.A., D.A. Kinner, and X. Ubeda. 2009. Linking hydraulic properties of fire-affected soils to infiltration and water repellency. *Journal of Hydrology* 379: 291–303.

Moritz, M.A., and S.L. Stephens. 2008. Fire and sustainability: Considerations for California's altered future climate. *Climatic Change* 87: S265–S271.

Naeem, S, ed. 1996. *Biodiversity and ecosystem functioning in restored ecosystems: Extracting principles for a synthetic perspective*. Washington: Island.

Neary, D.G., C.C. Klopatek, L.F. DeBano, and P.F. Ffolliott. 1999. Fire effects on belowground sustainability: A review and synthesis. *Forest Ecology and Management* 122: 51–71.

Norberg, J., and G.S. Cumming. 2008. *Complexity theory for a sustainable future*. New York: Columbia University Press.

Nystrom, M., and C. Folke. 2001. Spatial resilience of coral reefs. *Ecosystems* 4: 406–417.

O'Neill, K., D. Richter, and E. Kasischke. 2006. Succession-driven changes in soil respiration following fire in black spruce stands of interior Alaska. *Biogeochemistry* 80: 1–20.

Odum, H.T. 1960. Ecological potential and analogue circuits for the ecosystem. *American Scientist* 48: 1–8.

Odum, E.P. 1968. Energy flow in ecosystems–A historical review. *American Zoologist* 8: 11–18.

Odum, E.P. 1969. The strategy of ecosystem development. *Science* 164: 262–270.

Oppenheimer, C., P. Kyle, V. Tsanev, A. McGonigle, T. Mather, and D. Sweeney. 2005. Mt. Erebus, the largest point source of NO_2 in Antarctica. *Atmospheric Environment* 39: 6000–6006.

Parshall, T., and D.R. Foster. 2002. Fire on the New England landscape: Regional and temporal variation, cultural and environmental controls. *Journal of Biogeography* 29: 1305–1317.

Peters, D.P.C., R.A. Pielke Sr., B.T. Bestelmeyer, C.D. Allen, S. Munson-McGee, and K.M. Havstad. 2004. Cross-scale interactions, nonlinearities, and forecasting catastrophic events. *Proceedings of the National Academy of Sciences of the United States of America* 101: 15130–15135.

Peters, D.P.C., B.T. Bestelmeyer, and M.G. Turner. 2007. Cross-scale interactions and changing pattern-process relationships: consequences for system dynamics. *Ecosystems* 10: 790–796.

Peterson, G.D. 2002. Contagious disturbance, ecological memory, and the emergence of landscape pattern. *Ecosystems* 5: 329–338.

Pietikainen, J., R. Hiukka, and H. Fritze. 2000. Does short-term heating of forest humus change its properties as a substrate for microbes? *Soil Biology and Biochemistry* 32: 277–288.

Poage, N.J., P.J. Weisberg, P.C. Impara, J.C. Tappeiner, and T.S. Sensenig. 2009. Influences of climate, fire, and topography on contemporary age structure patterns of Douglas-fir at 205 old forest sites in western Oregon. *Canadian Journal of Forest Research* 39: 1518–1530.

Power, M.J., J. Marlon, N. Ortiz, et al. 2008. Changes in fire regimes since the Last Glacial Maximum: an assessment based on a global synthesis and analysis of charcoal data. *Climate Dynamics* 30: 887–907.

Prescott, C.E., J.P. Corbin, and D. Parkinson. 1989. Biomass, productivity, and nutrient-use efficiency of aboveground vegetation in four Rocky Mountain coniferous forests. *Canadian Journal of Forest Research* 19: 309–317.

Raffa, K.F., B.H. Aukema, B.J. Bentz, A.L. Carroll, J.A. Hicke, M.G. Turner, and W.H. Romme. 2008. Cross-scale drivers of natural disturbances prone to anthropogenic amplification: The dynamics of bark beetle eruptions. *Bioscience* 58: 501–517.

Raison, R.J. 1979. Modification of the soil environment by vegetation fires, with particular reference to nitrogen transformations: A review. *Plant and Soil* 51: 73–108.

Randerson, J.T., H. Liu, M.G. Flanner, S.D. Chambers, Y. Jin, P.G. Hess, G. Pfister, M.C. Mack, K.K. Treseder, L.R. Welp, F.S. Chapin, J.W. Harden, M.L. Goulden, E. Lyons, J.C. Neff, E.A.G. Schuur, and C.S. Zender. 2006. The impact of boreal forest fire on climate warming. *Science* 314: 1130–1132.

Randerson, J.T., F.M. Hoffman, P.E. Thornton, N.M. Mahowald, K. Lindsay, Y.H. Lee, C.D. Nevison, S.C. Doney, G. Bonan, R. Stockli, C. Covey, S.W. Running, and I.Y. Fung. 2009. Systematic assessment of terrestrial biogeochemistry in coupled climate-carbon models. *Global Change Biology* 15: 2462–2484.

Richardson, D.M. 1998. *Ecology and evolution of Pinus*. Cambridge: Cambridge University Press.

Rietkerk, M., S.C. Dekker, P.C. de Ruiter, and J. van de Koppel. 2004. Self-organized patchiness and catastrophic shifts in ecosystems. *Science* 305: 1926–1929.

Rogers, K.H. 2003. Adopting a heterogeneity paradigm: Implications for management of protected savannas. In *The Kruger experience: Ecology and management of savanna heterogeneity*, eds. S.R. Du Toit, K.H. Rogers, and H.C. Biggs, 41–58. Washington: Island.

Romme, W.H. 1982. Fire and landscape diversity in subalpine forests of Yellowstone National Park. *Ecological Monographs* 52: 199–221.

Romme, W.H., and D.G. Despain. 1989. Historical perspective on the Yellowstone fires of 1988. *Bioscience* 39: 695–699.

Romme, W.H., D.B. Tinker, G.K. Stakes, and M.G. Turner. 2009. Does inorganic nitrogen limit plant growth 3–5 years after fire in a Wyoming, USA, lodgepole pine forest? *Forest Ecology and Management* 257: 829–835.

Running, S.W. 2006. Climate change: Is global warming causing more, larger wildfires? *Science* 313: 927–928.

Running, S.W. 2008. Ecosystem disturbance, carbon, and climate. *Science* 321: 652–653.

Rustad, L.E. 2008. The response of terrestrial ecosystems to global climate change: Towards an integrated approach. *The Science of the Total Environment* 404: 222–235.

Scheffer, M., J. Bascompte, W.A. Brock, V. Brovkin, S.R. Carpenter, V. Dakos, H. Held, E.H. van Nes, M. Rietkerk, and G. Sugihara. 2009. Early-warning signals for critical transitions. *Nature* 461: 53–59.

Scheller, R.M., and D.J. Mladenoff. 2007. An ecological classification of forest landscape simulation models: tools and strategies for understanding broad-scale forested ecosystems. *Landscape Ecology* 22: 491–505.

Schlesinger, W.H. 1997. *Biogeochemistry: An analysis of global change*. San Diego: Academic.

Schoennagel, T., M.G. Turner, and W.H. Romme. 2003. The influence of fire interval and serotiny on postfire lodgepole pine density in Yellowstone National Park. *Ecology* 84: 2967–1978.

Schoennagel, T., T.T. Veblen, and W.H. Romme. 2004. The interaction of fire, fuels, and climate across Rocky Mountain forests. *Bioscience* 54: 661–676.

Schoennagel, T., E.A.H. Smithwick, and M.G. Turner. 2008. Landscape heterogeneity following large fires: Insights from Yellowstone National Park, USA. *International Journal of Wildland Fire* 17: 742–753.

Schultz, M.G., A. Heil, J.J. Hoelzemann, A. Spessa, K. Thonicke, J.G. Goldammer, A.C. Held, J.M.C. Pereira, and M. van het Bolscher. 2008. Global wildland fire emissions from 1960 to 2000. *Global Biogeochemical Cycles* 22, GB2002, doi:10.1029/2007GB003031.

Shakesby, R.A., and S.H. Doerr. 2006. Wildfire as a hydrological and geomorphological agent. *Earth Science Reviews* 74: 269–307.

Smithwick, E.A.H., M.E. Harmon, S.M. Remillard, S.A. Acker, and J.F. Franklin. 2002. Potential upper bounds of carbon stores in forests of the Pacific Northwest. *Ecological Applications* 12: 1303–1317.

Smithwick, E.A.H., M.G. Turner, M.C. Mack, and F.S. Chapin III. 2005. Post-fire soil N cycling in northern conifer forests affected by severe, stand-replacing wildfires. *Ecosystems* 8: 163–181.

Smithwick, E.A.H., D.M. Kashian, M.G. Ryan, and M.G. Turner. 2009a. Long-term nitrogen storage and soil nitrogen availability in post-fire lodgepole pine ecosystems. *Ecosystems* 12: 792–806.

Smithwick, E.A.H., M.G. Ryan, D.M. Kashian, W.H. Romme, D.B. Tinker, and M.G. Turner. 2009b. Modeling the effects of fire and climate change on carbon and nitrogen storage in lodgepole pine (*Pinus contorta*) stands. *Global Change Biology* 15: 535–548.

Sonderegger, D.L., H.N. Wang, W.H. Clements, and B.R. Noon. 2009. Using SiZer to detect thresholds in ecological data. *Frontiers in Ecology and the Environment* 7: 190–195.

Tinker, D.B., and D.H. Knight. 2000. Coarse woody debris following fire and logging in Wyoming lodgepole pine forests. *Ecosystems* 3: 472–483.

Tinker, D.B., W.H. Romme, W.W. Hargrove, R.H. Gardner, and M.G. Turner. 1994. Landscape-scale heterogeneity in lodgepole pine serotiny. *Canadian Journal of Forest Research* 24: 897–903.

Turner, M.G., and V.H. Dale. 2001. Comparing large, infrequent disturbances: What have we learned? *Ecosystems* 1: 493–496.

Turner, M.G., W.H. Romme, R.H. Gardner, R.V. O'Neill, and T.K. Kratz. 1993. A revised concept of landscape equilibrium: Disturbance and stability on scaled landscapes. *Landscape Ecology* 8: 213–227.

Turner, M.G., W.W. Hargrove, R.H. Gardner, and W.H. Romme. 1994. Effects of fire on landscape heterogeneity in Yellowstone National Park, Wyoming. *Journal of Vegetation Science* 5: 731–742.

Turner, M.G., W.H. Romme, R.H. Gardner, and W.H. Hargrove. 1997. Effects of fire size and pattern on early succession in Yellowstone National Park. *Ecological Monographs* 67: 411–433.

Turner, M.G., W.H. Romme, and D.B. Tinker. 2003. Surprises and lessons from the 1988 Yellowstone fires. *Frontiers in Ecology and the Environment* 1: 351–358.

Turner, M.G., D.B. Tinker, W.H. Romme, D.M. Kashian, and C.M. Litton. 2004. Landscape patterns of sapling density, leaf area, and aboveground net primary production in postfire lodgepole pine forests, Yellowstone National Park (USA). *Ecosystems* 7: 751–775.

Turner, M.G., E.A.H. Smithwick, K.L. Metzger, D.B. Tinker, and W.H. Romme. 2007a. Inorganic nitrogen availability after severe stand-replacing fire in the Greater Yellowstone ecosystem. *Proceedings of the National Academy of Sciences of the United States of America* 104: 4782–4789.

Turner, M.G., D.M. Turner, W.H. Romme, and D.B. Tinker. 2007b. Cone production in young post-fire *Pinus contorta* stands in Greater Yellowstone (USA). *Forest Ecology and Management* 242: 119–126.

van Nes, E.H., and M. Scheffer. 2005. Implications of spatial heterogeneity for catastrophic regime shifts in ecosystems. *Ecology* 86: 1797–1807.

Vitousek, P.M., and J.M. Melillo. 1979. Nitrate losses from disturbed forests: Patterns and mechanisms. *Forest Science* 25: 605–619.

Vitousek, P.M., and W.A. Reiners. 1975. Ecosystem succession and nutrient retention: A hypothesis. *Bioscience* 25: 376–381.

Wall, D.H. 2007. Global change tipping points: above- and below-ground biotic interactions in a low diversity ecosystem. *Philosophical Transactions of the Royal Society B-Biological Sciences* 362: 2291–2306.

Wan, S., D. Hui, and Y. Luo. 2001. Fire effects on nitrogen pools and dynamics in terrestrial ecosystems: a meta-analysis. *Ecological Applications* 11: 1349–1365.

West, D.C., H.H. Shugart, and J.W. Ranney. 1981. Population structure of forests over a large area. *Forest Science* 27: 701–710.

Westerling, A.L., H.G. Hidalgo, D.R. Cayan, and T.W. Swetnam. 2006. Warming and earlier spring increase western U.S. forest wildfire activity. *Science* 313: 940–943.

Williams, J.W., and S.T. Jackson. 2007. Novel climates, no-analog communities, and ecological surprises. *Frontiers in Ecology and the Environment* 5: 475–482.

Woodmansee, R.G., and L.S. Wallach. 1981. Effects of fire regimes on biogeochemical cycles. In *Terrestrial nitrogen cycles*, eds. F.E. Clark and T. Rosswall, 649–669. Stockholm: Ecological Bulletins.

Yang, J., H.S. He, and S.R. Shifley. 2008. Spatial controls of occurrence and spread of wildfires in the Missouri Ozark Highlands. *Ecological Applications* 18: 1212–1225.

Yermakov, Z., and D.E. Rothstein. 2006. Changes in soil carbon and nitrogen cycling along a 72-year wildfire chronosequence in Michigan jack pine forests. *Oecologia* 149: 690–700.

Zimmermann, J., S.I. Higgins, V. Grimm, J. Hoffmann, and A. Linstädter. 2010. Grass mortality in semi-arid savanna: The role of fire, competition and self-shading. *Perspectives in Plant Ecology, Evolution and Systematics* 12: 1–8.

Chapter 7
Reconstructing Landscape Pattern of Historical Fires and Fire Regimes

Tyson Swetnam, Donald A. Falk, Amy E. Hessl, and Calvin Farris

7.1 Introduction

Analysis of historical fire patterns of severity provides a view of fire regimes before they were altered by contemporary forest management practices such as logging, road-building, grazing, and fire suppression. Historical fire data can place contemporary observed fire data in a longer temporal context, and establish prior likelihoods to test outputs from predictive fire behavior and forest vegetation simulation models. When integrated with biophysical and remote-sensing data, fire-history data have been modeled to create both coarse scale (1 km^2, Schmidt et al. 2002) and fine scale (30 m^2, Rollins and Frame 2006) maps of fire regimes for the contiguous United States (LANDFIRE 2007). When joined with analysis of contemporary fires, the spatial properties of historical fires can provide a valuable perspective for fire and fuel management decisions (Schmidt et al. 2002). For these and other reasons, spatial reconstruction of historical fires is of both scientific and management interest.

The guiding scientific motivations for reconstructing landscape-scale spatio-temporal properties of fires include interests in:

- Reconstructing previously unrecorded fires, including perimeter estimates and internal patterns of heterogeneity in burn severity. By estimating the size and pattern of historical fires we can compare historical landscape-scale effects to modern fires at a given location. These site-specific reconstructions can help local managers improve fire management strategies.
- Estimating properties of fire regimes, such as fire frequency, fire-size distribution and fire rotation.
- Correlating landscape to regional spatiotemporal patterns of fire occurrence with other observations (proxy and instrumental), especially in relation to climate. By comparing the periodicity of historical fires to climate variability we may

T. Swetnam (✉)
School of Natural Resources and the Environment, University of Arizona,
Tucson, AZ 85721-0001, USA
e-mail: tswetnam@gmail.com

D. McKenzie et al. (eds.), *The Landscape Ecology of Fire*, Ecological Studies 213,
DOI 10.1007/978-94-007-0301-8_7, © Springer Science+Business Media B.V. 2011

narrow our inference about historical controls on the size, intensity, and frequency of fires, such as the role of inter-annual climate variation in modulating fuel accumulation (Swetnam and Betancourt 1992).

- Understanding the relative influence of top-down and bottom-up regulation of fires and fire regimes (Falk et al. 2007; Chap. 1; Chap. 3; Chap. 4). Spatial reconstruction can yield insights into these fundamental drivers of fire regimes through correlations with both stable (e.g. topographic) and time-varying (climate, fuels) factors.
- Identifying post-disturbance legacies, which can strongly influence ecological processes as well as subsequent fires, in a spatially explicit framework. Disturbance is a primary driver of establishment and succession in many ecosystems (Connell and Slatyer 1977; Turner et al. 1998). By comparing disturbance and establishment dates at specific locations to reconstructed climate and life history data, we can better understand the role of disturbance in demography and long-term forest dynamics.

Fire is an inherently spatiotemporal process, which progresses contagiously across landscapes (Rothermel 1972; Duarte 1997; Finney 1999; Chap. 1). Synchronous fire occurrence across large regions, at scales beyond the spread of individual events, reflects entrainment by synoptic weather patterns and multi-year climate variability (Johnson and Wowchuck 1993; Swetnam and Betancourt 1992). By contrast, fine-scale heterogeneity is the signature of bottom-up regulation, which operates primarily by modifying fire behavior and effects. Variable fire severities ramify postfire legacies and influence landscape fire dynamics by creating mosaics of vegetation and various successional stages that influence subsequent events (Collins et al. 2009). In turn, these landscape mosaics modify a wide range of ecological processes such as species distributions and biogeochemical cycling (Beeson et al. 2001; Wondzell and King 2003; Chap. 6).

In contemporary fires, we can observe fire behavior, intensity, and severity in real time, as well as antecedent fuel conditions, fuel loading, ignition locations, and weather. Processed multi-spectral high-resolution imagery from satellites, airplanes, and other remote-sensing platforms reveals effects of fires, both during and after the event, and enables highly detailed reconstruction of spatial pattern of multiple fire-related variables (Fig. 7.1; White et al. 1996; Keane et al. 2001; Miller and Yool 2002). The evolution of fire research in these areas has greatly advanced our understanding of fire as a spatial process, but the temporal coverage of these methods is relatively short. For example, satellite-based mapping of fire perimeters became widespread in the 1990s with the availability of image analysis software for personal computers. Satellite imagery for fire mapping is available from the AVHRR series of satellites (1981–Present), with better resolution data after the 1984 launch of LANDSAT 5; other systems such as MODIS (1999–), LANDSAT 7 (1999–), and QuickBird (2001–) represent additional sources of data. Landscape fire severity mosaics based on normalized-difference greenness from analysis of remotely derived images are increasingly available in many areas (MTBS 2010, Eidenshink et al. 2007).

For historical fires, human observer accounts and recorded station weather go back in time only as far as instrumental or written records. In western North

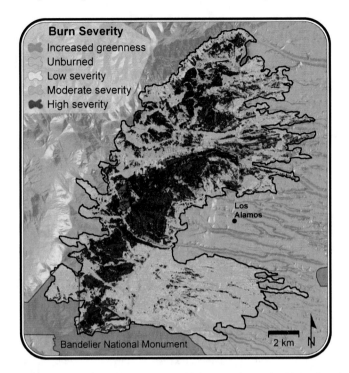

Fig. 7.1 Landsat Thematic Mapper (*TM*) image of fire severity in the Cerro Grande Fire near Los Alamos, New Mexico, 2002 (Data source: Miller and Yool (2002))

America prior to the twentieth century, low population density and a relative paucity of observers in rural areas limited the detail of historical fire records for most mountain ranges in the western United States. Proxy data prior to 1900 consist mainly of tree-ring based annually dated fire scars and tree recruitment data, which are the primary sources of information for reconstructing past disturbances.

In this chapter we review:

1. The essentials of the fire-scar proxy and fire-history reconstructions in both spatial and non-spatial contexts.
2. Analytical methods of spatial interpolation used to create likelihood surfaces from point records.
3. Case studies of fire history that have been used to reconstruct landscape spatial patterns in historic fires.

Many early fire-history analyses focused primarily on temporal properties of fire regimes, such as fire frequency and relationships to inter-annual climate. Over time, understanding spatial controls on fire regimes has become an important theme in fire history. Grissino-Mayer et al. (1996, 2006), Kaib et al. (1998); Barton (1999); Brown et al. (2001); and Iniguez et al. (2008) analyzed differences in fire occurrence along elevational gradients and across vegetation types in the "Sky Island" ranges of the Madrean Archipelago in southwestern U.S. and northern Mexico. Similar work in California, Oregon, and Utah has used spatial networks of fire-scarred trees to reveal patterns of fire occurrence reflecting topographic control of fire spread (Taylor and Skinner 2003; Taylor 2000; Beaty and Taylor 2001; Heyerdahl et al. 2001; Brown et al. 2008a; Kellogg et al. 2008). These studies represent important progress in understanding spatial attributes of fire *regimes*, although not necessarily at the level of individual fires or fire years.

High-resolution global positioning systems (GPS), field data recording, and the advancement of geographic information systems (GIS) have facilitated the construction and analysis of spatially explicit sets of point and polygon fire history data. These data represent multidimensional, multivariate predictor vectors (e.g., climate variables, disturbance, forest age structure, species distributions). Geospatial data and tools thus open the way for spatially explicit reconstructions of historical disturbance events, provided that a suitably accurate and georeferenced proxy record is available (Hessl et al. 2007; Farris et al. 2010).

7.2 Methods: Reconstructing Spatial Pattern of Fire

In this section we focus on techniques and proxy methods for determining the spatial extent of historical fires. These techniques involve probabilistic or geometric interpolations of fire occurrence between fire-scarred trees to reconstruct spatial pattern of historic fires. The basic approach, using georeferenced point data to reconstruct a landscape property, is similar regardless of the proxy used. The central problem in reconstructing spatial pattern in historic fires is thus one of interpretation. Solving this problem requires interpolating from point-based records such as tree recruitment dates and fire scars, to create a likelihood data surface. These methods vary in their accuracy and precision, depending on spatial scale and sample intensity.

7.2.1 Fire Scars

Fire scars are growth lesions caused by death of cambial cells along part of the circumference of a tree from exposure to extreme heat (Fritts 1976; Gutsell and Johnson 1996; Johnson and Miyanishi 2001). Heat energy from a fire must penetrate the bark in order to reach the cambium and cause cellular mortality, which appears later as a lesion. Tree species vary widely in the degree of thermal protection afforded by bark; many species in western North American forests have thick bark that provides substantial insulation against conduction of heat to the cambial layer (Vines 1968; Ryan and Reinhardt 1988). As a flaming front passes, heat flux through the bark causes injury to the cambium. Smoldering combustion of ground fuels after passage of the flaming front can also contribute to scarring by causing mortality of basal and cambial tissues (de Mestre et al. 1989). Consequently, the process of fire-scar formation reflects highly localized conditions of energy transfer, on the order of a few square meters around the base of a tree. Provided the tree is *in situ* when collected, a fire scar thus provides affirmative evidence of the occurrence of fire at a point location. Individual scarred trees can be georeferenced in the field using GPS and compiled in a GIS database.

The lack of a scar, however, does not necessarily indicate that fire was not present at the site (Fall 1998; Parsons et al. 2007). Fire scars may fail to form if heat penetration of the bark is insufficient to damage cambial cells. Once formed, scars may be destroyed by subsequent mechanical damage or rot or be burned off by subsequent fires. Consequently, a fundamental asymmetry exists in the record: for any given tree, a scar is affirmative evidence of an event, but the lack of a scar is ambiguous in its interpretation. Fire historians overcome this dilemma by compiling *composite fire records* (CFR), which are the union of the sets of fire dates recorded by multiple individual trees within a defined area (Dieterich 1980). As the number of samples in the CFR increase, its reliability as a representation of the true fire record for the site increases as a collector's curve (Falk 2004). In general, the first few records in the CFR capture the most widespread fires (those that tend to occur on multiple trees); successive samples added tend to record progressively smaller fires, until eventually the record is saturated (Falk 2004; Hessl et al. 2004).

Use of a CFR as the basis for spatial fire reconstruction increases the reliability of the record, but it also introduces an additional complication for spatial inference because trees that make up the CFR must be sampled over some definable area. The CFR converts an inherently "point" record (a scarred tree) to an area record that dictates the minimum mapping unit (MMU)—the area represented by all of the fire-scarred trees used to generate the CFR. The MMU may be a CFR compiled from trees within a small area (typically a stand of trees <1–10 ha). Definitive spatial inference cannot be made for an area larger or smaller than the MMU: we cannot know for certain whether fires burned between sample points outside the MMU sample unit, nor can we reliably infer pattern at scales smaller than the MMU. When fires are detected at two points that are close together compared to what we know of fire spread (e.g. 100 m apart) it is assumed likely that fire burned continuously between the points, but this assumption becomes less likely as distance increases (Kellogg et al. 2008).

The temporal resolution of fire-scar records is typically annual to sub-annual. This is achieved by *crossdating* the annual ring record, which compares the patterns of tree growth (as reflected in ring properties) to a growth chronology for the area (Fritts 1976; Schweingruber 1988), thereby calibrating the fire-scar record to calendar time with annual resolution. Once the rings are dated with annual precision, the lesion caused by fire damage can generally be placed within a single year of growth, and often to season (Dieterich and Swetnam 1984). If the lesion cannot be associated positively with a single ring, temporal errors of accuracy are generally within ± 2 years. For the purposes of spatial reconstruction of historic fires, the fire scar record thus provides a reliable proxy of fire occurrence with high spatiotemporal accuracy, especially when CFRs are used, material is crossdated, and the MMU is respected.

Current evidence suggests that well replicated, spatially distributed fire-scar CFRs reflect spatial properties of fire accurately at the temporal scale of a year (Falk 2004; Van Horne and Fulé 2006; Brown et al. 2008a, b). For example, Farris et al. (2010) demonstrated that observed twentieth-century fire perimeters in the Rincon Mountains of Arizona were recorded accurately by CFRs based on fire-scarred trees, especially for fires larger than a few hectares. CFRs are especially reliable when sample size is sufficient to capture the majority of fire events at each point (Falk and Swetnam 2003; Falk 2004; Farris et al. 2010).

7.2.2 Spatial Interpolation Techniques

"Everything is related to everything else, but near things are more related than distant things."- Tobler's 1st Law of Geography (1970).

Because fire spreads largely as a contagious process, spatial autocorrelation is expected at some scale (Heyerdahl et al. 2001; Peterson 2002; Kellogg et al. 2008). Consequently, two adjacent points recording fire in a single year are more likely to have been burned by the same fire than two points that are more distant from one another, which may have burned in different fires during the same year. Because the fire was not observed, we cannot say absolutely whether the area between points burned. However, we can calculate likelihoods of this being the case, using probabilistic interpolations. The primary fire-scar record can be thought of as a binary (occurrence or absence) array in space and time, with interpolations into the intervening spaces.

Hessl et al. (2007) demonstrated multiple techniques for creating likelihood surfaces of area burned by interpolating spatial point data. Here we review three techniques for spatial interpolation (Fig. 7.2), and discuss the strengths and weakness of each specifically for the task of reconstructing spatial pattern in historic fires.

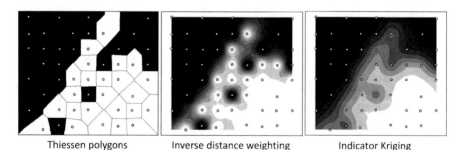

| Thiessen polygons | Inverse distance weighting | Indicator Kriging |

Fig. 7.2 Examples of Thiessen polygons (*TP*), Inverse Distance Weighting (*IDW*), and Indicator Kriging (*IK*) interpolation methods for the Monument Canyon study site, New Mexico. The year 1893 is shown in each of the panels. Black represents composite fire records (*CFRs*) that burned, white represents *CFRs* that have no record of fire in that year. Inverse distance weighting used a power parameter equal to two with nine neighbors; Indicator Kriging (*IK*) used a variable search radius with a spherical semi-variogram model

7.2.2.1   Thiessen Polygons

Thiessen polygons (TP), or Voronoi diagrams, are constructed by bisecting the vector between every pair of nearest-neighbor sample points p (Fig. 7.2, left). Polygons are created where these bisecting lines intersect, creating a space-filling array with areas, denoted by $V(p_i)$, associated with each sample point (Okabe et al. 1992). All bisecting lines on the plane are partitioned such that all lines around $V(p_i)$ are equidistant to two or more p. Thiessen polygons have the advantage that no parameterization (such as a regression slope or distance-decay rate) is required; polygons are delineated directly from the spatial array of sample points. A disadvantage is that polygon size is a function of sample point density. Sparse samples generate relatively large polygons and consequently coarser reconstruction of a spatial process (Watson 1981), whereas dense sampling generates smaller and more precise polygons. The Thiessen Polygon approach assumes that all the area within a Thiessen polygon experienced the same fire, so spatial reconstructions are typically binary (fire, no fire).

7.2.2.2   Inverse Distance Weighting

Inverse distance weighting (IDW) assigns values of a variable based on the distance from the nearest neighbor of a known point (Isaaks and Srivastava 1989):

$$\hat{Z}(s_0) = \frac{\sum_{i=1}^{n} \lambda_i Z(s_i)}{\sum_{i=1}^{n} \lambda_i}$$

where $\hat{Z}(s_0)$ is the value to be predicted for location s_0, n is the number of neighbors used for prediction, $Z(s_i)$ is the weighted value at location s_i, and λ_i is the assigned weight, defined as

$$\lambda i = \frac{1}{d^p},$$

where d is distance to the nearest neighbor and p is an exponent chosen by the modeler.

As its formulation suggests, IDW assigns greater weight to nearby sample points than to those that are more distant (Fig. 7.2, center). This is conceptually suitable for a contagious process such as a spreading fire, in that the probability is greater that areas near a fire-scarred tree were affected by the same fire than more distant locations on the landscape. However, parameterization of IDW is somewhat arbitrary, particularly the assignment of the power parameter p, which sets the rate of decay (Isaaks and Srivastava 1989; Burrough and McDonnell 1998).

7.2.2.3 Indicator Kriging

Indicator Kriging (IK) interpolates values of a binary variable between known sample points by constructing a functional relationship between distance and *semivariance*:

$$\hat{\gamma}(d) = \frac{1}{2n}\sum\nolimits_{i=1}^{n}\{z(x_i) - z(x_i + d)\}^2$$

where variables $z(x_i)$ and $z(x_i + d)$ are objects with values of interest separated by distance d, for n pairs of observations (Burrough and McDonnell 1998). A plot of $\hat{\gamma}(d)$ by d is the *empirical variogram*. Three components describe the shape of the variogram: the nugget, sill, and range. The nugget describes uncorrelated noise, the sill the between-point distance beyond which spatial covariance is negligible, and the range is the distance at which the sill is observed (Fig. 7.3). Using the variogram, IK gives a minimum-variance estimator of the interpolated likelihood surface between data points (Isaaks and Srivastava 1989; Fig. 7.2, right).

In application, IDW is often more conservative than IK, because with IK the probability of two events co-occurring decreases with distance (d) from an observed point more slowly than with typically chosen weighting parameters in IDW. For example, Hessl et al. (2007) found that IK was consistently more likely to assume that intervening areas had burned than IDW.

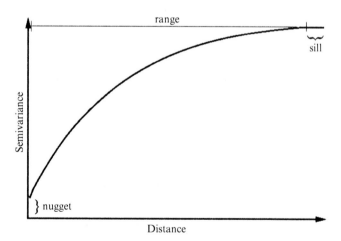

Fig. 7.3 Example of a variogram. The area nearest a composite fire record (*CFR*) or fire scarred tree has the highest probability of occurrence, the 'nugget' represents the area inside a *CFR*, the range is the distance from a *CFR* until the likelihood of the interpolation is small or equal to 0, and the sill is the scale at which variance stabilizes. Indicator Kriging uses the variogram to minimize likelihood in interpolation

7.2.3   Fire Regime Metrics

In addition to spatial patterns *per se*, spatial fire reconstruction allows estimation of important statistical properties of individual fires and fire regimes. These metrics can also be estimated using non-spatial data, but the availability of spatial fire reconstruction can provide other sources of data for these calculations. Two such metrics are:

7.2.3.1   Annual Area Burned

Annual area burned (AAB) can be estimated non-spatially from the percentage of trees or plots recording fire (with the assumption of uniform spacing) as:

$$A_i = (A_s)(P_i)$$

where A_i is the area burned in the ith year, A_s is the study area size, and P_i is the proportion of recording samples in the ith year that detected a fire scar (Morrison and Swanson 1990; Taylor 2000; Farris et al. 2010). Spatial reconstruction of annual fire perimeters provides an alternative means of estimating AAB, by using GIS functions to calculate polygon size A_i for each fire year. In either case, mean annual area burned A (in units of area year^{-1}) is then:

$$A = \frac{1}{T}\sum_{i=1}^{T} A_i$$

where T is the total length of the fire record in years. The fire size distribution can also be derived from the A_i values assuming that the study area is large enough to capture entire fire perimeters.

7.2.3.2   Natural Fire Rotation (NFR)

The natural fire rotation (NFR) is the time (usually calculated in years) required to burn a defined amount of area (A_s) at least once (Heinselman 1973). NFR is calculated using the total time of a sample record, T, and P_s, the proportion of the area burned during T, as:

$$NFR = T / P_s$$

NFR thus integrates mean fire frequency and mean annual area burned ($\bar{A}$) derived above. The crux of the calculation is the derivation of P_s, which can be estimated from non-spatial data in a spatially implicit manner by assuming that the proportion of recording sites with fire in a given year is correlated linearly with annual area burned, as:

$$P_s = \sum_{i=1}^{T} Pi$$

over i years. If fire perimeters are known, NFR (in years) can be derived in a spatially explicit manner as:

$$NFR = \frac{A_s}{\bar{A}}$$

For example, using a 100-year fire history record in a study area of $A_s = 10,000$ ha with reconstructed fire perimeters giving $\bar{A} = 100$ ha year^{-1},

$$NFR = \frac{10,000 \text{ ha}}{100 \text{ ha year}^{-1}} = 100 \text{ years}$$

It is worth noting that in his original exposition, Heinselman (1973) cautioned that NFR, like any single statistic, averages out a great deal of potentially ecologically meaningful variation in fire occurrence over space and time.

7.2.4 Case Studies

The geospatial methods described here may be sensitive to different confounding factors, depending on the spatial extent and resolution of sampling and the degree to which the sample represents the fire regime of the surrounding landscape. We explore these issues by comparing three case studies that used a variety of geospatial interpolation methods to estimate fire perimeters. The study sites are from different geographic regions (Fig. 1, in preface) and are different sizes (Fig. 7.4), but have similar fire regimes, tree species, and sampling techniques. These three sites also provide similar dependent (occurrence or absence of fire in a given year) and independent variables (distance and time between fire events).

Fire-scarred materials in the three studies were analyzed using standard dendrochronological techniques (Stokes and Smiley 1968; Dieterich and Swetnam 1984; Fritts and Swetnam 1989). The smallest site (Monument Canyon, Falk 2004) has the highest sampling resolution for fine scale (200 m between points) reconstruction. At the mid-scale site (500 m between points, Mica Mountain) Farris et al. (2010) compared fire-scar data to surveyed fire perimeters. At the largest site (100–1,000 m between points, Swauk Creek Watershed, Everett et al. 2000; Hessl et al. 2004), Hessl et al. (2007) compared multiple techniques for reconstructing landscape fires. We identify physical characteristics of the sites and the comparative project methodologies from each location (Table 7.1). Full details of the sampling methods at each study site are available in the cited references.

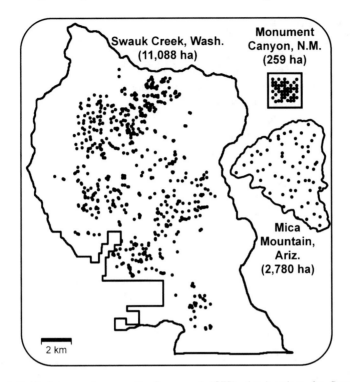

Fig. 7.4 Individual tree and composite fire record (*CFR*) plot locations for Swauk Creek Watershed (*SCW*), Mica Mountain (*MIC*), and Monument Canyon (*MCN*). *SCW* represents single trees, *MCN* and *MIC* used plot-based *CFRs*

Table 7.1 Site and sampling characteristics for each of the three case study location. The Swauk Creek study was based on individual trees rather than composite fire records

	Monument Canyon	Mica Mountain	Swauk Creek
Vegetation type	Rocky Mountain ponderosa pine	Madrean ponderosa pine	Pacific Northwest ponderosa pine
Historic fire severity	Low	Low	Low to mixed
Mean elevation (m)	2,500	2,400	800
Sample type	Plot	Plot	Targeted trees
Sample area (ha)	254	2,780	11,088
Plot size (ha) and # (n)	0.1 ha (45)	1.0 ha (52)	No plots (665)
Minimum mapping unit	Plot CFR[a]	Plot CFR	Individual tree
Number of trees sampled	198	405	665
Trees per ha sampled	0.78	0.15	0.06
Interpolation methods	IK, IDW	TP	TP, IDW, IK, expert
Verification	None	Independently mapped fire perimeters	Comparative

[a] Composite fire record

7.2.4.1 Case Studies

Monument Canyon Research Natural Area (MCN) is a 254-ha protected forested stand in the Santa Fe National Forest, in the Jemez Mountains of northern New Mexico, USA (35° 48′ N, 106° 37′ W; mean elevation 2,500 m) on the southwestern rim of the Valles Caldera (Falk 2004). Topographically, most of MCN is relatively level or gently sloping mesa tops and small drainages. Vegetation is primarily ponderosa pine (*Pinus ponderosa*); with Douglas-fir (*Pseudotsuga menziesii*), white fir (*Abies concolor*), and southwestern white pine (*Pinus strobiformis*) occurring along northern aspects and mesic drainages between mesas. Quaking aspen (*Populus tremuloides*) and Gambel oak (*Quercus gambelii*) occur on disturbed sites, Rocky Mountain juniper (*Juniperus scopulorum*) and piñon (*Pinus edulis*) on warm and rocky southwestern aspects. Historically, MCN experienced frequent low-severity fire (mean fire interval [MFI] ≤ 4 years for the 254 ha, Falk and Swetnam 2003; Falk 2004). MCN was designated as a research natural area (RNA) in 1935, and has been reserved from grazing and burning for nearly a century. Because of intentional fire suppression in surrounding areas, the RNA is now more departed from its HRV (historical range of variability, Morgan et al. 1994) than areas surrounding it.

To establish the site's fire history, Falk (2004) collected and analyzed demographic and fire-history data using georeferenced 0.5-ha plots in a 200 m grid design (~4 ha cells). A total of 45 grid plots containing 198 trees were dated to the calendar year (mean sample density = 0.78 trees ha^{-1}). Composite fire chronologies were compiled for each grid cell (μ = 4 trees/cell). Sample depth analysis indicated that the record is reliable beginning in 1598 (Falk 2004).

For spatial fire reconstruction, a binary *year* × *fire* matrix (fire, no fire) was compiled for all composite fire records over the period 1598–1900. IK and IDW algorithms were used to interpolate fire occurrence over the landscape between sampled grid points in ArcMap 9.3 (ESRI 2008). We used these interpolations to create surface maps of the probability of fire occurrence by year based on known fire occurrence at the grid sample points. To evaluate spatial pattern, we restricted our reconstructions to the period 1700–1910, the period during which all grid cells were recording (meaning that at least one tree had been scarred previously and was thus more likely to record a subsequent fire).

Mica Mountain (MIC) is a 2,780 ha study site located in the Rincon Mountains, Saguaro National Park, Arizona, USA (32° 12′ N, 110° 32′ W, mean elevation 2,400 m) (Farris et al. 2010). Terrain varies from gentle slopes near the peak to steep drainages in lower elevations. Vegetation at MIC is dominated by ponderosa pine; southwestern white pine is subdominant at higher elevations, and evergreen oaks (*Quercus hypoleucoides, Q. turbinella*) at lower elevations. Isolated stands of mixed conifers dominated by ponderosa pine, Douglas-fir, white fir, southwestern white pine, and Gambel oak exist on north aspects. Historically, MIC experienced frequent low-severity fire in ponderosa pine stands, and low to mixed severity in mixed-conifer stands (Baisan and Swetnam 1990). MIC has been managed by the National Park Service (NPS) since 1933 and has been not been logged commercially or heavily grazed; the entire range is roadless above 1,200 m. Prescribed and wildland fires were common throughout the twentieth century across MIC, with NPS recording fire perimeters since the 1940s.

To establish the site's fire history, Farris et al. (2010) collected and analyzed demographic and fire-history data using georeferenced 1-ha plots in a 1.2 km grid design. A total of 37 random and 25 grid plots containing 405 trees were dated to the calendar year (mean sample density = 0.15 trees ha^{-1}). Composite fire chronologies were compiled for each grid cell ($\mu = 8$ trees/cell).

For spatial reconstruction, a binary *year x fire* matrix (fire, no fire) was compiled for all composite records over the period 1937–2000. TP were used to interpolate fire occurrence over the landscape between the sampled grid points. To verify geospatial pattern reconstructed by interpolation from fire-scar data, the resulting polygons were compared to surveyed and mapped NPS fire perimeters (Fig. 7.5) from 1933-present, and cumulative spatial patterns of fire frequency were evaluated. During this period there were polygons with as many as 9 fires recorded in the MIC study area; Farris et al. (2010) limited their inferences to fires after 1700.

The *Swauk Creek Watershed* (SCW) is an 11,088-ha study site located on the Okanogan-Wenatchee National Forest of central Washington, USA (47° 15′ N, 120° 38′ W, mean elevation 800 m; Everett et al. 2000; Hessl et al. 2004). The area has steep-sided valleys on partially metamorphosed sedimentary rock and

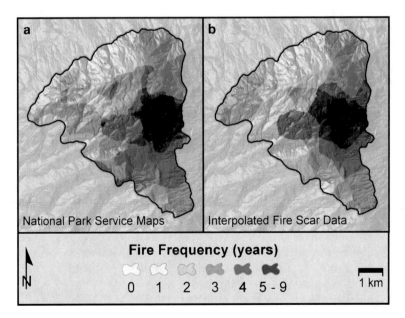

Fig. 7.5 Spatial patterns of fire frequency from 1937 to 2000 calculated from (**a**) National Park Service Atlas maps and (**b**) fire-scar data interpolated using *TP*. Farris et al. (2010) found a Pearson's cross correlation of 0.81 between surveyed and reconstructed perimeters (Based on Farris et al. (2010))

volcanic rock. Vegetation within SCW includes ponderosa pine along a narrow northeast- to-southwest band between 600–1,200 m elevation, and Douglas-fir, grand fir (*Abies grandis*), western larch (*Larix occidentalis*), and lodgepole pine (*Pinus contorta*) at higher elevations. Historically, SCW experienced low-severity fire in the ponderosa pine zone, low to mixed severity in the mixed-conifer zone, and high severity in the lodgepole pine zone.

To establish the site's fire history, Everett et al. (2000) and Hessl et al. (2004) used targeted sampling across the SCW to produce spatially explicit fire histories. Fire-scarred trees were georeferenced using triangulation and input into GIS. A total of 665 trees were dated to the calendar year (mean sample density = 0.06 trees ha^{-1}). Point fire-return intervals were calculated from single trees, but Hessl et al. (2004) also identified synchronous fire years with 10% and 25% of all trees scarred. The interpolations of Hessl et al. (2007) used only those years in which a minimum of four trees recorded fire. Sample-depth analysis indicated that the record is reliable beginning in 1700 (Hessl et al. 2004).

For spatial fire reconstruction, a binary *year x fire* matrix (fire, no fire) was compiled for all trees over the period 1700–1899. IK, IDW, TP, and an expert approach were used to interpolate fire occurrence over the landscape between sampled trees. The expert approach involved a local fire ecologist drawing perimeters by hand based on observed fire scars and topographic controlling features (Hessl et al. 2007). TP, IK, and IDW were conducted in a GIS.

To evaluate spatial pattern, Hessl et al. (2007) restricted reconstructions to the period 1700–1899; area-under-curve (AUC) was calculated from receiver operating characteristic plots (Fielding and Bell 1997). AUC is a measure of accuracy for presence-absence models (Manel et al. 2001).

7.3   Results: Spatially Reconstructed Fire Histories

7.3.1   Fine-Scale Spatial Fire History

The MCN study area was too small to detect entire large (>250 ha) fires, although it did capture possible boundary dynamics between some events in successive years (Fig. 7.6). Single and multiple plots also recorded occasional small fires. Multiple points recording fire tended to occur during years with low (negative) regional Palmer drought severity index (PDSI) reconstructed values, whereas single point fires tended to occur during years with high (positive) PDSI values (Falk 2004); PDSI derived from instrumental and tree-ring data available from the paleoclimatology program at the National Climatic Data Center (*http://www.ncdc.noaa.gov/paleo/*). Large fires tended to occur after antecedent wet years. These results are consistent with other fire-history studies that indicate an increase in fine fuels after several wet years, leading to widespread regional fires during dry years in fuel-limited systems (Grissino-Mayer et al. 2000; Schoennagel et al. 2004).

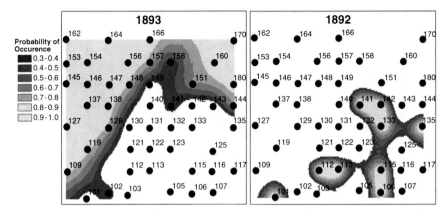

Fig. 7.6 Fire probability surface generated by *IK* for two successive fire years in Monument Canyon. *IK* used a variable search radius; interpolated values less than 0.30 were removed from the display. Most areas burned by the 1892 fire did not reburn in 1893, with the exception of plot 141. No single tree recorded both fires, however. Plot 122 burned in 1893 but its interpolated value is small because it is isolated between other unburned plots

The spatial resolution of the MCN reconstruction reveals patterns that may reflect within-perimeter heterogeneity. For example, between 1892 and 1893 two fires burned virtually the entire surface of MCN (the most recent large fire occurred 5 years prior in 1887). Only one point recorded both fires; several points record no fire both years, and most points burned in 1887 re-burned in 1892 or 1893. As a result, burned areas in the 2 years fit like jigsaw puzzle pieces (Fig. 7.6), suggesting a self-organizing limitation to fire spread by available fine fuels (Chap. 3).

7.3.2   Mid-Scale Spatial Fire History

Of the three spatial reconstructions in this comparison, the MIC study area was the only one that had the benefit of being validated with surveyed NPS fire perimeters. MIC also experienced less human-caused disturbance and fire suppression during the twentieth century and is therefore probably closest to its HRV (Morgan et al. 1994). Farris et al. (2010) used Pearson's cross-correlation coefficient and a Kolmogorov-Smirnov (KS) test (Zar 1984) to compare the NPS record and their reconstruction for spatial correspondence of frequency values, grouping fire frequency into classes and comparing predicted (interpolated) and actual (NPS) fire frequency. Pearson's cross-correlation coefficient (Zar 1984) between maps was $r = 0.81$ ($p = 0.001$) reflecting strong correspondence between high and low fire frequency surfaces (Fig. 7.6). The KS test showed no difference in frequency distribution ($p = 0.97$). Less than 15% of the total area in each map (fire-scar estimated vs. NPS atlas) differed by a frequency of more than one fire.

For twentieth century fires, Farris et al. (2010) calculated NFR to be 26.8 years based on NPS surveyed perimeters, compared to 29.6 years from interpolated perimeters. Pre-twentieth century NFRs were the same as previous fire-history studies in MIC (Baisan and Swetnam 1990) that used 'targeted' samples, and in nearby areas with similar species composition and climate (Grissino-Mayer et al. 1996; Kaib 1998; Swetnam and Baisan 1996; Van Horne and Fulé 2006). The high correlation of surveyed burned area to reconstructed burned area (Fig. 7.6) demonstrates that fire-scarred trees can be accurate and reliable recorders of the spatial properties of surface fire events at landscape scales.

7.3.3 Broad-Scale Spatial Fire History

The SCW data set detected both small and large fires in the period 1700–1899 used for reconstruction (Hessl et al. 2007). IK produced the largest estimated burned areas; IDW, TP, and the expert approach generated smaller estimates, respectively. Hessl et al. (2007) found that spatial interpolations can be replicated at landscape scales. Of the three quantitative methods, IK was determined to be the most accurate using AUC analysis (AUC mean = 0.92, SD = 0.10), followed closely by IDW (AUC mean = 0.89, SD = 0.13). TP was weakest (AUC mean = 0.78, SD = 0.13) (Hessl et al. 2007).

The SCW dataset detected entire fire perimeters during the historical period (Fig. 7.7). Multiple fires that appear to be spatially separate were detected in the same year (1895) (Hessl et al. 2007). The size of the SCW study area appears to be large enough to capture many complete fire perimeters, but many events appear to have burned beyond the study area boundary.

7.4 Insights from Spatial Reconstruction of Fire Histories

Reconstruction of the spatial properties of historic fires provides a unique window into how fire operated historically as a landscape process. As these case studies indicate, the landscape spatial pattern of historic fires can be reconstructed in considerable detail. In this section we summarize the potential contribution of spatial fire reconstruction for understanding fire regimes and properties of individual fires, the potential for applying these methods to other ecosystem types with different fire regimes, issues and caveats in spatial reconstruction, and the application of such work for fire and ecosystem management.

7.4.1 Basic Insights from Case Studies

At each spatial scale, geospatial interpolations had specific advantages and drawbacks associated with sample density, edge effects, and the degree to which the sample area represented the landscape or regional fire regime. At MCN, reconstructed fire

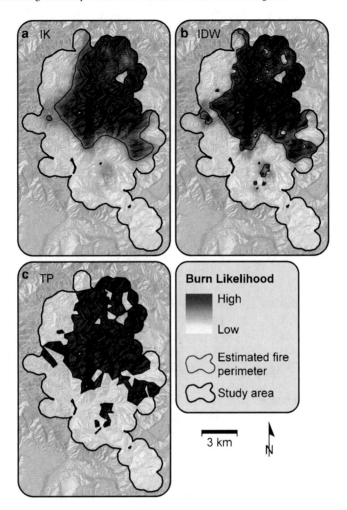

Fig. 7.7 Indicator Kriging (*IK*) (**a**), inverse distance weighting (*IDW*) (**b**), Thiessen polygons (*TP*) (**c**), from the 1776 fire in Swauk Creek Watershed (*SCW*). One large fire is observed in the northeastern part of the study area. What may be a separate fire is visible south of the main fire (Source: Hessl et al. (2007))

perimeters appeared to follow topographic features, suggesting that there were no biases in the global predictions of fire perimeters even if there might be inaccuracies in individual reconstructions. MCN occasionally recorded small fires (<100 ha, 1–4 CFR plots), typically in positive PDSI years. From the mid-scale reconstruction Farris et al. (2010) observed 12 twentieth-century fire years greater than 100 ha (up to 1,600 ha) with numerous perimeters overlapping at different dates and locations. Farris et al. (2010) also found that fire scars were reliable recorders of fire when compared to historical surveys. The broad-scale reconstruction (SCW, Hessl et al. 2007) employed a distributed record from individual scarred trees. The SCW study area appears to have been large enough to contain the perimeters of both individual large fires and multiple large fires within a single year. The case studies in this chapter suggest that historical fires in the MIC and SCW study sites were typically larger than 100 ha and often larger than 1,600 ha. The observed large fire years in MCN typically involved >100 ha, but their edges often extended outside the study area.

7.4.2 Understanding Topographic Control of Fire Spread

Spatial pattern reconstruction can provide insights into landscape elements that regulate fire regimes. In the Klamath Mountains of Oregon, Taylor and Skinner (2003) used cluster analysis to identify landscape segments with coherent patterns of synchronous fire occurrence. Segment boundaries corresponded generally to landscape features such as ridgelines and drainages. The effectiveness of these features varied with the severity of fire weather. Under severe conditions, fire would overrun features that could control the spread of fire under more moderate conditions. In this case, the spatial pattern of fire was revealed primarily by the spatiotemporal relationships of annual fire records.

Reconstructions at MCN, MIC, and SCW also reflect the controlling influence of topography on fire spread. For example, the 1892–1893 fires in MCN (Fig. 7.6) are limited along mesa tops on the north and south sides of the study area. Farris et al. (2010) recorded the highest frequency of fires in an area below a large rock escarpment above the east slope of Mica Mountain, and the reconstructions at SCW reflect the limiting effects of topography in some years near the watershed margin (Fig. 7.7).

7.4.3 Reconstructing Spatial Heterogeneity in Fire Occurrence and Burn Severity

All three case studies captured spatial heterogeneity in fire occurrence at some scale. The MCN site was small in spatial extent, but the closely spaced sampling grid (200 m) and high sample density (0.78 trees ha^{-1}) used to create local composite fire records captured likelihoods about fire perimeters and heterogeneity in burn severity at a much finer scale than the other studies (Fig. 7.5). SCW was sampled at a lower sample density (0.06 scarred trees ha^{-1}), but the large number and spatial dispersion of scarred trees facilitated reconstruction of entire (and possibly multiple) fires at landscape scales. In some respects, the MIC site combined the best attributes of both, using composite fire records in gridded study plots, while covering an entire forested landscape. MIC's reconstruction also used documented fire perimeters from the twentieth century to test the faithfulness of fire-scarred trees as recorders of fire events. This provided strong validation that spatial patterns of historical fires can be reconstructed accurately from fire scars. Other recent studies (Parsons et al. 2007; Shapiro-Miller et al. 2007) also have found strong correlation between fire scars and fire atlases.

Depending on the desired application, spatial fire reconstruction could be used to estimate the historical fire patch size distribution, or estimate historical spatial patterns of fire severity. In contemporary landscape fire analysis, patches are typically classified by fire severity (Kotliar et al. 2003). Patch sizes can strongly influence postfire succession, hydrology, soil stability, and wildlife use (Agee 1998; Bonnet et al. 2005). Fire scars do not indicate severity precisely, but the occurrence of a scar generally can be interpreted as evidence of locally low- to moderate-severity fire, intense enough to scar the tree but not so intense as to kill it. Reconstruction of the patch size distributions of historical fires, particularly the proportion of a landscape in low- and moderate-severity patches, would be a valuable reference point for interpreting the patch mosaics produced by contemporary fires (Morgan et al. 2001).

7.4.4 Reconstructing Landscape Patterns of Fire across Multiple Years

Prior to fire suppression, fires in much of western North America may have burned for weeks or even months before being extinguished by rain or by cooler weather, or by reaching the limit of flammable vegetation. Depending on fire severity and frequency, the limit of fire extent in 1 year would alter the spatial pattern of fuels in subsequent years. The time during which fuel conditions would influence the spread and effects of subsequent fires varied with the vegetation type and climate.

Fire behavior theory predicts that over time, the spatial footprint of successive fires created and maintained complex landscape mosaics dependent on fire severity and time since fire (Heinselman 1973; Minnich and Dezzani 1991). In ecosystems with less frequent (10^2–10^3 years) moderate- to high-severity events, fuel complexes and stand structure would be altered substantially immediately postfire, including extensive mortality, consumption of canopy and understory fuels, and altered soil and hydrology. Such conditions would represent a very different fire environment in the years immediately following a high-severity event, and would modify the behavior and effects of subsequent events (Finney et al. 2005; Kulakowski and Veblen 2007; Collins and Stephens 2008). If fire intervals were typically long, the fuel matrix could become re-established by the time climate conditions were favorable for another widespread event, and the effects on fire spread would decline with time.

Such relationships have been observed in a few contemporary landscapes where fires have been allowed to retain at least some element of their original role and dynamics. Collins et al. (2009) found that the perimeters of fires in the Sierra Nevada reflected different fire behavior in areas that had burned in the previous fire. This "self-limiting" property of fire regimes in a given place and time would create a dynamic landscape equilibrium similar to a shifting mosaic model of stand age and composition, though with no assumption of stationarity (Scholl and Taylor 2010; Chap. 1; Chap. 3). Reconstructing the spatiotemporal patterns of historic fires with sufficiently fine resolution could provide a window into past landscape dynamics beyond what can be observed in contemporary conditions, especially where both fire suppression and forest management have altered the relationships of fire to landscapes (Morgan et al. 2001).

Interpolation techniques can be used to infer spatial properties of a landscape process such as fire from point records such as fire scars. However, an important element in fire spatial reconstruction is the evaluation of the accuracy of interpolated fire pattern, by comparing it to maps derived from other methods (Hessl et al. 2007; Farris et al. 2010). Note that two basic types of calibration are possible: comparison with actual historical fire perimeters (where these are known), and comparison with the properties of contemporary fires. These comparisons can be based on a variety of sources, including:

1. Field-mapped fire data (e.g. fire atlases) maintained by many land-managing agencies (Shapiro-Miller et al. 2007; Collins and Stephens 2008; Farris et al. 2010).
2. Aerial photographs or remotely sensed imagery such as normalized burn severity maps (Miller and Yool 2002; Cocke et al. 2005; Lentile et al. 2006; Monitoring Trends in Burn Severity, mtbs.gov).
3. Field reconnaissance and georeferencing of individual points on postburn landscapes, and comparison with the location of fire-scarred trees (Parsons et al. 2007).
4. Written records describing the location and spatial distribution of burned areas, which is sometimes the only source of correlative information for historical fires (Shapiro-Miller et al. 2007).

In some cases, fire-scar evidence may be more accurate at the point scale than a CFR, because of the inherently fine-scale processes that scar trees. Mapped fire perimeters are useful primarily for verifying the accuracy of interpolation of spatial fire pattern between sampled points. In most cases the present range of variability can be estimated more accurately than the historical range of variability (HRV) (Morgan et al. 1994; Wong et al. 2003).

7.4.5   Estimation of Statistical Properties of Fire Regimes

Spatial reconstruction of multiple events in time and space offers insights into statistical properties of fire regimes, beyond the landscape pattern of individual fire events or years. Falk (2004), McKenzie et al. (2006), and Farris et al. (2010) used spatially explicit reconstruction of fire history to estimate MFI at multiple spatial scales. By re-sampling fire occurrence data in moving windows of varying size, they showed that MFI is a scale-dependent parameter, generally following a power-law distribution analogous to the species-area theorem. Such cross-scale analysis, which requires spatially explicit and georeferenced fire history data, may indicate the interaction of top-down and bottom-up regulation of fire regimes (Moritz 2003; Falk et al. 2007; Kellogg et al. 2008; Chaps. 1 and 3).

Spatially explicit reconstruction of past fires could enable calculation of NFR for historical fire regimes, a property that has been difficult to estimate without recorded fire perimeters. Unlike MFI, NFR is a self-scaling (or scale-independent) statistic derived from the product of annual area burned and fire frequency. In the past, tree-ring studies were not sufficiently well distributed to allow accurate estimation of fire sizes. However, the development of interpolated fire perimeters using historical records could provide fire size estimations, and thus facilitate estimation of NFR. The MIC and SCW study areas here were large enough to capture full fire perimeters (Hessl et al. 2007; Farris et al. 2010).

Fire regimes are often grouped into three general severity classes: low, moderate, and high (Brown and Smith 2000). However, many fire regimes are actually mixed in both space and time. Fire scars are most frequently associated with low- and moderate severity fire regimes, where trees may be scarred but tend to survive the event. Reconstructing spatial pattern is inherently more difficult in mixed- and high-severity fire regimes where the nature of the disturbance event tends to destroy some or all of the annual record.

Mixed-severity fire regimes can be described either as spatial mosaics of severity classes (spatial mixtures), or variation in time of fire severity in a given location or forest type (temporal mixtures) (Heyerdahl et al. 2001). Because many mixed-severity fire regimes include a high-severity component, the records produced in these forest types tend to be incomplete. Proxy techniques include convergent lines of evidence such as fire-scars on trees bordering chaparral, and death dates from killed trees and scars on survivor trees adjacent to high-severity patches (Margolis et al. 2007; Margolis and Balmat 2009). Age-structure analysis, including death and recruitment data, can also help identify areas when and where mixed-severity fires may have taken place (Brown et al. 2008a,b). When these data types are well replicated on the landscape, and dated with sufficient accuracy and precision, they can be used to reconstruct historic fire spatial pattern (Brown et al. 2008a,b). The spatial resolution of these methods can be fairly high (10^3–10^4 m^2), but temporal resolution of age reconstructions (10–50 years) is lower than for fire scar evidence.

High-severity fires typically involve extensive mortality of overstory trees (leading to stand replacement), so multiple surviving trees with fire scars are rare. The primary proxy technique in this regime type is age-structure analysis, which includes finding recruitment ages of trees, cohort identification, and comparing death dates of remnant material to living trees (Romme and Despain 1982; Johnson 1992). Aspen cohorts, which can rapidly re-colonize a burned site after fire, provide a higher temporal resolution than other recruitment species and can be used to date high-severity events to annual or near-annual resolution. Consequently, spatial mapping of age structure of aspen clones within mixed conifer or spruce-fir stands can reveal historical stand-replacing fires (Turner et al. 2003; Margolis et al. 2007).

7.4.6   Temporal Considerations in Interpreting Landscape Patterns of Historical Fire

The landscape composition that we see today is not the one observed by Nativ Americans, European colonists, or even the first U.S. or Canadian Forest Servic employees. Vegetation change within the last 50–100 years has been dramatic i some locations, as many time-series photos show (Turner et al. 2003). Defining reference conditions for the range species composition and landscape structure under historical climatic conditions is paramount in determining the degree o departure from the historical range of variability (HRV) (Morgan et al. 1994).

Absence of evidence, it is often said, is not evidence of absence (Altman and Bland 1995). In the case of fire regimes, historical fire evidence can be destroyed or degraded by subsequent fires, weathering, physical removal, or decomposition. The further back in time that a fire has occurred, the less likely it is to leave suffi-cient evidence for reconstruction. There are therefore temporal limits to accurate landscape fire reconstruction, depending on the fire regime and persistence o proxy evidence. In much of western North America, this limit is reached abou 500–700 years ago (i.e., AD 1300–1500), and in some areas much later.

7.4.7   Applications of Spatial Fire History Reconstructions in Ecosystem Management

Every year fires burn forests and rangelands in western North America that have ecological, economic, and social value. The desire to live in exurban homes tha border forested areas, i.e. the "wildland-urban interface" (WUI), has increased th risk to human lives and property greatly over the last 25 years (Pyne 2001; Fa et al. 2010; Chap. 11). Recent regional climate change has increased the length the fire season in some areas and amplified the number of ignitions and hectar burned over the last decade. As a result, "super fires" may be burning at great intensities than did historical events (Westerling et al. 2006,). The coupled effe of fire suppression over the last century and rising annual temperatures have led an environment with the potential for rapid and unpredictable change (Chap. 1). It h been estimated that the majority of existing western forests are at risk of underg ing vegetation type-conversion after stand-replacing fire due to a combination factors including changing climate, invasive species, and land use (McKenzie et 2004; Savage and Mast 2005). Similar patterns are evident in fire regimes globa (Krawchuk et al. 2009).

In volatile political and uncertain climatic environments, land managers respon sible for fire on public lands have a critical need for information that will help then understand how fire behaves at landscape scales. For example, the increasing exten and complexity of the WUI zone increases the risk and potential damage to human infrastructure from escaped fires, while decreasing the flexibility of land managers who wish to return fire to the landscape as a natural process and to reduce the probability of catastrophic fires (Pyne 2001; Chap. 11). In such a context, better understanding of spatial properties of fire as a landscape process is significant both for designing management strategies and for communicating to the public that fire is a natural process.

As spatial fire datasets become more common and publicly available via the internet and other media, managers have access to an unprecedented level of information. Some of the potential benefits of historical fire reconstruction to forest management include:

1. *Recognition of fire as a complex landscape process.* Contemporary fires in many areas are burning under anomalous fuel conditions. Moreover, most large fires occur under extreme weather in which control measures are inadequate, whereas historically most fires probably burned under more moderate conditions (Swetnam et al. 1999). As a result, the public image of wildfire is of a highly destructive process, when in fact historically many fires were relatively more benign in their ecological effects. Improved maps of historical fire mosaics would demonstrate that fire and ecosystems can coexist on the landscape, and in fact have done so for thousands of years. Illustrations of the spatial extent of past fires could also be useful for designing appropriately scaled management regimes, and for public understanding of the large role that fire should be playing in many western landscapes.

2. *Understanding feedbacks and self-limiting properties in landscape dynamics.* The few parts of North America where landscape fires are not always suppressed have valuable lessons to teach fire science. Areas such as the Gila Wilderness (New Mexico), Selway-Bitterroot Wilderness (Montana and Idaho), subalpine forests of the Central Rocky Mountains (Colorado), and the central Sierra Nevada wilderness (California) are living laboratories for the landscape spatial dynamics of fire (Miller and Urban 2000; Rollins et al. 2001; Kulakowski and Veblen 2007; Collins et al. 2009). Large spreading fires often drop to the surface or go out entirely when they reach the footprint of fires from recent past, which still reflect the legacy of altered fuel complexes (Agee and Skinner 2005; Collins et al. 2009). Even in more highly managed landscapes with extensive fire suppression, such as northern Arizona and southern California, past fires alter the behavior of subsequent events (Finney et al. 2005; Moritz 2003; Scholl and Taylor 2010). These natural experiments complement the lessons of spatial fire reconstruction, which suggest that historical landscapes were resilient because fire maintained a dynamic relationship with vegetation. Although large severe fires undoubtedly occurred at some spatial scale in forests (especially at higher elevations where such events were typical), the legacy of many centuries of low-severity fire offers support for fire policies such as wildland fire use for resource benefit. Careful reconstruction of inter-annual patterns in historic fires, such as those illustrated here, can place these landscape dynamics in an even longer temporal perspective.

3. *Fire severity mosaics and landscape configuration.* Many recent large fires have created very large high-severity patches. For example, the proportion of the

2001 Cerro Grande fire Los Alamos, New Mexico classified as high-severity was not unusually high for this vegetation type, but the largest high-severity patch was larger than any other known natural fire (Fig. 7.1; Miller and Yool 2002). Similarly, the largest high-severity patches in the 2003 Aspen Fire and 2004 Nutall-Gibson fires in Arizona were caused in part by management actions taken during suppression efforts. Burnout operations conducted as a defensive strategy caused these high-severity patches. An understanding of how postfire mosaics were created prior to the suppression era would help to place these novel landscape features in context, and allow managers to assess whether some occurrence of high-severity fire would fall within the historic range of landscape variability.

Spatial reconstruction of historic fires and fire regimes is just one part of an overall growth in the field of fire science toward embracing complexity. Combined with detailed observations and monitoring of contemporary fires, and computer modeling of fire spread on complex landscapes, spatial reconstruction will help build a richer understanding of fire as a landscape process.

References

Agee, J.K. 1998. The landscape ecology of western forest fire regimes. *Northwest Science* 72: 24–34.

Agee, J.K., and C.N. Skinner. 2005. Basic principles of forest fuel reduction treatments. *Forest Ecology and Management* 211: 83–96.

Altman, D.G., and J.M. Bland. 1995. Absence of evidence is not evidence of absence. *British Medical Journal* 311: 485.

Baisan, C.H., and T.W. Swetnam. 1990. Fire history on a desert mountain range: Rincon Mountain Wilderness, Arizona, USA. *Canadian Journal of Forest Research* 20: 1559–1569.

Barton, A.M. 1999. Pines *versus* oaks: Effects of fire on the composition of Madrean forests in Arizona. *Forest Ecology and Management* 120: 143–156.

Beaty, R.M., and A.H. Taylor. 2001. Spatial and temporal variation of fire regimes in a mixed conifer forest landscape, southern Cascades, California, USA. *Journal of Biogeography* 28:955–966.

Beeson, P.C., S.N. Martens, and D.D. Breshears. 2001. Simulating overland flow following wildfire: Mapping vulnerability to landscape disturbance. *Hydrological Processes* 15: 2917–2930.

Bonnet, V.H., A.W. Schoettle, and W.D. Shepperd. 2005. Postfire environmental conditions influence the spatial pattern of regeneration for *Pinus ponderosa*. *Canadian Journal of Forest Research* 35: 37–47.

Brown, J.K., and J.K. Smith, eds. 2000. *Wildland fire in ecosystems: Effects of fire on flora*. General Technical Report RMRS-GTR-42-2. Ft. Collins: U.S. Forest Service.

Brown, P.M., M.W. Kaye, L.S. Huckaby, and C.H. Baisan. 2001. Fire history along environmental gradients in the Sacramento Mountains, New Mexico: Influences of local patterns and regional processes. *Ecoscience* 8: 115–126.

Brown, P.M., E.K. Heyerdahl, S.G. Kitchen, and M.H. Weber. 2008a. Climate effects on historical fires (1630–1900) in Utah. *International Journal of Wildland Fire* 17: 28–39.

Brown, P.M., C.L. Wienk, and A.J. Symstad. 2008b. Fire and forest history at Mount Rushmore. *Ecological Applications* 18: 1984–1999.

Burrough, P.A., and R.A. McDonnell. 1998. *Principles of geographical information systems*. Oxford: Oxford University Press.

Cocke, A.E., P.Z. Fule, and J.E. Crouse. 2005. Comparison of burn severity assessments using differenced normalized burn ratio and ground data. *International Journal of Wildland Fire* 14: 189–198.

Collins, B.M., and S.L. Stephens. 2008. Managing natural wildfires in Sierra Nevada wilderness areas. *Frontiers in Ecology and the Environment* 5: 523–527.

Collins, B.M., J.D. Miller, A.E. Thode, M. Kelly, J.W. van Wagtendonk, and S.L. Stephens. 2009. Interactions among wildland fires in a long-established Sierra Nevada natural fire area. *Ecosystems* 12: 114–128.

Connell, J.H., and R.O. Slatyer. 1977. Mechanisms of succession in natural communities and their role in community stability and organization. *The American Naturalist* 111: 1119–1144.

Dieterich, J.H. 1980. The composite fire interval–A tool for more accurate interpretation of fire history. In *Proceedings of the fire history workshop*, eds. M. Stokes and J. Dieterich, 8–14. General Technical Report RM-81. Ft. Collins: U.S. Forest Service.

Dieterich, J.H., and T.W. Swetnam. 1984. Dendrochronology of a fire-scarred ponderosa pine. *Forest Science* 30: 238–247.

de Mestre, N.J., E.A. Catchpole, D.H. Anderson, and R.C. Rothermel. 1989. Uniform propagation of a planar fire front without wind. *Combustion Science Technology* 65: 231–244.

Duarte, J.A.M.S. 1997. Fire spread in natural fuel—computational aspects. In: D. Stauffer, Editor, *Annual Reviews of Computational Physics*, World Scientific, Singapore.

Eidenshink, J., B. Schwind, K. Brewer, Z. Zhu, B. Quayle, and S. Howard. 2007. A project for monitoring trends in burn severity. *Fire Ecology* 3: 3–21.

ESRI. 2008. ArcMap 9.3. ESRI, Redlands.

Everett, R., J. Townsley, and D. Baumgartner. 2000. Inherent disturbance regimes: A reference for evaluating the long-term maintenance of ecosystems. In *Mapping wildfire hazards and risks*, eds. R.N. Sampson, R.D. Atkinson, and J.W. Lewis, 265–288. New York: Food Products Press/ Haworth Press.

Fall, J.G. 1998. Reconstructing the historical frequency of fire: A modeling approach to developing and testing methods. M.S. thesis, Simon Fraser University, Burnaby.

Falk, D.A. 2004. Scaling rules for fire regimes. Ph.D. dissertation, University of Arizona Tucson.

Falk, D.A., and T.W. Swetnam. 2003. Scaling rules and probability models for surface fire regimes in Ponderosa pine forests. In *Fire, fuel treatments, and ecological restoration*, Proceedings RMRS-P-29. eds. P.N. Omi and L.A. Joyce, 301–317. Fort Collins: U.S. Forest Service, Rocky Mountain Research Station.

Falk, D.A., C. Miller, D. McKenzie, and A.E. Black. 2007. Cross-scale analysis of fire regimes. *Ecosystems* 10: 809–826.

Falk, D.A., C. Cox, D. Hill, T. McKinnon, E. Rosenberg, K. Siderits, and T.W. Swetnam. 2010. *Living with fire: Land-use planning for forest health and safe communities*. Technical Report of the Arizona Forest Health Advisory Council, Office of the Governor.

Farris, C.A., C.H. Baisan, D.A. Falk, S.R. Yool, and T.W. Swetnam. 2010. Spatial and temporal corroboration of a fire-scar fire history reconstruction in a frequently burned ponderosa pine forest in Arizona. *Ecological Applications* 20: 1598–1614.

Fielding, A.H., and J.F. Bell. 1997. A review of methods for the assessment of prediction errors in conservation presence/absence models. *Environmental Conservation* 24: 38–49.

Finney, M.A. 1999. Mechanistic modeling of landscape fire patterns. In *Spatial modeling of forest landscape change*, eds. D.J. Mladenoff and W.L. Baker, 186–209. Cambridge: Cambridge University Press.

Finney, M.A., C.W. McHugh, and I.C. Grenfell. 2005. Stand- and landscape-level effects of prescribed burning on two Arizona wildfires. *Canadian Journal of Forest Research* 35: 1714–1722.

Fritts, H.C. 1976. *Tree rings and climate*. London: Academic.

Fritts, H.C., and T.W. Swetnam. 1989. Dendroecology: A tool for evaluating variations in past and present forest management. *Advances in Ecological Research* 19: 111–189.

Grissino-Mayer, H.D., and T.W. Swetnam. 2000. Century-scale climate forcing of fire regimes in the American Southwest. *Holocene* 10: 213–220.

Grissino-Mayer, H.D., C.H. Baisan, and T.W. Swetnam 1996. Fire history in the Pinaleño Mountains of southeastern Arizona: Effects of human-related disturbances. In *Biodiversity and management of the Madrean Archipelago: The sky islands of the southwestern United States and northwestern Mexico*, eds. L.B. DeBano and G.G. Gottfried, 399–407. General Technical Report RM-GTR-264. Ft. Collins: U.S. Forest Service.

Gutsell, S.L., and E.A. Johnson. 1996. How fire scars are formed: Coupling a disturbance process to its ecological effect. *Canadian Journal of Forest Research* 26: 166–174.

Heinselman, M.L. 1973. Fire in the virgin forests of the boundary waters canoe area, Minnesota. *Quaternary Research* 3: 329–382.

Hessl, A.E., D. McKenzie, and R. Schellhaas. 2004. Drought and Pacific decadal oscillation linked to fire occurrence in the inland Pacific Northwest. *Ecological Applications* 14: 425–442.

Hessl, A.E., J. Miller, J. Kernan, D. Keenum, and D. McKenzie. 2007. Mapping paleo-fire boundaries from binary point data: Comparing interpolation methods. *Professional Geographer* 59: 87–104.

Heyerdahl, E.K., L.B. Brubaker, and J.K. Agee. 2001. Spatial controls of historical fire regimes: A multiscale example from the interior West USA. *Ecology* 82: 660–678.

Iniguez, J.M., T.W. Swetnam, and S.R. Yool. 2008. Topography affected landscape fire history patterns in southern Arizona, USA. *Forest Ecology and Management* 256:295–303.

Isaaks, E.H., and R.M. Srivastava. 1989. *Applied geostatistics*. New York: Oxford University Press.

Johnson, E.A. 1992. *Fire and vegetation dynamics*. Cambridge: Cambridge University Press.

Johnson, E.A., and K. Miyanishi, eds. 2001. *Forest fires: Behavior and ecological effects*. San Diego: Academic.

Johnson, E.A., and D.R. Wowchuck. 1993. Wildfires in the southern Canadian Rocky Mountains and their relationship to mid-tropospheric anomalies. *Canadian Journal of Forest Research* 2: 1213–1222.

Kaib, J.M. 1998. Fire history in riparian canyon pine-oak forests and the intervening desert grasslands of the southwest borderlands: a dendroecological, historical, and cultural inquiry. M.S. thesis, University of Arizona, Tucson.

Kaib, M., C.H. Baisan, H.D. Grissino-Mayer, and T.W. Swetnam 1996. Fire history in the gallery pine-oak forests and adjacent grasslands of the Chiricahua Mountains of Arizona. In *Effects of fire on Madrean Province ecosystems*. General Technical Report RM-GTR-289. eds. P. F. Ffolliott, L. F. DeBano, and M. B. Baker et al., 253–264. Fort Collins: U.S. Forest Service.

Keane, R.E., R. Burgan, and J. van Wagtendonk. 2001. Mapping wildland fuels for fire management across multiple scales: Integrating remote sensing, GIS and biophysical modeling. *International Journal of Wildland Fire* 10: 301–319.

Kellogg, L.-K.B., D. McKenzie, D.L. Peterson, and A.E. Hessl. 2008. Spatial models for inferring topographic controls on low-severity fire in the eastern Cascade Range of Washington, USA. *Landscape Ecology* 23: 227–240.

Kotliar, N.B., S.L. Haire, and C.H. Key. 2003. Lessons from the fires of 2000: Post-fire heterogeneity in ponderosa pine forests." In *Fire, fuel treatments, and ecological restoration: Conference proceedings*. Proceedings RMRS-P-29. *2002 16–18 April.*, eds. P.N. Omi and L.A. Joyce, 277–280. Fort Collins: U.S. Forest Service.

Krawchuk, M.A., M.A. Moritz, M.-A. Parisien, J. Van Dorn, and K. Hayhoe. 2009. Global pyrogeography: The current and future distribution of wildfire. *PLoS ONE* 4(4): e5102.

Kulakowski, D., and T.T. Veblen. 2007. Effect of prior disturbances on the extent and severity of wildfire in Colorado subalpine forests. *Ecology* 88(3): 759–769.

LANDFIRE. 2007. Homepage of the LANDFIRE Project, U.S. Department of Agriculture, Forest Service; U.S. Department of Interior. Available: http://www.landfire.gov/index.php. Accessed 26 Feb 2010.

Lentile, L.B., Z.A. Holden, A.M.S. Smith, M.J. Falkowski, A.T. Hudak, P. Morgan, S.A. Lewis, P.E. Gessler, and N.C. Benson. 2006. Remote sensing techniques to assess active fire characteristics and post-fire effects. *International Journal of Wildland Fire* 15: 319–345.

Manel, S., H.C. Williams, and S.J. Ormerod. 2001. Evaluating presence-absence models in ecology: The need to account for prevalence. *Journal of Applied Ecology* 38: 921–931.

Margolis, E.Q., and J.B. Balmat. 2009. Fire history and fire-climate relationships along a fire regime gradient in the Santa Fe Municipal Watershed, NM, USA. *Forest Ecology and Management* 258: 2416–2430.

Margolis, E.Q., T.W. Swetnam, and C.D. Allen. 2007. A stand-replacing fire history in upper montane forests of the southern Rocky Mountains. *Canadian Journal of Forest Research* 37: 2227–2241.

McKenzie, D., S.J. Prichard, A.E. Hessl, and D.L. Peterson. 2004. Empirical approaches to modeling wildland fire in the Pacific Northwest, USA: Methods and applications to landscape simulations. In *Emulating natural forest landscape disturbances: Concepts and applications*, eds. A.J. Perera, L.J. Buse, and M.G. Weber. New York: Columbia University Press. Chapter 7.

McKenzie, D., A.E. Hessl, and L.-K.B. Kellogg. 2006. Using neutral models to identify constraints on low-severity fire regimes. *Landscape Ecology* 21: 139–152.

Miller, C., and D.L. Urban. 2000. Connectivity of forest fuels and surface fire regimes. *Landscape Ecology* 15: 145–154.

Miller, J.D., and S.R. Yool. 2002. Mapping forest post-fire canopy consumption in several overstory types using multi-temporal Landsat TM and ETM data. *Remote Sensing of Environment* 82: 481–496.

Minnich, R. A., and R. J. Dezzani. 1991. Suppression, fire behavior, and fire magnitudes in Californian chaparral at the urban/wildland interface. Pages 67 – 83 in J. J.DeVries, editor. California watersheds at the urban interface. Report 75. University of California, Water Resources Center, Davis.

Monitoring Trends in Burn Severity (MTBS). 2010. Available: http://www.mtbs.gov. Accessed 27 Aug 2010.

Morgan, P., G. Aplet, J. Haufler, H. Humphries, M. Moore, and W. Wilson. 1994. Historical range of variability: A useful tool for evaluating ecosystem change. *Journal of Sustainable Forestry* 8: 87–112.

Morgan, P., C. Hardy, T.W. Swetnam, M.G. Rollins, and D.G. Long. 2001. Mapping fire regimes across time and space: Understanding coarse and fine-scale patterns. *International Journal of Wildland Fire* 10: 329–342.

Moritz, M.A. 2003. Spatiotemporal analysis of controls on shrubland fire regimes: Age dependency and fire hazard. *Ecology* 84: 351–361.

Morrison, P.H., and F.J. Swanson. 1990. *Fire history and pattern in a Cascade Range landscape*. General Technical Report PNW-GTR-254. Portland: U.S. Forest Service.

Okabe, A., B. Boots, and K. Sugihara. 1992. *Spatial tessellations concepts and applications of Voronoi diagrams*. New York: Wiley.

Parsons, R.A., E.K. Heyerdahl, R.E. Keane, B. Dorner, and J. Fall. 2007. Assessing accuracy of point fire intervals across landscapes with simulation modeling. *Canadian Journal of Forest Research* 37: 1605–1614.

Peterson, G.D. 2002. Contagious disturbance, ecological memory, and the emergence of landscape pattern. *Ecosystems* 5: 329–338.

Pyne, S.J. 2001. *Fire: A brief history*. Seattle: University of Washington Press.

Rollins, MG. and C.K. Frame, tech. eds. 2006. T*he LANDFIRE prototype project: Nationally consistent and locally relevant geospatial data for wildland fire management*. General Technical Report RMRS-GTR-175. Fort Collins: U.S. Forest Service.

Rollins, M.G., T.W. Swetnam, and P. Morgan. 2001. Evaluating a century of fire patterns in two Rocky Mountain wilderness areas using digital fire atlases. *Canadian Journal of Forest Research* 31: 2107–2123.

Romme, W.H., and D. Despain. 1982. Fire and landscape diversity in subalpine forests of Yellowstone National Park. *Ecological Monographs* 52: 199–221.

Rothermel, R. 1972. *A mathematical model for predicting fire spread in wildland fuels*. Research Paper INT-115. Ogden: U.S. Forest Service.

Ryan, K.C., and E.D. Reinhardt. 1988. Predicting postfire mortality of seven western conifers. *Canadian Journal of Forest Research* 18: 1291–1297.

Savage, M., and J.N. Mast. 2005. How resilient are southwestern ponderosa pine forests after crown fires? *Canadian Journal of Forest Research* 35: 967–977.

Schmidt, K.M., Menakis, J.P., Hardy, C.C., Hann, W.J., and D.L. Bunnell, 2002. *Development of coarse-scale spatial data for wildland fire and fuel management.* General Technical Report RMRS-GTR-87. Fort Collins: U.S. Forest Service.

Schoennagel, T., T.T. Veblen, and W.H. Romme. 2004. The interaction of fire, fuels, and climate across Rocky Mountain forests. *Bioscience* 54: 661–676.

Scholl, A.E., and A.H. Taylor. 2010. Fire regimes, forest change, and self-organization in an old-growth mixed-conifer forest, Yosemite National Park, USA. *Ecological Applications* 20: 362–3809.

Schweingruber, F.H. 1988. *Tree rings: Basics and applications of dendrochronology.* Dordrecht: Kluwer Academic Publishers.

Shapiro-Miller, L.B., E.K. Heyerdahl, and P. Morgan. 2007. Comparison of fire scars, fire atlases, and satellite data in the northwestern United States. *Canadian Journal of Forest Research* 37: 1933–1943.

Stokes, M.A., and T.L. Smiley. 1968. *An introduction to tree-ring dating.* Chicago: University of Chicago Press.

Swetnam, T.W., and C.H. Baisan. 1996. Historical fire regime patterns in the southwestern United States since AD 1700. In *Fire effects in southwestern forests: The second La Mesa fire symposium.* General Technical Report RM-GTR-286. ed. C. D. Allen, 11–32. Fort Collins: U.S. Forest Service.

Swetnam, T.W., and J.L. Betancourt. 1992. Temporal patterns of El Nino/Southern Oscillation – Wildfire teleconnections in the southwestern United States. In *El Nino: Historical and paleoclimatic aspects of the Southern Oscillation*, eds. H.F. Diaz and V. Markgraf, 259–270. Cambridge: Cambridge University Press.

Swetnam, T.W., C.D. Allen, and J.L. Betancourt. 1999. Applied historical ecology: Using the past to manage the future. *Ecological Applications* 9: 1189–1206.

Taylor, A.H. 2000. Climatic influences on fire regimes in the Lake Tahoe Basin, California and Nevada. In *Fire conference 2000: National congress on fire ecology, prevention, and management*, eds. K.E.M. Galley and T.P. Wilson. Tallahassee: Tall Timbers Research Station.

Taylor, A.H., and C.N. Skinner. 2003. Spatial patterns and controls on historical fire regimes and forest structure in the Klamath Mountains. *Ecological Applications* 13: 704–719.

Tobler, W. 1970. A computer movie simulating urban growth in the Detroit region. *Economic Geography* 46(2): 234–240.

Turner, M.G., W.L. Baker, C.J. Peterson, and R.K. Peet. 1998. Factors influencing succession: Lessons from large, infrequent natural disturbances. *Ecosystems* 1: 511–523.

Turner, M.G., W.H. Romme, and D.B. Tinker. 2003. Surprises and lessons from the 1988 Yellowstone fires. *Frontiers in Ecology and the Environment* 1: 351–358.

Van Horne, M.L., and P.Z. Fulé. 2006. Comparing methods of reconstructing fire history using fire scars in a southwestern United States ponderosa pine forest. *Canadian Journal of Forest Research* 36: 855–867.

Vines, R.G. 1968. Heat transfer through bark and the resistance of trees to fire. *Australian Journal of Botany* 16: 499–514.

Watson, D.F. 1981. Computing the n-dimensional tessellation with application to Voronoi polytopes. *The Computer Journal* 2: 167–172.

Westerling, A.L., H.G. Hidalgo, D.R. Cayan, and T.W. Swetnam. 2006. Warming and earlier spring increase western U.S. forest wildfire activity. *Science* 313: 940–943.

White, J.D., K.C. Ryan, C.C. Key, and S.W. Running. 1996. Remote sensing of forest fire severity and vegetation recovery. *International Journal of Wildland Fire* 6: 125–136.

Wondzell, S.M., and J.G. King. 2003. Postfire erosional processes in the Pacific Northwest Rocky Mountain regions. *Forest Ecology and Management* 178: 75–87.

Wong, C.M., H. Sandmann, B. Dorner. 2003. *Historical variability of natural disturbances in British Columbia: A literature review.* Kamloops: FORREX-Forest Research Extension Partnership FORREX Series 12.

Zar, J.H. 1984. *Biostatistical analysis.* Englewood Cliffs: Prentice-Hall.

Chapter 8
Fire and Invasive Plants on California Landscapes

Jon E. Keeley, Janet Franklin, and Carla D'Antonio

8.1 Introduction

Throughout the world, the functioning of natural ecosystems is being altered by invasions from nonnative plants and animals. Disturbances that alter ecosystem processes often initiate species invasions. Increasingly it is evident that fire-prone ecosystems can be highly vulnerable both to invasion during the immediate postfire period and to alterations of fire regime by altered fuel bed properties after invasion. Here we explore how temporal and spatial patterns of burning affect invasion and the prevalence of nonnative species, and how fundamental variation in fire regime characteristics pose challenges for articulating unifying principles of the relationship between fire and the invasion process at the landscape scale.

Many landscapes in the western United States are dominated by ecosystems where fire is a natural and necessary process for long-term sustainability of those systems. However, despite the obvious resilience of many ecological communities to periodic fire, it is misleading to think of species in these systems as being fire-adapted. Rather, they are adapted to a particular temporal and spatial pattern of burning. This is captured in the concept of a fire regime, which includes the fuel types consumed, frequency and timing of burning, intensity of the fire, and the spatial distribution of individual fire events (Keeley et al. 2009a). Fires are often referred to as disturbances but in many communities fire has been a historic, routine process, and the real "disturbances" to the system are perturbations to the fire regime that lie outside the historic realm. Such disturbances include increased fire frequency, as well as suppression and exclusion of fire, and these can create conditions conducive to species invasions.

J.E. Keeley (✉)
Western Ecological Research Center, U.S. Geological Survey, Sequoia National Park,
47050 Generals Hwy, Three Rivers, CA 93271, USA
and
Department of Ecology and Evolutionary Biology, University of California, Los Angeles,
CA 90095, USA
e-mail: jon_keeley@usgs.gov

D. McKenzie et al. (eds.), *The Landscape Ecology of Fire*, Ecological Studies 213,   193
DOI 10.1007/978-94-007-0301-8_8, © Springer Science+Business Media B.V. 2011

We address the complex relationship between fire and nonnative species invasions, by focusing on a complex landscape with diverse fire regimes and ecosystems that pose different challenges with respect to alien invasions. Our goal is to understand the features of the fire regime critical to thresholds that influence system susceptibility to invasion and how these thresholds vary between different ecosystems. In the context of future global changes in climate, human population growth, and landscape use, we will attempt to define the potential trajectories for these systems and how future fire regime characteristics may affect biological invasions.

8.2   The Setting: California

Fire is a prominent ecosystem process over much of the California landscape, including diverse plant communities of grasslands, shrublands and forests. These communities broadly sort out along a moisture gradient, although other factors, including substrate, land use history, and fire, can play important roles in determining their distribution. The Mediterranean climate of this region has characteristics that contribute to making it a fire-prone landscape. This includes a wet and cool winter with growing conditions sufficient to generate moderate primary productivity followed by a long, dry, and hot summer that converts much of this production into available fuel for wildfires. Many plant communities in California are resilient to fire, and species exhibit traits apparently selected for by fire. Thus, it is somewhat surprising that fire can be an important driver behind alien plant invasions in this landscape. Conversely, with the wide breadth of species introduced into California, it is not surprising that some would be fire responsive.

Fire regimes are markedly different across California both within and across vegetation types (Sugihara et al. 2006). Since settlement by Euro-Americans in the late eighteenth and nineteenth centuries, fire regimes have been greatly altered and fire regimes are outside the historical range of variability, although in very different ways in forests, shrublands, and grasslands.

Fire suppression has proven effective at excluding fire from conifer forests in mountainous regions of the state. As a result many of these forests have accumulated fuels in excess of what occurred historically. Higher elevation conifer-dominated forests have historically burned in high-frequency (scaled at approximately 10-year intervals) lightning-ignited fires that consumed predominantly surface understory fuels, or they have burned in a mixed pattern of low intensity surface and high intensity crown fires. Frequent fires contributed to landscape fuel mosaics conducive to small patchy burns.

Fuel loads in these forests include both surface fuels and increased density of saplings or ladder fuels. Fire suppression is only one of the factors contributing to these fuels. In ponderosa pine forests with herbaceous understory intense livestock grazing has reduced surface fuels, which have worked to exclude fires and allow increased ladder fuels. Logging has contributed to both increased surface fuels as well as increased density of even-aged trees (Odion et al. 2004; Stephens and Collins 2004; van Wagtendonk and Fites-Kaufman 2006). This homogenization of stand structure has affected fire regimes. Efforts to restore high-frequency low-severity fire to these forested ecosystems (Keeley and Stephenson 2000) are greatly complicated by restrictions related to increased human settlement in mountain areas, the volume and homogeneity of the fuels, and nonnative plant species not present prior to Euro-American settlement.

By contrast, lower elevation shrublands retain most fuels in the canopy and historically had less frequent lightning ignitions resulting in longer intervals (closer to 100-year intervals) between high-intensity crown fires. Although fire sizes likely varied, these ecosystems would have burned periodically in massive landscape scale fires, probably much larger than fires in forested ecosystems (Moritz 2003; Keeley 2006a). Lower elevation ecosystems, particularly shrublands, have experienced increased fire frequency since the middle of the 20th century (Keeley et al. 1999), which is quite unlike the recent history in conifer forests (Fig. 8.1).

In contrast to both forests and shrublands in California, fire regimes in grasslands are poorly understood (Wills 2006). Since Euro-American settlement, grasslands have undergone profound changes in composition and today are dominated by nonnative annual grasses. These species are presumed to have displaced native bunchgrasses on some landscapes, native forbs on other landscapes, and shrublands on still other landscapes (Keeley 1990; Hamilton 1997). Prior to Euro-American settlement the distribution of native grasslands would have been most strongly influenced by edaphic factors and Native American settlement and burning patterns (Wells 1962; Huenneke and Mooney 1989; Keeley 2002). Due to the combination of fuel characteristics (ready ignition, rapid spread rates) and anthropogenic burning (e.g. Anderson 2005a), fires were likely frequent and of low intensity.

Although California's forests, woodlands, shrublands, and grasslands tend to occupy different elevational zones, the topographically diverse landscape often produces a fine-gain a mosaic of vegetation types. This mosaic is fine-scaled in the coastal ranges and xeric southern California mountains (Franklin and Woodcock 1997). Plant formations with very different physiognomy, fuel structure, fire regime types (crown fire, surface fire, mixed), and fire response of dominant woody species (resistance versus resilience) often occur in close proximity. Fire regimes in this landscape can also vary at fine spatial scales most likely as a function of fuel characteristics (Stephens et al. 2009). However, such patterns are often obliterated under extreme fire conditions as evident in some of the recent megafires that have consumed all fuels in their wake (Keeley et al. 2004, 2009b).

Nonnative plant species have been present in California for several centuries, primarily introduced during early Euro-American settlement (Klinger et al. 2006). The following sections will consider the interactions of fire, climate change, and invasive species on California's landscapes for each of three major terrestrial vegetation types: forest, shrublands and grasslands.

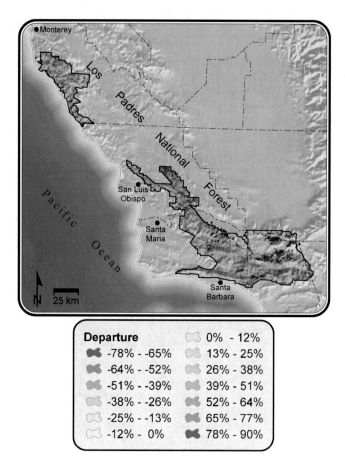

Fig. 8.1 Percentage departure of current mean fire return interval (1910–2006) from reference mean fire return interval (pre-Euro American settlement). Areas with negative departures (e.g., lowland chaparral and sage scrub) are experiencing more frequent fire today than in the presettlement period. Areas with positive departures (e.g., high elevation ponderosa pine) are experiencing less frequent fire today than in the presettlement period (Data courtesy of Dr. Hugh Safford, regional ecologist for U.S. Forest Service, Pacific Southwest Region)

8.3 Forests

Montane conifer forests in California generally have fewer alien plant species, a lower abundance of aliens, and a different collection of alien species than lower elevation foothill oak savanna or grassland (Keeley 2001; Keeley et al. 2003; Klinger et al. 2006). This follows a general pattern of decreased alien presence with increasing elevation in California (Mooney et al. 1986; Rejmanek and Randall 1994; Schwartz et al. 1996) and could involve numerous factors such as shorter growing seasons, forests with lower surface light levels, different disturbance regimes, fewer alien propagules, or fewer potential alien invaders adapted to conditions at higher elevations.

Present problems with invasive species are the result of repercussions from past management practices as well as unintended side effects of present management practices. The anomalously high woody fuel accumulation resulting from logging, livestock grazing, and fire suppression has put many forests on a trajectory away from low- or mixed-severity fire to larger, more intensive crown fires. California has experienced several of these anomalously high-intensity fires in the past decades. These have created crown gaps that appear to be outside the historical range of variability.

One such example occurred recently in southern California mountains in 2003 and eliminated most all of the conifer forests in Cuyamaca State Park (Franklin et al. 2006). Forest recovery has been slow, hindered by establishment of the early seral stage shrub layer as well as high cover of native and nonnative herbs around dry meadows (former homesteads and ranches) (Franklin et al. 2006). This is even though the dominant conifer, Coulter pine (*Pinus coulteri*), is partially cone serotinous (Borchert 1985), suggesting that it has potential for rapid reestablishment following fire. The slow recovery suggests that pine establishment, which is dependent on seed dispersal from burned cones or nearby adult trees, has been affected by the extent and severity of the fire. Also, cheatgrass (*Bromus tectorum*) was found in greater abundance in the second (wetter) postfire year than in the first (drier) (Franklin et al. 2006), and had increased further in abundance by the fourth growing season, even though it was a dry year (Franklin, submitted). However, we do not know if this nonnative grass will persist or spread as the forest regenerates and affect the future fire regime in this region.

To put this recent event into historical perspective, extensive stand replacing fires driven by easterly winds were reported in these mountains in fall 1899, in the Los Angeles Times (29 September, 1899) and the Julian Sentinel (4 October, 1899), where it was lamented "...above all we mourn the loss of our forests." It is not possible to glean details (size, severity, location) about a fire event from this kind of historical description, but it suggests that the recent crown-fire events in these forests may not have been entirely outside the historical range of variability.

Current fire management practices have the potential to influence alien plant invasions at nearly all stages, including both prefire treatments as well as postfire restoration responses (Keeley 2006b). Agencies are increasingly aware of this and are taking steps to minimize these impacts even during fire events, e.g., by checking firefighting equipment for alien propagules prior to entering wildland areas during fires.

Prefire fuel treatments pose one of the biggest risks for alien plant invasion largely because the treatments reduce surface fuels as well as open the forest canopy, both of which promote the growth of herbaceous species. Globally, forest management in western North America is unique in its focus on restoring historical conditions of forest structure and processes. The philosophy is that these forests persisted under such conditions prior to intensive contemporary land management, and thus returning forests to those conditions will bring us closer to ensuring sustainability (Millar 1997). One of the primary impacts of reintroducing historical fire frequencies is that it greatly reduces surface and ladder fuels to the point where forests retain their historical fire regime of low- or mixed-severity fires and are less vulnerable to high-intensity crown fires. Alien species in western North America are recent introductions, however, and historical landscapes were allowed to recover after fire in an environment largely free from threat of invasion. Today the alien presence interferes with expected responses to fuel management. For example, following a series of prescription burns in ponderosa pine forests to reintroduce historical fire frequencies to the lower elevations of Kings Canyon National Park sites (Keeley and McGinnis 2007) were heavily invaded by cheatgrass (*Bromus tectorum*). When historical fire frequencies were applied through prescription burning, they were too frequent to allow canopy closure and enough litter accumulation to inhibit cheatgrass. They were also too frequent to generate sufficient fire intensity to destroy cheatgrass seedbanks. The conclusion from that experience is that future fire management may want to set its goals to some middle ground between short historical fire frequencies, which favor cheatgrass, and very long fire free periods, which lead to hazardous fuel loads.

Postfire management has also played a role in the spread of plant invaders. On USFS lands in the Sierra Nevada it has long been common silvicultural practice to utilize herbicides to eliminate the natural seral stage of ceanothus (*Ceanothus*) (Fig. 8.2a) and other shrubs in order to grow "better" ponderosa pine plantations by reducing competition between pine seedlings and shrubs. However, *Ceanothus* species are nitrogen-fixing (Delwiche et al. 1965; Conard et al. 1985) and important to ecosystem recovery following nitrogen volatilization by fires (Hellmers and Kelleher 1959; Binkley et al. 1982). The result of shrub removal is to increase the dominance of annual aliens, in particular various species of brome grass (*Bromus*) (Fig. 8.2b). Not only does this alter the native to nonnative understory composition, but it also affects habitat and seed sources for small mammals and greatly alters the fuel structure of young forests. The greater proportion of fine fuels increases the probability of fires spreading in these young stands. One example from the central Sierra Nevada is the Cleveland Fire in El Dorado County (Fig. 8.3). The extensive red brome (*Bromus madritensis*) and cheatgrass (*B. tectorum*) invasion after several

Fig. 8.2 The 2006 Star Fire in the northern Sierra Nevada burned across two national forests, which applied different postfire treatments: (**a**) The Tahoe National Forest did not use herbicides to eliminate early-seral stage shrubs such as Ceanothus, whereas (**b**) the Eldorado National Forest used repeated herbicide treatments to eliminate the shrub layer and replace it with herbaceous native and non-native species (Photos by (**a**) Tom McGinnis and (**b**) by Jon Keeley)

Fig. 8.3 The 1992 Cleveland Fire area was sprayed with herbicides and replanted with pine seedlings. Annual alien grass invasion fueled a repeat fire in 2001, the St. Pauli Fire, which destroyed a significant portion of the 8-year-old plantation (Photo by Tom McGinnis)

herbicide treatments to destroy shrubs, produced grass fuels sufficient to carry a fire at 8 years (St. Pauli Fire), which destroyed a substantial portion of the plantation.

8.4 Shrublands

Chaparral and California sage scrub are typically closed-canopy shrublands that are relatively resistant to invasion by nonnative species. The most common disturbance that sets them on a trajectory of invasion is a perturbation in the fire regime, in particular increases in fire frequency (Figs. 8.4 and 8.5). These ecosystems are highly resilient to fires at frequencies of more than 20 years but as the interval between fires decreases, more and more native species are lost due to insufficient time between fires for recovery (Zedler et al. 1983; Haidinger and Keeley 1993; Keeley et al. 2005). The lowest fire frequency threshold of tolerance varies with vegetation type and landscape position. Chaparral is generally not resilient to return intervals shorter than 20 years, whereas sage scrub in the interior cannot tolerate return intervals shorter than 10 years, although coastal versions of sage scrub can sometimes tolerate shorter return intervals. In general, these tolerances decrease with increasing site aridity; for example, pole-facing exposures are resilient to more frequent fires than those facing the equator. Other changes in fire regime that may have an impact on invasions include fire intensity (Fig. 8.6), fire season and fire size. Land management practices such as grazing and mechanical disturbance may also enhance invasion (Stylinski and Allen 1999).

Fig. 8.4 The entire chaparral scene shown here was burned in the 1970 Laguna Fire. The background has not re-burned since 1970 and this mature stand is largely free of alien species. The middle and front parts of the scene were burned in the 2001 Viejas Fire and the middle section is an early seral stage of chaparral dominated by short-lived natives and relatively few alien species. The foreground was burned a third time by the 2003 Cedar Fire and is dominated by aliens (Fig. 8.5). The high frequency of fires has also reduced the ability of natives to recover and placed the community on a trajectory that favors alien persistence (Photo by Richard Halsey)

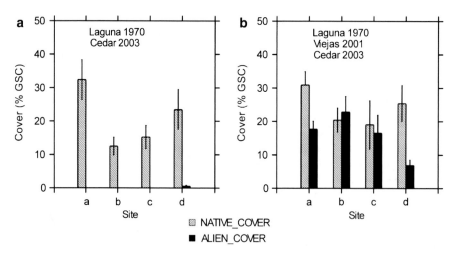

Fig. 8.5 Sites with 33 years between fires recover with little or no alien species whereas sites burned at shorter intervals have a substantial alien species load (Data from Tess Brennan and Jon Keeley)

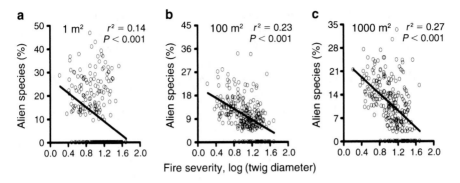

Fig. 8.6 In contrast to conifer forests where higher fire severity often increases alien invasion (e.g., Keeley et al. 2003), in chaparral increased fire severity, observed at three spatial scales, is associated with decreased alien invasion (Keeley et al. 2008)

Ozone pollution (Westman 1979) and nitrogen pollution (Allen et al. 1998) have been postulated to be involved in alien invasions of these shrublands, but probably only when coupled with some other disturbance that first opens the closed canopy shrubland and allows for annual grass and forb invasion (Keeley et al. 2005). Regional patterns of pollution, fire, and invasion are consistent with this model. Although there is a strong gradient of increasing pollution from the coast to the interior (Padgett et al. 1999), loss of native shrublands and invasion by alien grasses and forbs happens frequently in the unpolluted coastal plain and foothills (Keeley et al. 2005). Studies that have reported landscape patterns of invasion consistent with pollution have not adequately considered fire history. The frequency of fire is extremely high on interior polluted landscapes, and one of the commonly cited examples of pollution-driven type conversion from shrubland to grassland, Box Springs Mountain in Riverside County, has burned and reburned repeatedly in the last 50 years (Cal Fire 2007). Westman (1979) used landscape patterns of ozone pollution to demonstrate pollution-driven type conversion, but without any consideration of fire return intervals at his study sites. Recently Talluto and Suding (2008) found evidence for both fire and nitrogen as factors in alien invasion, and suggested that nitrogen was likely important on those parts of the landscape that were unburned. However, it is doubtful that very much of that landscape was unburned because their study used a fire history database that excluded fires less than 40 ha, and these smaller fires generally constitute more than 95% of all the fires in that region (Keeley, n.d.). Smaller fires on these landscapes likely play a crucial role in creating a patchwork mosaic of type conversion that increases alien propagule availability throughout the region.

One of the primary limitations to the pollution model is a lack of a clear mechanistic basis for how grasses displace shrubs. Numerous experimental studies have failed to detect a competitive advantage of grasses over shrubs under elevated nitrogen (e.g., Allen et al. 1998; Padget and Allen 1999; Yoshida and Allen 2001). Recently it has been proposed that interactions between nitrogen pollution and mychorrizae may inhibit shrub seedling recruitment and give invasive grasses a competitive edge (Siguenza et al. 2006). Such a model could not account for displacement of intact shrublands since recruitment is largely restricted to open sites after fire. The mechanism for how this would work remains to be determined since most all experiments have used soil nitrogen levels characteristic of the late summer dormant season whereas soil nitrogen levels during the winter growing season are often indistinguishable between polluted and unpolluted sites. In summary, we do not rule out the possibility that nitrogen pollution plays some role in nonnative plant invasion in coastal California, but there is no evidence this can occur without physical disturbance such as fire, mechanical disturbance, or livestock grazing first opening up the shrub canopy. Once this occurs, grasses promote further burning and these landscapes may develop into a more open mixed shrubland/grassland, depending on the availability of alien seed, potentially contributing to an acceleration of the invasion process (Fig. 8.7).

Fire is a necessary ecosystem process for the sustainability of California shrublands, but postfire conditions provide a window of opportunity for alien plants to invade. Generally, the shrub canopy regenerates rapidly and most aliens are excluded during the early seral stages. Most of these aliens are annual species. An exception appears to be occurring in chaparral and sage scrub along the southern and central coast

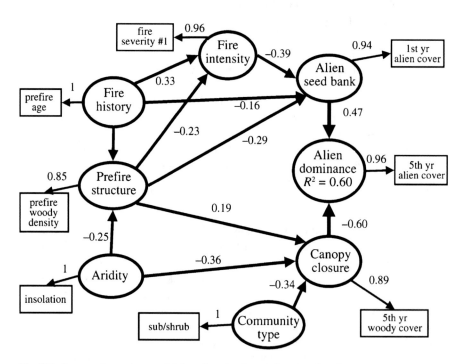

Fig. 8.7 Structural equation model for alien dominance 5 years after fire using a path analysis that separates latent (*ellipses*) and measured (*rectangles*) variables. Two latent variables in the original model, nitrogen deposition and landscape position, were not significant and were excluded. Path strength is indicated with standardized regression coefficients, and R^2 is given for the response variable of alien dominance (Keeley et al. 2005)

where an alien perennial grass, *Ehrharta calycina*, aggressively invades burned shrublands and appears to inhibit regeneration of native species (Roye 2004; D'Antonio personal observations).

When multiple fires occur within the time frame of one decade, shrub regeneration is compromised and often leads to permanent invasion. What is critical is the shortest interval between fires (Zedler et al. 1983; Haidinger and Keeley 1993; Jacobson et al. 2004). Thus, any increase in fire frequency, through either wildfires or prescription burning, should be viewed as a potential contributor to alien invasion.

Determining alien invasion is a multi-faceted problem and there are many factors that determine alien success (Fig. 8.7). One of the key factors is propagule availability (D'Antonio et al. 2001). After fire, alien propagule presence is determined by both temporal and spatial factors. Site history plays a major role and young seral stage stands are more likely to have aliens and alien seed banks than more mature stands.

Fuel structure also affects the invasion process. Heterogeneity of fuel distribution changes fire intensity (Odion and Davis 2000), and this can affect survival of alien propagules (e.g. D'Antonio et al. 1993). The invasion process can alter the dominant plant functional types, which in turn alters fire intensity. For example, invasive grasses, particularly annuals, reduce fire intensity and enhance seed survivorship in the soil (Keeley 2006b). In one case, evergreen shrublands were invaded after fire by a succulent (Zedler and Schied 1988) that altered fuel structure, potentially suppressing spread of future fires (D'Antonio 2000).

Fire management activities likewise may alter fire regimes in ways that favor alien species, both by providing suitable habitat and altering fire intensities. For example, trails or fuel breaks may promote alien invasions (Merriam et al. 2006). Although many fuel breaks have contributed to fire operations, doubtlessly many have not, and there is a need for careful evaluation of the benefits relative to the financial and resource costs of these activities.

Sites that have been highly disturbed from frequent short interval fires or other disturbances such as construction activities or livestock grazing in open shrubland/ grassland associations will have a greater presence of alien species and hence seed prior to the fire than mature closed canopy systems (Stylinski and Allen 1999). Aliens tend to be annual grasses and forbs, and they alter the fire regime from an active crown fire to a mixture of surface and crown fires. This tends to increase fire frequency in younger shrubland stands. Soil temperatures during grass fires are lower than in shrub fires, favoring survival of alien propagules (Keeley 2006b).

We would expect that this phenomenon is not unique to chaparral and sage scrub but also applies to other ecosystems, particularly those closed-canopy systems that typically burn in crown fires. The level of disturbance is likely to depend on the system. As a general rule, in crown-fire ecosystems the historical fire return interval was much greater than the time to canopy closure, and this seems to hold for shrubland as well as crown-fire forested systems (Baker 2006).

Disturbances that lead to alien invasion are those that reduce the ratio, current fire-return interval : time-to-canopy closure, and when this ratio drops to less than 1, alien invasion is very likely.

Spatial patterns at many scales may affect the invasion process. One of the critical factors is the regeneration mode of the dominant shrub species. Typically mesic north facing slopes in California shrublands are dominated by vigorous resprouting species. Recovery from fire is not dependent on seedling recruitment; basal resprouts recover very rapidly. In contrast, drier sites are often dominated by non-sprouting obligate seeding species that require up to two decades of regrowth to replenish the soil seed bank. On these sites a fire interval of less than a decade hinders shrub recruitment and creates an ecological vacuum that is readily filled by alien species. However, on more mesic slopes much more frequent fires may be required to effect such a change.

The extent to which the conversion of native shrublands to alien dominated grasslands has already occurred on the California landscape is unknown, although early observations suggested that it was substantial and that Native Americans had played a significant role in the process (Cooper 1922). Nonnative-dominated annual grasslands cover a substantial portion of the wildland landscape in the state and are derived from shrublands, native grasslands or forblands. Wells (1962) examined substrate preferences for a portion of San Luis Obispo County in the central coast region and concluded that a significant portion of annual grasslands were formerly shrublands that had been displaced by frequent fires, perhaps long before Euro-Americans arrived. He suggested that substrate was an important determinant of native shrubland and native grassland distribution but not for alien-dominated grasslands, whose distribution was largely influenced by disturbance. Although fire frequency plays a major role in driving alien invasions in shrubland landscapes, other perturbations in the fire regime likewise play a role. For example, fire intensity, often measured as fire severity, is generally high in most shrubland fires. These systems are very resilient to high fire intensity and variations in intensity have little impact on native plant recovery (Keeley et al. 2008). However, as fire intensity decreases, alien invasion increases due to a variety of correlated factors. Lower fire intensity occurs in more open stands with a mixture of grasses and shrubs; thus, they are likely to have more alien propagules in the soil at the time of fire. These stands also generally have burned in the recent past and thus reburning presents an obstacle for the regeneration of the native species. Lastly, lower intensity likely enhances survivorship of alien propagules.

Fire season may have a profound role in affecting alien invasions. For example, reduced native recovery has been reported for out-of-season prescribed burns (Keeley 2006b) and this vacuum is always filled with alien species. The mechanism by which out-of-season burning decreases native plant recovery is unknown, but it is commonly attributed to prescribed burns during winter or spring that cause heating of seed banks with moist heat, which is often lethal (Parker 1987). Perhaps more important though is that winter burning greatly decreases the length of the first growing season. For most seedlings having the growing season reduced from a typical 6 months (following summer or fall burns) to perhaps as little as 1 month (following a winter burn) could limit survival during the ensuing dry summer.

We expect fire size to play a role because it increases the chances of reburning sites that were recently burned. In the 2007 wildfire season, more than 30,000 ha that burned in southern California overlapped fires from 2002 and 2003 (Keeley et al. 2009b). Such a short return interval increases the likelihood that alien species were present, reduces fire intensity in reburned areas, and decreases native shrub survival. It is unclear if this in turn increases continuity of alien dominated sites. Fires on these landscapes burn through a mosaic of native shrublands and alien-dominated grasslands and as fire size increases the mixtures become more diverse. It is to be expected that as the boundary between burned shrublands and alien-dominated grasslands increases the latter would provide a source of alien propagules and increase the chance that adjacent shrublands would act as a sink for these aliens.

8.5   Grasslands

Although fire has long been considered to be a dominant structuring force in grass-lands around the world, its history and role in California's Mediterranean grass-lands are murky. Early Euro-American visitors to California reported that the indigenous peoples were burning frequently, presumably to keep landscapes free of woody species and thus creating open forblands or grasslands (Keeley 2002; Anderson 2005a,b; Minnich 2008). Reports of indigenous burning exist from coastal, Central Valley and Sierra Nevada foothill grassland-type habitat suggest-ing that grassland fires were an important landscape feature across the state (Anderson 2005a, 2007). Indeed, records of individual tribes suggest that at least 35 tribes used fire to manage grassland (here defined broadly to include forblands) for particular plants or for hunting (Reynolds 1959; Anderson 2007). Historical journal observations and ethnographic records present a reasonable basis for infer-ring that fire played a significant role in the expansion of pre-Euro-American grasslands, and this is supported by pollen records (Anderson 2005b) and phyto-liths (Hopkinson 2003).

After Euro-American colonization, it is likely that the role of fire changed. The initially "open" (grassland or forbland) habitats created by indigenous burning likely were maintained by intensive livestock grazing during the mission era (Minnich 2008). During the last 100 years, fire was apparently used to convert shrublands to annual grasslands as the expansion of agriculture in the late 1800s reduced available open lands for grazing (Tyler et al. 2007). As these newly formed and existing grasslands were utilized for livestock and crop production, burning appears to have become an uncommon activity although records about this are poor. Indeed, Greenlee and Langenheim (1990) estimate that prior to widespread live-stock grazing, fire frequencies in the central coast near to human settlements were potentially every 1–5 years. After the cessation of indigenous burning and with the advent of widespread crop agriculture and livestock grazing they estimate that fire frequencies dropped to every 20–30 years. Although these numbers are largely speculative, it is conceivable that there have been dramatic changes in fire occur-rence in grass-like habitats over the past two centuries.

The invasion of European annual grasses into California's grasslands that apparently occurred beginning in the late 1700s is thought to be due to a combination of intensive year round livestock grazing and conversion to crop agriculture followed by land abandonment (Jackson 1985). The frequent use of fire by indigenous peoples to manipulate composition likely aided the establishment of opportunistic annual grasses that arrived with Euro-American settlers (Keeley 2002). It has been hypothesized that frequent burning of shrublands to create grasslands resulted in landscapes susceptible to rapid establishment of alien grasses and forbs without any grazing or crop agriculture (Keeley 2002). Although the hypothesis that frequent indigenous burning contributed to the vulnerability of California grassland to the initial invasions cannot be tested, it also cannot be discounted since it is now known that many nonnative annual species can tolerate or increase with grassland fires. Thus, it is reasonable to assume that the common occurrence of "grassland" fires in the 1700s and early to mid 1800s contributed to the rapid spread of some nonnative grassland species.

In contrast to the other ecosystems discussed in this chapter, California grasslands are already very heavily invaded by nonnative species and it is not clear that fire will make them more susceptible to further invasion. Since European annual grasses dominate sites that have not burned for decades, it is also clear that they do not rely on fire in any way to maintain their dominance. Hypotheses regarding the persistent dominance of annual grasses are reviewed elsewhere (e.g. Corbin et al. 2007). Also in contrast to forests and shrublands, managers of grassland are trying to use fire to manipulate composition away from nonnative species or to "tip the balance" in favor of native species (Corbin et al. 2004). Given the highly diverse nonnative flora in California grasslands, and that on many landscapes annual grasslands occupy former shrubland sites, a goal of eliminating nonnative species and reestablishing native grasslands is unrealistic. Nevertheless, fire is a useful tool for manipulating composition in some areas and under some circumstances.

In a meta-analysis of the outcome of fire management treatments across California grasslands, Bainbridge and D'Antonio (in prep.; reanalysis of Corbin et al. 2004) found that fire can depress the abundance of European annual grasses, but only for the immediate season after fire. Whether a single fire tips composition towards natives is site-dependent and generalities are difficult to find. Generally, single fire events slightly depress alien annual grasses but may increase exotic forbs, depending on the site and species pool. For example, in annual grass-dominated sites in the Carrizo plains, experimental burning in spring promoted the exotic forb *Erodium cicutarium* and the native forb *Phacelia ciliata* (Meyer and Schiffman 1999) but the dominant invasive annual grasses were unaffected by fire. Likewise, Reiner et al. (2006) and Reiner (2007) report an increase in both native and nonnative forbs with fire. They used repeated fires (up to 13) to reduce nonnative grass dominance on their sites in the California Central Valley.

Efforts to reduce specific invasive species with fire may be more successful than general efforts to shift composition, but only if fire is repeated often. For example, DiTomaso et al. (1999) used repeated fire to reduce the abundance of yellow star thistle (*Centaurea solstitialis*), a noxious weed in California grasslands. Because the seeds of this species live for many years (Callihan et al. 1993), at least three fires are necessary to obtain reductions in star thistle. However, these treatments were never sustainable, as within a few years of ending burning, starthistle populations rebounded (Kyser and DiTomaso 2002). Clearly, prescribed burning provides only temporary reduction and does not affect sustainable control of this alien, and may even exacerbate the alien situation (Fig. 8.8). Likewise, DiTomaso et al. (2001) and Betts (2003) used multiple fires to reduce the abundance of the invasive grasses medusahead (*Taeniatherum caput-medusae*) and barbed goat grass (*Aegilops triuncialis*), but what happens once fire is removed is unknown.

Several studies point to the important role of preburn composition in determining the relative effects of fire on native vs. nonnative species. Native species tend to

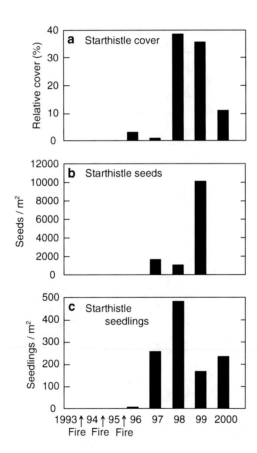

Fig. 8.8 Yellow starthistle (*Centaurea solstitialis*) cover and seed and seedling production following three consecutive annual burns applied to extremely dense populations of this noxious alien weed. Immediate postfire results were very promising (DiTomaso et al. 1999), but follow-up studies indicate that burning destabilized these grasslands and allowed subsequent reinvasion once burning was stopped (Kyser and DiTomaso 2002)

increase in sites where they were already reasonably abundant (Harrison et al. 2003), but decrease in sites where they are rare to begin with. DiTomaso and Johnson (2006) reported that the invasive black mustard (*Brassica nigra*) increased to almost complete dominance after management fires in a Sierra Nevada foothill grassland site. The species was present as a dormant seed bank prior to fire and can respond rapidly after fire. As a consequence, this species can present surprises when fire is used to target other species. For example, Moyes et al. (2005) used prescription burning with the goal of targeting ripgut brome (*Bromus diandrus*). The treatment was successful in nearly eliminating this grass, but the space was readily occupied by fire-stimulated germination of the mustard seed bank (Fig. 8.9).

It is widely believed that cattle grazing or targeted grazing by goats can reduce fuel accumulations, and create discontinuities in fuel. Conceptually then, grazers should be a useful tool for reducing the spread of fire over landscapes. However, every year fires ignite in grassy vegetation and grazed grasslands are observed to burn. To evaluate how grazing influences the probability of successful ignition or fire spread would require a landscape analysis of where fires start, the patterns they follow as they spread, and a knowledge of the grazing regime at the local scale across the region. Fire-spread modeling does support the notion that fuel discontinuities can reduce fire spread rates and fireline intensities (Finney 2001), but strongly wind-driven fires have been shown to burn across a range of vegetation types, fuel ages, and fuel structures (Moritz 2003). These fires may be indifferent to past grazing in grasslands, as was observed in October 2007 when a Santa Ana-wind-driven fire burned through heavily grazed pastureland (Fig. 8.10) that fire behavior models would predict should not carry fire. Indeed, a study of fire spread

Fig. 8.9 Postfire shift in cover from the non-native annual grass *Bromus diandrus* (lower left foreground), which was targeted in a prescribed burn, to mustard (upper right foreground) in the Santa Monica Mountains National Recreation Area of Ventura County, California (Photo by Andrew Moyes)

Fig. 8.10 Heavily grazed grasslands in Pamo Valley, San Diego County, burned in the 2007 Witch Fire (Photo by Richard Halsey)

rates in 121 grass-fueled fires suggested that wind speed is more important than grass type or fuel structure (e.g. standing versus cut, dense versus thinned) in driving fire spread rates (Cheney et al. 1993). Nevertheless, fire intensity and spread rates are demonstrably lower in grazed grassland under some conditions (Diamond et al. 2009) and grazing, if it leaves little residual dry matter, can reduce standing fuel and fireline intensity (Davison 1996; Diamond et al. 2009). Thus, grazing is being used to create a landscape scale fuel mosaic that under the right conditions (not extreme) could reduce fire spread rates and intensity in grass-dominated vegetation (McAdoo et al. 2007). Many of the large fires in California, however, burn under extreme weather conditions (Mensing et al. 1999; Moritz 2003), and under these conditions, grazing is unlikely to be important in modifying fire behavior (e.g. Launchbaugh et al. 2008).

8.5.1   Fire and the Grassland/Shrubland Matrix

At the time of Euro-American settlement, it is possible that vegetation heterogeneity was less than today due to the widespread use of fire to keep vegetation "open." It has been documented today that fire contributes to the current landscape mosaic of grassland, shrubland, and woodland (e.g., Callaway and Davis 1993). On grassland sites prone to shrub colonization, a reduction in fire frequency gradually leads to conversion of grassland to shrubland (Keeley 2002) with accompanying increases in standing fuel and the potential therefore for higher-intensity fires. The trend of increased presence of woody vegetation on landscapes previously supporting extensive grassland is particularly apparent in the San Francisco Bay area. Contrary to conventional wisdom, this trend is not related to disruption of the natural fire regime by fire suppression, but rather due to a reduction in anthropogenic ignitions and cessation of intensive livestock grazing (Keeley 2005). Thus, this so-called shrubland invasion is perhaps better viewed as a recolonization following the cessation of anthropogenic disturbance, at least for the dominant native shrub, coyote bush (*Baccharis pilularis*). However, non-native invasive shrubs such as scotch, french, and spanish brooms (*Cytissus scoparius*, *Genista monspessulana* and *Spartium junceum* respectively) and gorse (*Ulex europaeus*) are spreading widely into grasslands in the northern and central coast regions. Germination of the brooms is stimulated by fire, and recurrent short return interval fires can be used to reduce the broom seedbank and reduce standing adult plants (Alexander and D'Antonio 2003a, b). Nevertheless, using fire to control these nonnative invaders presents a challenge in increasingly urban or suburbanized landscapes and with rising societal concern about the efficacy, safety, and environmental impacts of controlled burning.

Over the past decade, the potential role of atmospheric nitrogen (N) deposition in influencing California vegetation has been recognized (Weiss 2006) and detailed mapping of nitrogen plumes suggests that a substantial fraction of coastal habitats and portions of the California deserts could be affected by N deposition. Nitrogen deposition promotes invasive annual grasses in open scrub vegetation of the Mojave Desert (Brooks 2003), and in nutrient-stressed serpentine grasslands near the coast (Weiss 1999). These alien grasses in turn either increase, or have the potential to increase, fire frequency (Beatley 1966; D'Antonio and Vitousek 1992). Although this grass/fire cycle can happen independent of N deposition (D'Antonio and Vitousek 1992; Keeley 2006b), N deposition could accelerate the onset of such a cycle in some habitats.

8.6   Future Directions for Fire and Invasive Species Interactions in California

Alien species presence on California landscapes will likely increase under future conditions. Two issues that need to be given serious consideration are the continuation of certain fire management practices and changes in fire regimes induced by global changes in human demography and climate. The most critical stressors are likely to vary with vegetation type.

8.6.1   Forests

In understanding future impacts on forests, we distinguish between two fire regime types in forests of western North America: those driven by herbaceous understory fuels (e.g., ponderosa pine forest) and those with dead litter as the fuel source (e.g., mixed conifer forest). Each is likely to respond differently to climate change and they may not be equally susceptible to invasion by alien species under altered fire regimes. A major paleoecological reconstruction of the American Southwest based on tree ring analysis showed that major fire years coincided with extreme drought. However, antecedent (1–3 years prior) wet conditions affecting grass (fine fuel) production were significantly correlated with fire extent only in ponderosa pine forest. The lack of any significant lag effect of antecedent moisture on fire activity in mixed-conifer forest was explained in terms of the greater persistence of snow pack in spring and lack of importance of fine fuels in fire dynamics (Swetnam and Betancourt 1998). Ponderosa pine forests may be more susceptible than other forest types to altered fire regimes under global warming, and to invasion by non-native herbaceous species that could, in turn, affect the fire regime.

In conifer forests throughout the mountainous western USA, the interactions of fire, a warming climate, and invasive herbaceous vegetation are of great concern (Dale et al. 2001; Brooks et al. 2004; Keeley 2006a; Millar et al. 2007). Fewer but larger higher-severity fires resulting from warmer windier conditions predicted under climate-warming scenarios could homogenize the forest landscape mosaic into large even-aged patches. On the other hand, reduced fuels due to higher decomposition rates and reduced tree seedling establishment, or to the establishment of annual grasses, could lead to lower fire intensity and higher frequency under global warming. How might changes in the patch mosaic size, in conjunction with spatially varying pressure from invasive plant species, affect forest succession on these landscapes? Will it lead to increasing "type conversion"? How big a role will exotic invasive species play in forested areas? Will nonnative grass cover increase in forested areas leading to feedbacks in the fire cycle, promoting high–frequency low-severity fire?

Predicted changes in fire regime, driven by climate warming, are likely to reduce the extent of old-growth fire-affected forests. Kaufmann et al. (2007) describe a conceptual model of late successional or old-growth forest structure in fire-prone forest types. An important factor is the scale of patchiness. Increase in very large fires will erase fine-scale mosaics of differently aged stands, especially if they are severe.

Increased spring and summer temperatures, projected by virtually all climate models for future decades owing to anthropogenic climate change, could reinforce the trend toward longer fire seasons and large wildfires. A study of northern California also projected similar trends of more fires and larger burned areas in some regions of the state under the warmer windier conditions that are projected by climate models (Fried et al. 2004). However, ecological feedbacks are likely to complicate these projections, and will be discussed in the following sections. More frequent larger fires in recent decades in mid- to high-elevation western forests coincide with warmer temperatures and earlier spring snowmelt (Westerling et al. 2003, 2006), although the number of fires *per se* and snowpack are not generally correlated in the west (Medler et al. 2002).

Extensive areas of the California Floristic Province comprise broadleafed evergreen forest (Douglas-fir–tanoak–madrone, ponderosa pine–black oak, and canyon live oak–Coulter pine)—about 20% of the land area of California or about equal to the conifer forest area (Lenihan et al. 2003). The mixed evergreen forest formation is predicted to expand northward and upslope at the expense of subalpine conifer forest types under global warming scenarios. Lenihan et al. (2003) also suggested that anthropogenic climate change in the coming century might be less likely to affect fire weather (summer drought will persist) and more likely to affect California's fire regimes by changing the amount and character of fuels. This might increase inter-annual variability in the area burned—i.e., with more extreme events. Therefore, fire-prone forests, extensive and perhaps severe forest fires, and invasive species are all likely to expand in California forests under global warming.

In addition to nonnative grass invasions, there is also growing concern about the outbreaks of insects (such as pine beetles) and emerging infectious diseases on the landscape. For example, historic fire suppression in mixed-evergreen forest of northern coastal California has led to land cover change, the expansion of broadleaf woodland at the expense of chaparral, reduction in spatial heterogeneity of plant communities, and subsequent increased susceptibility of oak woodlands to lethal infection of *Phytophthora ramorum*, an introduced plant pathogen causing Sudden Oak Death (Moritz and Odion 2005). Somewhat more subtle changes in forest structure due to fire suppression are considered to be factors contributing to the effect of white pine blister rust (*Cronartium ribicola*) on *Pinus lambertiana* in the Sierra Nevada (van Mantgem et al. 2004). Despite disturbing trends in losses of this tree, models suggest managers still have time to alter the outcome of this invasion and prevent complete demise of this key forest species.

8.6.2 Shrublands

During the Zaca Fire of 2007, over 800 km of fire breaks were bulldozed through relatively pristine old-growth chaparral. Although restoration is planned, these areas have the potential for greatly exacerbating the alien species problem in this region. Particularly troubling is that they surround and dissect a wilderness area that previously was relatively isolated from aliens. The strategy behind this extensive use of clear-cutting chaparral is akin to what political commentator Ron Suskind (2006) describes as the 1% doctrine. Namely, in some circles if there is a 1% chance that an action will pay off, it is considered a legitimate course of action. In other words, there is a hierarchy of values that preempt any sort of cost–benefit analysis. Although fire management has never had such an extreme policy, if a treatment had any potential benefit in reducing fire hazards it was considered a legitimate strategy. However, agencies are increasingly faced with having to balance numerous issues other than just fire hazard reduction, and often conservation of natural resources comes in conflict. Balancing these issues will necessitate a more thorough cost-benefit analysis to fire management decisions so that the impacts on resources such as alien plant invasion are considered in the equation. One might expect that a cost-benefit analysis that considered potential alien invasions might have resulted in a more judicious use of fire breaks in the Zaca Fire. Of course, these decisions cannot be made during major fire events, but rather must be thoroughly considered and incorporated into a fire management plan with specific analyses that consider all resource costs in association with perceived benefits of reduced fire hazard.

Application of a cost benefit analysis to fire management will be easier in some ecosystems than in others. Besides the economics, the issues to be balanced are fire hazard and resources. Forested ecosystems with a history of frequent surface fire regimes are assembled from both overstory and understory species well adapted to frequent fires. A frequent fire regime is also compatible with reducing fire hazard, and thus balancing the resource costs and fire hazard–reduction benefits of prescribed burning is relatively easy compared to many other ecosystems such as chaparral.

Crown-fire ecosystems like chaparral are a different story. The historical fire frequency was significantly longer than forested ecosystems and the vegetation comprises many species that are rather vulnerable to frequent fire. Fire management strategies designed to incorporate frequent prescription burning as a cure for fire hazard are often at odds with resource conservation. Because humans share much of this landscape it is inevitable that in some cases fire hazard reduction will preempt resource issues. When fuel treatments such as frequent prescribed burning or mechanical crushing (Fig. 8.11) are applied to this landscape they have the potential for doing resource damage. Thus, a cost-benefit analysis might result in a more judicious use of such treatments.

Of course, with no end in sight for population growth, the fire management and alien plant problems are destined to become worse unless some changes occur in development patterns. Smart growth that promotes infilling within the development footprint, rather than continuing urban sprawl, has some potential for slowing this problem.

Fig. 8.11 Fuel treatment of chaparral through crushing. Such sites will invariably become dominated by alien species and native communities are not likely to regain this site for an extremely long time. These treatments are designed solely for fire hazard reduction and can be viewed as sacrificing natural resources (Photo by Wendy Boes)

8.6.3   Grasslands

The future frequency and impact of fire in California grasslands will depend on climate, nitrogen deposition, and grazing, all of which influence the density and nature of the fuel bed. Development and human presence within the landscape will also influence the nature of fuels, frequency of ignitions, control efforts, and subsequent burning patterns. The regional species pool will also determine which species are present to respond to fire events.

Models of future climate vary in terms of the direction and amount of rainfall change projected for California grasslands. Substantial increases in California precipitation were predicted by early general circulation models (Hadley Centre HadCM2) and CCM1 (Dukes and Shaw 2007). Increased precipitation within the growing season, however, may have little effect on productivity in California grassland settings (Pitt and Heady 1978; Reever-Morghan et al. 2007; Dukes and Shaw 2007). Thus, it may have little effect on fire frequency or intensity. The recent Hadley Centre model version 3 (HadCM3) and the Geophysical Fluids Dynamic Laboratory model (GFDL) predict decreases in total annual precipitation in California (Pope et al. 2000; Cayan et al. 2006). Reduced growing season precipitation events could lead to a decline in grassland productivity. A reduction in fuel density and biomass could reduce fire intensities when fires burn through grassy vegetation.

Recent Global Change Models that predict a decrease in growing season precipitation for California grasslands also predict that precipitation will be packaged into fewer more extreme events. To date there have been no field experiments that recreate these climate conditions. We suggest here that such repackaging could favor native perennial species because individual events will saturate the soil beyond the ability of introduced annual grasses to transpire water. Native perennial species, by contrast, with their deeper roots and longer period of activity, could access this water over a wider range of depths and time periods. Native perennial shrubs such as coyote brush (*Baccharis pilularis*) readily colonize grassland under wet conditions (Williams et al. 1987). Conversion to coyote brush shrublands should decrease fire frequency due to the change from fine summer–dry fuel to evergreen shrub fuel (though not everywhere, c.f. Keeley 2005). Fire intensity might increase, however, due to the greater biomass accumulation in such a shrubland and the tendency for shrublands to burn as crown fires. If the beneficiaries of altered precipitation events are native perennial grasses, it is more difficult to predict their impact on fire regimes. In tropical savannas, some recent studies have shown that introduced grasses can increase fuel loads by up to seven-fold greatly increasing fire intensities (D'Antonio 2000; Rossiter et al. 2003), but these systems contain both native and nonnative grasses that are substantially larger than typical grasses in California. No studies have yet compared fire frequency and intensity between native perennial bunchgrass-dominated vs. alien annual grass-dominated California grassland.

8.7   Conclusions

The California Floristic Province spans a large latitudinal range in a topographically diverse area of western North America. The entire region experiences a Mediterranean climate with warm dry summers, but the fire regimes in the grasslands, shrublands, woodlands and forests of California are as diverse as the terrain. In the historical period the landscape mosaic was affected by half a millennium of grazing, anthropogenic burning, logging, fire suppression, urban expansion, and the deliberate or accidental introduction of invasive plants and other alien species. The future of the regional ecosystems facing combined impacts of climate change, land use change, nitrogen deposition, altered fire regimes, and species invasions must be considered in light of the history of the landscape.

Montane forests were subjected historically to logging, grazing, and fire suppression, leading to fire exclusion. Restoration of fire in these landscapes must be implemented cautiously in light of the threat of invasive herbaceous plant species. This is particularly urgent because although forested areas at higher elevations have historically been less threatened by alien plants than lower elevation shrublands and grasslands, this threat is likely to increase, as is the likelihood of large or severe fire, in association with anthropogenic climate warming. Native shrublands have been altered in the distant and recent past by frequent human-ignited fire. The greatest future threat to this ecosystem again lies in the combined impacts of increased fire frequency due to human population pressure and climate change, and the subsequent spread of invasive plant species. Exotic species probably became well established in native herbaceous plant formations several centuries ago as a result of the deliberate use of fire by people. In some cases, a specific prescription of repeated fire treatments has been used successfully to control non–native species, but in other cases prescribed fire has actually promoted them. Anthropogenic global warming may reduce fire frequency or intensity in native grasslands, but the altered precipitation regime that is predicted to occur may actually favor native plant species. A systematic approach is required to understand the complex and potentially synergistic impacts of fire, invasive species, climate change, and land use change on the landscape mosaic.

References

Alexander, J.M., and C.M. D'Antonio. 2003a. Effects of stand age and fire on the seed bank of french broom (*Genista monspessulana*) in California. *Restoration Ecology* 11: 185–197.

Alexander, J.M., and C.M. D'Antonio. 2003b. Control methods for the removal of French and Scotch Broom tested in coastal California. *Ecological Restoration* 21: 191–198.

Allen, E.B., P.E. Padget, A. Bytnerowicz, and R. Minnich. 1998. Nitrogen deposition effects on coastal sage vegetation of southern California. In *Proceedings of the international symposium on air pollution and climate change effects on forest ecosystems*. General Technical Report

PSW–GTR–166, eds. A. Bytnerowicz, M.J. Arbaugh, and S.L. Schilling, 131–139. Albany: U.S. Forest Service.

Anderson, M.K. 2005a. *Tending the wild: Native American knowledge and the management of California's natural resources*. Berkeley: University of California Press.

Anderson, R.S. 2005b. Contrasting vegetation and fire histories on the Point Reyes Peninsula during the pre–settlement and settlement periods: 15,000 years of change. Final report. http://www.nps.gov/pore/parkmgmt/upload/firemanagement_fireecology_research_anderson_0506.pdf. Accessed 25 Jan 2010.

Anderson, M.K. 2007. Native American uses and management of California's grasslands. In *California grasslands: Ecology and management*, eds. M. Stromberg, J. Corbin, and C.M. D'Antonio, 57–66. Berkeley: University of California Press.

Baker, W.L. 2006. Fire and restoration of sagebrush ecosystems in the western United States. *Wildlife Society Bulletin* 34: 177–185.

Beatley, J. 1966. Ecological status of introduced Brome grasses (*Bromus* spp) in desert vegetation in southern Nevada. *Ecology* 47: 548–554.

Betts, A.D.K. 2003. Ecology and control of goatgrass (*Aegilops triuncialis*) and medusaehead (*Taeniatherum caput–medusae*) in California annual grasslands. Ph.D. dissertation. Berkeley: University of California.

Binkley, D., K. Cromack, and R.L. Fredriksen. 1982. Nitrogen accretion and availability in some snowbrush ecosystems. *Forest Science* 28: 720–724.

Borchert, M. 1985. Serotiny and cone–habit variation in populations of *Pinus coulteri* (Pinaceae) in the southern Coast Ranges of California. *Madroño* 32: 29–48.

Brooks, M.L. 2003. Effects of increased soil nitrogen on the dominance of alien annual plants in the Mojave Desert. *Journal of Applied Ecology* 40: 344–353.

Brooks, M.L., C.M. D'Antonio, D.M. Richardson, J.B. Grace, J.E. Keeley, J.M. DiTomaso, R.J. Hobbs, M. Pellant, and D. Pyke. 2004. Effects of invasive alien plants on fire regimes. *Bioscience* 54: 677–688.

Cal Fire. 2007. FRAP fire history database. California Department of Forestry and Fire Protection. http://frap.cdf.ca.gov. Accessed June 2007.

Callaway, R.M., and F.W. Davis. 1993. Vegetation dynamics, fire, and the physical environment in coastal central California. *Ecology* 74: 1567–1578.

Callihan, R.H., T.S. Prather, and F.E. Northam. 1993. Longevity of yellow starthistle (*Centaurea solstitialis*) achenes in soil. *Weed Technology* 7: 33–35.

Cayan, D., E. Maurer, M. Dettinger, M. Tyree, K. Hayhoe, C. Bonfils, P. Duffy, and B. Santer. 2006. Climate scenarios for California. California Climate Change Center. http://www.energy.ca.gov/2005publications/CEC-500-2005-203/CEC-500-2005-203-SF.PDF. Accessed 25 Jan 2010.

Cheney, N.P., J.S. Gould, and W.R. Catchpole. 1993. The influence of fuel, weather and fire shape variables on fire–spread in grasslands. *International Journal of Wildland Fire* 3: 31–44.

Conard, S.G., A.E. Jaramillo, K. Cromack Jr., and S. Rose. 1985. *The role of the genus Ceanothus in western forest ecosystems*. General Technical Report PNW–GTR–182. Portland: U.S. Forest Service.

Cooper, W.S. 1922. *The broad–sclerophyll vegetation of California. An ecological study of the chaparral and its related communities. Publication No. 319*. Washington: Carnegie Institution of Washington.

Corbin, J.C., C.M. D'Antonio, and S.J. Bainbridge. 2004. Tipping the balance in the restoration of native plants: Experimental approaches to changing the exotic: native ratio in California grassland. In *Experimental approaches to conservation biology*, eds. M. Gordon and L. Bartol, 154–179. Los Angeles: University of California Press.

Corbin, J.C., A. Dyer, and E.W. Seabloom. 2007. Competitive interactions. In *California grasslands: Ecology and management*, eds. M. Stromberg, J. Corbin, and C.M. D'Antonio, 156–168. Berkeley: University of California Press.

D'Antonio, C.M. 2000. Fire, plant invasions and global changes. In *Invasive species in a changing world*, eds. H. Mooney and R. Hobbs, 65–94. Covelo: Island.

D'Antonio, C.M., and P. Vitousek. 1992. Biological invasions by exotic grasses, the grass–fire cycle and global change. *Annual Review of Ecology and Systematics* 23: 63–88.

D'Antonio, C.M., D. Odion, and C. Tyler. 1993. Invasion of maritime chaparral by the introduced succulent, *Carpobrotus edulis*: The roles of fire and herbivory. *Oecologia* 95: 14–21.

D'Antonio, C.M., J. Levine, and M. Thomsen. 2001. Ecosystem resistance to invasion and the role of propagules supply: A California perspective. *Journal of Mediterranean Ecology* 2: 233–245.

Dale, V.H., L.A. Joyce, S. McNulty, R.P. Neilson, M.P. Ayres, M.D. Flannigan, P.J. Hanson, L.C. Irland, A.E. Lugo, C.J. Peterson, D. Simberloff, F.J. Swanson, B.J. Stocks, and B.M. Wotton. 2001. Climate change and forest disturbances. *Bioscience* 51: 723–734.

Davison, J. 1996. Livestock grazing in wildland fuel management programs. *Rangelands* 18: 242–245.

Delwiche, C.C., P.J. Zinke, and C.M. Johnson. 1965. Nitrogen fixation by *Ceanothus*. *Plant Physiology* 40: 1045–1047.

Diamond, J.M., C.A. Call, and N. Devoe. 2009. Effects of targeted cattle-grazing on fire behaviour of cheatgrass-dominated rangeland in the northern Great Basin, USA. *International Journal of Wildland Fire* 18: 944–950.

DiTomaso, J.M., and D.W. Johnson eds. 2006. *The use of fire as a tool for controlling invasive plants*. Cal–IPC Publication 2006-01. Berkeley: California Invasive Plant Council.

DiTomaso, J.M., G.B. Kyser, and M.S. Hastings. 1999. Prescribed burning for control of yellow starthistle (*Centaurea solstitialis*) and enhanced native plant diversity. *Weed Science* 47: 233–242.

DiTomaso, J.M., K.L. Heise, G.B. Kyser, A. Merenlender, and R.J. Keiffer. 2001. Carefully timed burning can control barb goatgrass. *California Agriculture* 55: 47–53.

Dukes, J.S., and M.R. Shaw. 2007. Responses to changing atmosphere and climate. In *California grasslands: Ecology and management*, eds. M. Stromberg, J. Corbin, and C.M. D'Antonio, 218–232. Berkeley: University of California Press.

Finney, M.A. 2001. Design of regular landscape fuel treatment patterns for modifying fire growth and behavior. *Forest Science* 47: 219–228.

Franklin, J. 2010. Vegetation dynamics and exotic plant invasion following high severity crown fire in a southern California confier forest. *Plant Ecology* 207: 281–295.

Franklin, J., and C.E. Woodcock. 1997. Multiscale vegetation data for the mountains of Southern California: spatial and categorical resolution. In *Scale in remote sensing and GIS*, eds. D.A. Quattrochi and M.F. Goodchild, 141–168. Boca Raton: CRC/Lewis.

Franklin, J., L.A. Spears–Lebrun, D.H. Deutschman, and K. Marsden. 2006. Impact of a high–intensity fire on mixed evergreen and mixed conifer forests in the Peninsular Ranges of southern California, USA. *Forest Ecology and Management* 235: 18–29.

Fried, J.S., M.S. Torn, and E. Mills. 2004. The impact of climate change on wildfire severity: A regional forecast for northern California. *Climatic Change* 64: 169–191.

Greenlee, J.M., and J.H. Langenheim. 1990. Historic fire regimes and their relation to vegetation patterns in the Monterey Bay area of California. *American Midland Naturalist* 124: 239–253.

Haidinger, T.L., and J.E. Keeley. 1993. Role of high fire frequency in destruction of mixed chaparral. *Madroño* 40: 141–147.

Hamilton, J.G. 1997. Changing perceptions of pre–European grasslands in California. *Madroño* 44: 311–333.

Harrison, S., B.D. Inouye, and H.D. Safford. 2003. Ecological heterogeneity in the effects of grazing and fire on grassland diversity. *Conservation Biology* 17: 837–845.

Hellmers, H., and J.M. Kelleher. 1959. *Ceanothus leucodermis* and soil nitrogen in Southern California mountains. *Forest Science* 5: 275–278.

Hopkinson, P.J.M. 2003. Native bunchgrass diversity patterns and phytolith deposits as indicators of fragmentation and change in a California Coast Range grassland. Ph.D. dissertation, University of California, Berkeley.

Huenneke, L.F., and H.A. Mooney (eds.). 1989. *Grassland structure and function: California annual grasslands*. Dordrecht: Kluwer Academic.

Jackson, L.E. 1985. Ecological origins of California's Mediterranean grasses. *Journal of Biogeography* 12: 349–361.

Jacobson, A.L., S.D. Davis, and S.L. Babritius. 2004. Fire frequency impacts non–sprouting chaparral shrubs in the Santa Monica Mountains of southern California. In *Ecology, conservation and management of mediterranean climate ecosystems*, eds. M. Arianoutsou and V.P. Panastasis. Rotterdam: Millpress.

Kaufmann, M.R., D. Binkley, P.Z. Fule, M. Johnson, S.L. Stephens, and T.W. Swetnam. 2007. Defining old growth for fire–adapted forests of the Western United States. *Ecology and Society* 12: 15.

Keeley, J.E. [N.d.]. Unpublished data. Three Rivers: Western Ecological Research Center, U.S. Geological Survey (On file with: Jon Keeley).

Keeley, J.E. 1990. The California valley grassland. In *Endangered plant communities of southern California*. Special Publication No. 3, ed. A.A. Schoenherr, 3–23. Claremont: Southern California Botanists.

Keeley, J.E. 2001. Fire and invasive species in mediterranean–climate ecosystems of California. In *Proceedings of the invasive species workshop: The role of fire in the control and spread of invasive species*, eds. K.E.M. Galley and T.P. Wilson, 81–94. Tallahassee: Tall Timbers Research Station Misc. Publ. No. 11.

Keeley, J.E. 2002. Native American impacts on fire regimes in California coastal ranges. *Journal of Biogeography* 29: 303–320.

Keeley, J.E. 2005. Fire history of the San Francisco East Bay region and implications for landscape patterns. *International Journal of Wildland Fire* 14: 285–296.

Keeley, J.E. 2006a. South coast bioregion. In *Fire in California's ecosystems*, eds. N.G. Sugihara, J.W. van Wagtendonk, K.E. Shaffer, J. Fites-Kaufman, and A.E. Thoede, 350–390. Berkeley: University of California Press.

Keeley, J.E. 2006b. Fire management impacts on invasive plants in the western United States. *Conservation Biology* 20: 375–384.

Keeley, J.E., and T.W. McGinnis. 2007. Impact of prescribed fire and other factors on cheatgrass persistence in a Sierra Nevada ponderosa pine forest. *International Journal of Wildland Fire* 16: 96–106.

Keeley, J.E., and N.L. Stephenson. 2000. Restoring natural fire regimes to the Sierra Nevada in an era of global change. In *Wilderness science in a time of change*, comps. D.N. Cole, S.F. McCool, W.T. Borrie, and J. Loughlin, 255–265. Proceedings RMRS-P-15-VOL-5. Fort Collins: U.S. Forest Service.

Keeley, J.E., C.J. Fotheringham, and M. Morais. 1999. Reexamining fire suppression impacts on brushland fire regimes. *Science* 284: 1829–1832.

Keeley, J.E., D. Lubin, and C.J. Fotheringham. 2003. Fire and grazing impacts on plant diversity and alien plant invasions in the southern Sierra Nevada. *Ecological Applications* 13: 1355–1374.

Keeley, J.E., C.J. Fotheringham, and M.A. Moritz. 2004. Lessons from the 2003 wildfires in southern California. *Journal of Forestry* 102: 26–31.

Keeley, J.E., M. Baer–Keeley, and C.J. Fotheringham. 2005. Alien plant dynamics following fire in mediterranean–climate California shrublands. *Ecological Applications* 15: 2109–2125.

Keeley, J.E., T. Brennan, and A.H. Pfaff. 2008. Fire severity and ecosystem responses following crown fires in California shrublands. *Ecological Applications* 18: 1530–1546.

Keeley, J.E., H. Safford, C.J. Fotheringham, J. Franklin, and M. Moritz. 2009a. The 2007 Southern California wildfires: Lessons in complexity. *Journal of Forestry* 107: 287–296.

Keeley, J.E., G.H. Aplet, N.L. Christensen, S.G. Conard, E.A. Johnson, P.N. Omi, D.L. Peterson, and T.W. Swetnam. 2009b. *Ecological foundations for fire management in North American forest and shrubland ecosystems*. General Technical Report PNW–GTR–779. Portland: U.S. Forest Service.

Klinger, R.C., M.L. Brooks, and J.M. Randall. 2006. Fire and invasive plant species. In *Fire in California's ecosystems*, eds. N.G. Sugihara, J.W. van Wagtendonk, K.E. Shaffer, J. Fites–Kaufman, and A.E. Thoede, 499–519. Berkeley: University of California Press.

Kyser, G.B., and J.M. DiTomaso. 2002. Instability in a grassland community after the control of yellow starthistle (*Centaurea solstitialis*) with prescribed burning. *Weed Science* 50: 648–657.

Lenihan, J.M., R. Draper, D.B. Bachelet, and R.P. Neilson. 2003. Climate change effects on vegetation distribution, carbon and fire in California. *Ecological Applications* 13: 1667–1681.

McAdoo, K., B. Schultz, S. Swanson, and R. Orr. 2007. Northeastern Nevada wildfires part 2: Can livestock be used to reduce wildfires? University of Nevada Cooperative Extension Fact Sheet 07–21. http://www.unce.unr.edu/publications/files/nr/2007/fs0721.pdf. Accessed 25 Jan 2010.

Medler, M.J., P. Montesano, and D. Robinson. 2002. Examining the relationship between snowfall and wildfire patterns in the western United States. *Physical Geography* 23: 335–342.

Mensing, S.A., J. Michaelsen, and R. Byrne. 1999. A 560–year record of Santa Ana fires reconstructed from charcoal deposited in the Santa Barbara Basin, California. *Quaternary Research* 51: 295–305.

Merriam, K.E., J.E. Keeley, and J.L. Beyers. 2006. Fuel breaks affect nonnative species abundance in California plant communities. *Ecological Applications* 16: 515–527.

Meyer, M.D., and P.M. Schiffman. 1999. Fire season and mulch reduction in a California grassland: a comparison of restoration strategies. *Madroño* 46: 25–37.

Millar, C.I. 1997. Comments on historical variation and desired condition as tools for terrestrial landscape analysis. In *Proceedings of the sixth biennial watershed management conference*. ed. S. Sommarstrom, 105–131. Water Resources Center Report No. 92, Davis: University of California.

Millar, C.I., N.L. Stephenson, and S.L. Stephens. 2007. Climate change and forests of the future: Managing in the face of uncertainty. *Ecological Applications* 17: 2145–2151.

Minnich, R.A. 2008. *California's fading wildflowers*. Los Angeles: University of California Press.

Mooney, H.A., S.P. Hamburg, and J.A. Drake. 1986. The invasions of plants and animals into California. In *Ecology of biological invasions of North America and Hawaii*, eds. H.A. Mooney and J.A. Drake, 250–327. New York: Springer.

Moritz, M.A. 2003. Spatiotemporal analysis of controls on shrubland fire regimes: Age dependence and fire hazard. *Ecology* 84: 351–361.

Moritz, M.A., and D.C. Odion. 2005. Examining the strength and possibly causes of the relationship between fire history and sudden oak death. *Oecologia* 144: 106–114.

Moyes, A.B., M.S. Witter, and J.A. Gamon. 2005. Restoration of native perennials in a California annual grassland after prescribed spring burning and solarization. *Restoration Ecology* 13: 659–666.

Odion, D.C., and F.W. Davis. 2000. Fire, soil heating and the formation of vegetation patterns in chaparral. *Ecological Monographs* 70: 149–169.

Odion, D.C., E.J. Frost, J.R. Strittholt, H. Jiang, D.A. Dellasala, and M.A. Moritz. 2004. Patterns of fire severity and forest conditions in the western Klamath Mountains, California. *Conservation Biology* 18: 927–936.

Padgett, P.E., E.B. Allen, A. Bytnerowicz, and R.A. Minnich. 1999. Changes in soil inorganic nitrogen as related to atmospheric nitrogenous pollutants in southern California. *Atmospheric Environment* 33: 769–781.

Parker, V.T. 1987. Effects of wet–season management burns on chaparral vegetation: implications for rare species. In *Conservation and management of rare and endangered plants*, ed. T.S. Elias, 233–237. Sacramento: California Native Plant Society.

Pitt, M.D., and H. Heady. 1978. Responses of annual vegetation to temperature and rainfall patterns in northern California. *Ecology* 59: 336–350.

Pope, V.D., M.L. Gallani, P.R. Rowntree, and R.A. Stratton. 2000. The impact of new physical parameterizations in the Hadley Centre climate model: HadAM3. *Climate Dynamics* 16: 123–146.

Reever–Morghan, K., J. Corbin, and J. Gerlach. 2007. Water relations. In *California grasslands: Ecology and management*, eds. M. Stromberg, J. Corbin, and C.M. D'Antonio, 87–93. Berkeley: University of California Press.

Reiner, R.J. 2007. Fire in California grasslands. In *California grasslands: Ecology and management*, eds. M. Stromberg, J. Corbin, and C.M. D'Antonio, 207–217. Berkeley: University of California Press.

Reiner, R.J., P. Hujik, and O. Pollock. 2006. Predicting vegetation response to fire in California annual grassland. In *Assumptions used to justify prescribed fire as a restoration tool in California annual grasslands*, eds. R. Schlising and D. Alexander, 167–174. Chico: Butte Environmental Council.

Rejmanek, M., and J.M. Randall. 1994. Invasive alien plants in California: 1993 summary and comparison with other areas in North America. *Madroño* 41: 161–177.

Reynolds, R.D. 1959. The effect upon forest of natural fire and aboriginal burning in the Sierra Nevada. M.S. thesis, University of California, Berkeley.

Rossiter, N.A., S.A. Setterfield, M.M. Douglas, and L. Huntley. 2003. Testing the grass–fire cycle: Alien grass invasion in the tropical savannas of northern Australia. *Diversity and Distributions* 9: 169–176.

Roye, C.L. 2004. Plant assessment form. http://www.cal-ipc.org/ip/inventory/PAF/Ehrharta%20calycina.pdf. Accessed 25 Jan 2010.

Schwartz, M.W., D.J. Porter, J.M. Randall, and K.E. Lyons. 1996. Impact of nonindigenous plants. In *Sierra Nevada Ecosystem Project: Final report to Congress, Volume II, assessments and scientific basis for management options*, 1203–1218. Davis: University of California, Centers for Water and Wildland Resources.

Siguenza, C., L. Corkidi, and E.B. Allen. 2006. Feedbacks of soil inoculum of mycorrhizal fungi altered by N deposition on the growth of a native shrub and an invasive annual grass. *Plant and Soil* 286: 153–165.

Stephens, S.L., and B.M. Collins. 2004. Fire regimes of mixed conifer forests in the north–central Sierra Nevada at multiple spatial scales. *Northwest Science* 78: 12–23.

Stephens, S.L., J.J. Moghaddas, C. Edminster, C.E. Fiedler, S. Hasse, M. Harrington, J.E. Keeley, E. Knapp, J.D. McIver, K. Metlen, C. Skinner, and A. Youngblood. 2009. Fire treatment effects on vegetation structure, fuels, and potential fire severity western forests. *Ecological Applications* 19: 305–320.

Sugihara, N.G., J.W. van Wagtendonk, K.E. Shaffer, J. Fites-Kaufman, and A.E. Thoede, eds. 2006. *Fire in California's ecosystems*. Los Angeles: University of California Press.

Suskind, R. 2006. *The one percent doctrine*. New York: Simon and Schuster.

Swetnam, T.W., and J.L. Betancourt. 1998. Mesoscale disturbance and ecological response to decadal climatic variability in the American Southwest. *Journal of Climate* 11: 3128–3147.

Stylinski, C.D., and E.B. Allen. 1999. Lack of native species recovery following severe exotic disturbance in southern California shrublands. *Journal of Applied Ecology* 36: 544–554.

Talluto, M.V., and K.N. Suding. 2008. Historical change in coastal sage scrub in southern California, USA in relation to fire frequency and air pollution. *Landscape Ecology* 23: 803–815.

Tyler, C.M., D. Odion, and R.M. Callaway. 2007. Dynamics of woody species in the California grassland. In *California grasslands: Ecology and management*, eds. M. Stromberg, J. Corbin, and C.M. D'Antonio, 169–179. Berkeley: University of California.

van Mantgem, P.J., N.L. Stephenson, M.B. Keifer, and J.E. Keeley. 2004. Effects of an introduced pathogen and fire exclusion on demography of sugar pine. *Ecological Applications* 14: 1590–1602.

van Wagtendonk, J.W., and J. Fites-Kaufman. 2006. Sierra Nevada bioregion. In *Fire in California's ecosystems*, eds. N. Sugihara, J.W. van Wagtendonk, K.E. Shaffer, J. Fites-Kaufman, and A.E. Thode, 264–294. Berkeley: University of California.

Weiss, S. 1999. Cows, cars and checkerspot butterflies: nitrogen deposition and management of nutrient poor grasslands for a threatened species. *Conservation Biology* 13: 1478–1486.

Weiss, S. 2006. *Impacts of nitrogen deposition on California ecosystems and biodiversity*. CEC–500–2005–165. Sacramento: California Energy Commission, PIER Energy–related Environmental Research.

Wells, P.V. 1962. Vegetation in relation to geological substratum and fire in the San Luis Obispo quadrangle, California. *Ecological Monographs* 32: 79–103.

Westerling, A.L., A. Gershunov, T.J. Brown, D.R. Cayan, and M.D. Dettinger. 2003. Climate and wildfire in the western United States. *Bulletin of the American Meteorological Society* 84: 595–604.

Westerling, A.L., H.G. Hidalgo, D.R. Cayan, and T.W. Swetnam. 2006. Warming and earlier spring increase western US forest wildfire activity. *Science* 313: 940–943.

Westman, W.E. 1979. Oxidant effects on Californian coastal sage scrub. *Science* 205: 1001–1003.

Williams, K., R.J. Hobbs, and S.P. Hamburg. 1987. Invasion of an annual grassland in northern California by *Baccharis pilularis* ssp. *consanguinea*. *Oecologia* 72: 461–465.

Wills, R.D. 2006. Central Valley bioregion. In *Fire in California's ecosystems*, eds. N.G. Sugihara, J.W. van Wagtendonk, K.E. Shaffer, J. Fites-Kaufman, and A.E. Thoede, 295–320. Berkeley: University of California Press.

Yoshida, L.C., and E.B. Allen. 2001. Response to ammonium and nitrate by a mycorrhizal annual invasive grass and native shrub in southern California. *American Journal of Botany* 88: 1430–1436.

Zedler, P.H., and G.A. Scheid. 1988. Invasion of *Carpobrotus edulis* and *Salix lasiolepis* after fire in a coastal chaparral site in Santa Barbara County, California. *Madroño* 35: 196–201.

Zedler, P.H., C.R. Gautier, and G.S. McMaster. 1983. Vegetation change in response to extreme events: the effect of a short interval between fires in California chaparral and coastal scrub. *Ecology* 64: 809–818.

Chapter 9
Modeling Landscape Fire and Wildlife Habitat

Samuel A. Cushman, Tzeidle N. Wasserman, and Kevin McGarigal

9.1 Introduction

Global climate is expected to change rapidly over the next century (Thompson et al.
1998; Houghton et al. 2001; IPCC 2008). This will affect forest ecosystems both
directly by altering biophysical conditions (Neilson 1995; Neilson and Drapek
1998; Bachelet et al. 2001) and indirectly through changing disturbance regimes
(Baker 1995; McKenzie et al. 1996; Keane et al. 1999; Dale et al. 2001; McKenzie
et al. 2004; Westerling et al. 2006). Changes in biophysical conditions could lead
to species replacement in communities and latitudinal and altitudinal migrations
(Iverson and Prasad 2002; Neilson et al. 2005). Expected increases in the frequency, size, and severity of wildfires (Mearns et al. 1984; Overpeck et al. 1990;
Solomon and Leemans 1997; IPCC 2008), and other disturbances such as insect
outbreaks, may further amplify changes in vegetation structure, species composition, and diversity (Christensen 1988; McKenzie et al. 2004). These shifts in distributions of plant species may have large impacts on many aspects of ecological
diversity and function (Peters and Lovejoy 1992; Miller 2003).

Despite the magnitude of these potential ecosystem changes, relatively little
attention has been given to the effects of interactions between climate and natural
disturbance regimes on wildlife populations. Wildlife populations are critically
dependent on sufficient amount, quality, and spatial distribution of habitat.
The environmental conditions that provide habitat for each species in turn are a
dynamic product of the prevailing disturbance regime, in interaction with regional
climate. In the western United States, fire, as arguably the dominant landscape-
scale disturbance process, plays a keystone role in establishing and maintaining
habitat conditions for wildlife. Some species evolved in fire-dominated ecosystems
and consequently may have been negatively impacted by fire exclusion, whereas by

S.A. Cushman (✉)
Rocky Mountain Research Station, U.S. Forest Service, Missoula, MT 59801-5801, USA
e-mail: scushman@fs.fed.us

D. McKenzie et al. (eds.), *The Landscape Ecology of Fire*, Ecological Studies 213,   223
DOI 10.1007/978-94-007-0301-8_9, © Springer Science+Business Media B.V. 2011

contrast others may be sensitive to loss of habitat from fire. If climate change drives more frequent and extensive fires in forest ecosystems (McKenzie et al. 2004; Flannigan et al. 2005; Westerling et al. 2006), species dependent on extensive late-successional forest may decline while species dependent on early-successional habitat may increase. Furthermore, if a warmer climate increases disturbance by insects, changes to habitat could be accentuated by the interactions between fire and insect disturbance regimes.

In this context of changing climate and associated changes in natural disturbance regimes, managers need to decide whether to suppress fires, or manage vegetation, or both. From a management perspective, interactions of climate-induced changes in disturbance regimes with fire suppression and fuels treatment programs are difficult to anticipate based on a simplistic understanding of each factor acting independently. Consequently, it is important to evaluate the interactions among climate change, fire and insect disturbance processes, fire suppression, and fuels treatment more formally. Landscape dynamic simulation models are the most appropriate tool to conduct such evaluations. In this chapter, we illustrate this approach by evaluating the changes in natural disturbance regimes that could result from changes in climate, fire suppression, and vegetation management on the habitat capability of two wildlife species of concern, the American marten (*Martes americana*) and flammulated owl (*Otus flammeolus*) in the northern Rocky Mountains. Recent research on landscape habitat relationships in the study region has found these two species have contrasting ecological relationships. Marten is associated with late-successional closed-canopy mesic forests, whereas flammulated owl is associated with open-canopy, large size-class, dry forest types. These two ecological conditions may be expected to respond differently to changes in fire regime and vegetation management.

9.2 Methods

9.2.1 The Study Landscape

Prospect Creek Basin is a 47,058 ha watershed in the Lolo National Forest of western Montana (Fig. 9.1). We chose this landscape because a regional landscape analysis of biophysical characteristics identified it as highly representative of the surrounding 1,827,400 ha comprising three subsections (Coeur d'Alene Mountains, St. Joe-Bitterroot Mountains, and Clark Fork Valley and Mountains) of the Bitterroot Mountains Ecosection (Table 9.1). We classified the sample landscape into land cover classes based on the LANDFIRE project (http://www.landfire.gov). Specifically, land cover classes represent unique biophysical settings (BpS) or potential vegetation types (PVT). The only significant change we made to this classification scheme was to combine three separate BpS classes corresponding to

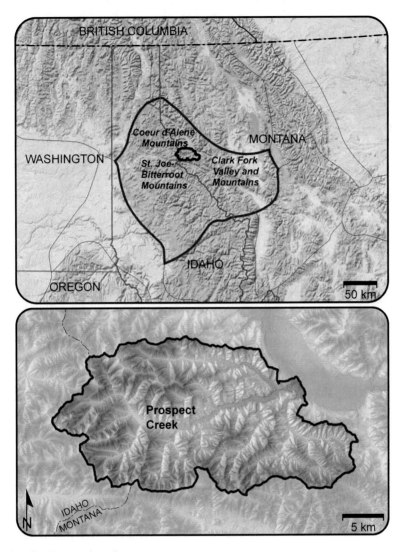

Fig. 9.1 Study area orientation map

"riparian" settings into a single "riparian" class. Full documentation of how these BpS classes were derived is available at the LANDFIRE website.

The spatial resolution of the landscape was set at 30 m, consistent with that of the data sources used in the LANDFIRE project. The spatial extent of the landscape was based on the hydrological watershed of Prospect Creek, a tributary of Clark Fork River, but for simulation purposes we included a 2-km wide buffer zone around the basin, bringing the total extent of the simulation landscape to 69,293 ha.

Table 9.1 Comparison of biophysical composition in the Prospect Creek basin study area to that of the surrounding ecosection. The table reports percentages of each landscape in eight biophysical types (http://www.landfire.gov)

Type	Percentage of landscape	
	Ecosection	Prospect creek
Mesic-Wet Spruce Fir	23.323	25.654
Mixed-Conifer Ponderosa Pine Douglas-fir	22.779	33.518
Western Hemlock Western Redcedar	20.599	11.269
Mixed Conifer Grand Fir	16.633	17.804
Riparian	5.326	4.649
Mixed Conifer Western Larch	4.382	3.269
Water	1.742	0.443
Subalpine Park	0.897	0.783
Total Area (ha)	1,827,400	47,058

9.2.2   Landscape Simulation Model

We used the Rocky Mountain Landscape Simulator (RMLANDS) (http://www.umass.edu/landeco/research/rmlands/rmlands.html) to simulate a variety of disturbance scenarios representing fire and insect disturbance regimes under current and future climate, two fire management strategies, and two vegetation management strategies. RMLANDS is a grid-based, spatially explicit, stochastic landscape model that simulates disturbance and succession processes affecting the structure and dynamics of Rocky Mountain landscapes. RMLANDS simulates two key processes: succession and disturbance. These processes are fully specified by the user (i.e., via model parameterization) and are implemented sequentially within 10-year time steps for a user-specified period of time.

9.2.2.1   Succession

RMLANDS simulates succession using a state-based transition approach in which discrete vegetation states are defined for each cover type. Each cover type has a separate transition model that uniquely defines its successional stages. Succession involves the probabilistic transition from one state to another over time and it occurs at the beginning of each time step in response to gradual growth and development of vegetation. Transition probabilities are typically based on the age of the stand (i.e., the time since the last stand-replacing event), but they can be based on any number of parameters, such as the abiotic setting (e.g., topographic setting) or disturbance history.

Succession is entirely patch-based. Specifically, each cell belongs to a patch, defined as contiguous (touching based on the eight-neighbor rule) cells sharing the same values for each of the attributes used to define succession probabilities. For example, age, time since low-mortality fire, and aspect are all used to define transition probabilities of a particular cover type transition model; contiguous cells with the same values for these three attributes will be treated as a patch and undergo probabilistic succession transitions together. Successional patches are not static; they change throughout the simulation in response to disturbance events, which can act to break up single patches into several new patches or to coalesce several patches into a single patch by changing the disturbance history at the cell level. This patch-based approach for succession avoids the salt-and-pepper effect that can occur with stochastic cell-based succession.

9.2.2.2 Disturbance Processes

RMLANDS simulates both natural and anthropogenic disturbances. Natural disturbances include wildfire and a variety of insect or pathogen outbreaks (e.g., mountain pine beetle). Each natural disturbance process is implemented separately, but can affect and be affected by other disturbance processes to produce changes in landscape conditions. For example, trees killed by mountain pine beetle can affect the local probability of ignition and spread of wildfires.

Climate plays a significant role in determining the temporal and spatial characteristics of the natural disturbance regime. RMLANDS uses a global parameter, as a proxy for climate, which affects initiation, spread, and mortality of all disturbances within a time step. This parameter can be specified as a constant mean or median with a user-specified level of temporal variability, a trend over time (with a specified variability), or as a user-defined trajectory reflecting the climate conditions during a specific reference period.

Disturbance events are initiated at the cell level, in contrast to succession. In each time step, each cell has a probability of initiation that is a function of its susceptibility to disturbance, and optionally, its spatial or temporal proximity to previous disturbance events or landscape features (e.g., roads). Susceptibility to wildfire, for example, is a function of factors that influence fuel mass and fuel moisture including: cover type, stand condition, time since last fire, time since last insect outbreak, elevation, aspect, and slope. Wildfire susceptibility is also a function of road proximity, which influences the risk of human-caused ignition.

Once initiated, the disturbance spreads to adjacent cells probabilistically. Each cell has a probability of spread that is a function of its susceptibility to disturbance (as above), which is modified by its topographic position relative to a burning cell (i.e., fires can burn more readily upslope), wind direction, and the influence of potential barriers (e.g., roads and streams). The probability of spread is further modified to reflect variable weather conditions associated with the disturbance event. This event modifier affects the final size of the disturbance and is specified as a user-defined size distribution. There is also an optional provision for the spotting of disturbances during spread so that disturbances are not constrained to contiguous spread only. The spotting feature as used in this analysis for both fire and insect disturbances.

Following disturbance spread, each cell is evaluated to determine the magnitude of ecological effect (i.e., severity) of the disturbance. Each cell can exhibit either high or low mortality of the dominant plants. High mortality occurs when all or nearly all (>75%) of the dominant plant individuals are killed; low mortality is assigned when less than 75% individuals are killed. Cells are aggregated into vegetation patches for purposes of determining mortality response, where patches are defined as spatially contiguous cells having the same cell attributes (e.g., identical disturbance history and age). So-called mixed severity fires produce a heterogeneous mixture of low- and high-mortality cells.

Following mortality determination, each vegetation patch is evaluated for potential immediate transition to a new stand condition (state). Transition pathways and rates of transition between states are defined uniquely for each cover type and are conditional on several attributes at the patch level. These disturbance-induced transitions are different from the successional transitions that occur at the beginning of each time step that represent the gradual growth and development of vegetation over time.

RMLANDS can also simulate a variety of vegetation treatments that result in immediate transition to a new state. These treatments are implemented via management regimes defined by the user. Management regimes are uniquely specified within management zones, or user-defined geographic units (e.g., urban-wildland interface vs. interior). Management zones are further divided into one or more management types based on cover type. Each cover type can be treated separately or it can be combined with other cover types to form aggregate management types. Each management type is then subject to a unique management regime, which consists of one or more treatment types and associated spatial and temporal constraints.

9.2.3   Wildlife Habitat Capability Model

We used HABIT@ (http://www.umass.edu/landeco/pubs/pubs.html) to quantify the habitat capability of the simulated landscapes for American marten and flammulated owl. HABIT@ is a multi-scale GIS-based system for modeling wildlife habitat capability (Fig. 9.2). We define habitat capability as the ability of the environment to provide the local resources (e.g, food, cover, nest sites) needed for survival and reproduction in sufficient quantity, quality, and distribution to meet the life-history requirements of individuals and local populations. Habitat capability is synonymous with habitat suitability.

HABIT@ models use GIS grids representing environmental variables such as cover type, stand age, canopy density, slope, hydrological regime, roads, and development. Input grids can represent anything pertinent to the species being modeled at any scale, depending only upon the availability of data. Complex derived grids representing specialized environmental variables (such as stream channel constraints, cliffs suitable for nesting, or rainfall patterns) can also be incorporated in HABIT@ models.

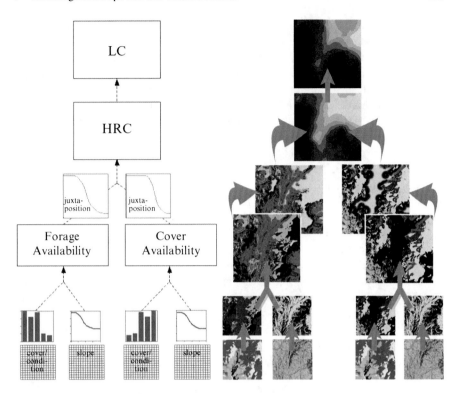

Fig. 9.2 HABIT@ is hierarchically organized into three primary levels. The lowest level comprises one or more local resources, each derived from one or more local resource indices based on GIS data. These are combined and summarized within home range equivalent areas at the second level – Home Range Capability (HRC). Home Range Capability is evaluated over an area much larger than the home range to produce an index of Landscape Capability (LC)

HABIT@ is spatially explicit and models habitat capability at three scales, corresponding to three levels of biological organization:

1. Local Resource Availability (LRA): the availability of resources important to the species life history, such as food, cover, or nesting, at the local, finest scale (a single cell or pixel).
2. Home Range Capability (HRC): the capability of an area corresponding to an individual's home range to support an individual, based on the quantity and quality of local resources, configuration and accessibility of those resources, and condition as determined by intrusion from roads and development.
3. Landscape Capability (LC): the capability of an area to support multiple home ranges; i.e., the ability of an area to support not only a single individual, but a local population.

HABIT@ returns a real number between 0 (no habitat value) and 1 (prime habitat) that represents the relative habitat capability for each cell. At the local resource level, the cell value indicates LRA, or the resources available at that cell (e.g., value as nesting habitat). At the home range level, the cell value indicates HRC, the resources available in a circular home range of a fixed size centered on that cell. Although the assumption of circular fixed size home ranges is seldom strictly true, it is not an unreasonable generalization when used consistently for comparative purposes. At this level, the HRC value will reflect if there are impediments to movement (e.g., food and nesting resources are across a road from one another). At the LC level, the resulting cell value indicates the value of a home range centered on that cell given that there is habitat in the neighborhood sufficient to support a local population. Specifically, the LC value of a cell reflects the quality of that location within a home range weighted by the sufficiency of the surrounding landscape to support additional proximal home ranges.

HABIT@ models are static; HABIT@ does not model population dynamics nor population viability. The results are relative measures of habitat capability, and do not necessarily correspond to animal density or fitness. HABIT@ is not an individual-based model, in that it does not explicitly model animal movement (although movement is accounted for implicitly in the assumption of home range size). HABIT@ models are, of course, limited by the availability, scale, and accuracy of available data, and the applicability of these data to the species being modeled. As with all habitat modeling, the greatest limitation is usually lack of knowledge of the habitat requirements of the species being modeled, and HABIT@ models are only as good as the biological information used to build them.

9.2.4   The Simulation Experiment

We established a full $2 \times 2 \times 2$ factorial design to evaluate the effect on habitat capability of three factors (climate, fire management, and vegetation management) and their interactions (Table 9.2).

Table 9.2 Factors in the modeling experiment

Factor	Levels	Description
Climate	Historical (HC)	Frequency, size, and severity of fire and insect disturbances calibrated to historical (1600–1900) regimes.
	Future (FC)	Frequency and probability of spread of fire and insect disturbances $= 1.1 \times HC$
Fire management	No suppression (NOSUP)	Frequency and size of fires calibrated to historical range of variability.
	Suppression (SUP)	Frequency of fires same as NOSUP; size of fires as in Table 9.3.
Vegetation management	No treatment (NOTRT)	No active vegetation management.
	Treatment (TRT)	WUI fuel reduction treatments with 3,000 ha per decade target; non-WUI post-disturbance salvage treatments up to 2,000 ha per decade.

9.2.4.1 Climate Factor

The two levels for the climate factor represented a contrast between natural disturbance regimes that have occurred under historical climate conditions and those that might be expected under future climate conditions.

Historical climate (HC). To represent natural disturbance regimes that have occurred under historic climate, we set the climate parameter in RMLands for the two dominant disturbance processes as follows:

- *Wildfire*—based on the historical record as represented by the mean Palmer Drought Severity Index (PDSI) (source: National Climatic Data Center), averaged over five sample locations in the vicinity of the study landscape for each 10-year interval for the period 1600–1900 (this sequence was repeated to create the 1,000 year time series needed for the simulation).The climate parameter affected the frequency and spread (i.e., size) of fires, which in combination affected the total area burned (Westerling and Swetnam 2003).
- *Pine beetle*—based on the historical record as represented by the *cumulative threshold* PDSI, averaged over five sample locations in the vicinity of the study landscape for each 10-year interval for the period 1600–1900, as above. The cumulative threshold PDSI is based on the maximum cumulative consecutive years of drought within each 10-year interval, but timesteps with an index < 1 are set to 0, preventing pine beetle disturbances from occurring. This results in periodic or episodic outbreaks (or epidemics) against a background of endemic levels of disturbance. This parameterization was based on local expert opinion and was consistent with published knowledge on beetle-climate interactions (e.g., Rogers 1996).

Future climate (FC). To represent natural disturbance regimes that might occur under future climate conditions, we set the climate parameter as above except increased the mean climate value by 10% (from 1 to 1.1). As implemented, the frequency and probability of spread (at the cell level) of disturbances both increased by 10%. This did not necessarily increase total area disturbed by 10%, however, because the climate parameter is only one of several variables affecting the disturbance processes. This factor level represents the case in which the frequency and severity of climate conditions conducive to burning and bark beetle outbreaks are increased by 10%. Although it is only one of many possible alternative future climate scenarios, we considered it to be within expectations of GCM predictions.

9.2.4.2 Fire Management

Two fire management factor levels represented a contrast between a "no suppression" policy and an "aggressive fire suppression" policy.

No suppression (NOSUP). To represent no suppression fire management policy, the frequency, size and severity of wildfires are based solely on the estimated historic range of variability (http://www.landfire.gov; Table 9.3).

Table 9.3 Percentage of fires in each size class under suppression and no suppression scenarios. Percentages in each size category are estimates from expert opinion

Size (ha)	Percentage of fires	
	No suppress	Suppress
1	76.25	92.25
10	12.00	4.00
100	6.00	2.00
1,000	3.00	1.00
10,000	1.50	0.50
100,000	0.75	0.25

Suppression (SUP). There are many possible ways to emulate the effect of fire suppression on fire frequency, size, and severity. For this factor level, we assumed that fire suppression per se does *not* change the frequency of ignitions or the severity of fires—although indirectly it will likely increase both over the long term if the vegetation becomes more flammable with age. Instead, we assumed that fire suppression directly affects the probability of fire spread, and thus directly influences the distribution of fire sizes. To emulate this effect, we modified the size distribution of fires in the spread parameters for wildfire (Table 9.2).

The settings here are designed to emulate a fire suppression policy that is reasonably but not perfectly effective in preventing the spread of fires. Consequently, while the distribution has many more small fires, very large fires (including the maximum fire size) still occur under the suppression scenario, but with a reduced probability.

9.2.4.3 Vegetation Management

Two vegetation management factor levels represented a contrast between "no treatment" and "aggressive vegetation treatment" management strategies.

No treatment (NOTRT). We simulated no active vegetation management to represent a "do nothing" management strategy.

Vegetation treatment (TRT). To represent an aggressive vegetation management policy, we attempted to emulate the current National Forest management focus on: (1) fuels reduction in the wildland-urban interface and (2) salvage of timber following large-scale disturbance events in the non-wildland-urban interface. Other management objectives, such as ecosystem restoration and timber stand improvement, were addressed only indirectly as by-products of the vegetation treatments aimed at fuels reduction and postfire salvage. Treatments were excluded from private lands, unsuitable timberland (as designated), riparian zones, and roadless areas. All other lands were considered eligible for treatments. The parameterization of vegetation treatments in RMLANDS can be quite complex. Rather than try to describe the detailed parameterization, a summary of the important distinctions is given below:

1. Wildland-urban interface (WUI) zone:

   Objective. WUI treatments were designed primarily to reduce fuels, thus reduce the risk of high-severity fire and improve the likelihood of effective fire suppression.

   Treatment intensity. The goal was to treat all eligible lands within the WUI (approximately 15,000 ha). Recognizing that even under the best circumstances, it is highly unlikely—and may not be desirable—to treat 100% of the eligible land, we instead assumed a target of 3,000 ha of land treated per decade on a 40-year treatment interval, with a goal of treating 12,000 ha every 40 years. However, numerous factors conspire against meeting this target in the simulation. For example, although closed-canopy forest conditions were targeted for treatment, previous occurrence of wildfire can leave considerably less closed-canopy forest to treat, and in some cases, less than the target. If we constrain treatments to sufficiently large contiguous areas of eligible lands containing suitable forest conditions for logistical and economic reasons, there will be locations and times when patches of eligible forest are simply too small or too scattered for efficient treatment. Thus, the targeted "maximum treatment area per timestep" was not necessarily met, but rather was a flexible target that varied depending on the vegetation conditions and due to other constraints.

   Treatment regime. Two silvicultural treatments were simulated: (1) "restoration" treatments, which involve the combination of individual tree removal (i.e., basal area reduction) and prescribed underburning (i.e., low mortality fire); and (2) "individual tree selection", which involves individual tree removal without underburning. Treatments were aggregated in units of 4–200 ha.

2. Non-wildland-urban interface (non-WUI) zone:

   Objective. Non-WUI treatments were designed primarily to salvage timber following major wildfires and insect outbreaks.

   Treatment Intensity. Given the spatial constraints outlined above, approximately 17,000 ha of land were eligible for treatments in the non-WUI zone. The goal was to salvage up to a *maximum* of 2,000 ha in any decade experiencing extensive wildfires or insect outbreaks.

   Treatment regime. Silvicultural treatments included a combination of "clearcut" and "individual tree selection". Both treatments were single-entry treatments without follow-up. Treatments were aggregated in units of 4–40 ha for clearcut and 10–200 ha for individual tree selection.

9.2.4.4 Landscape Capability Analysis

Each of the eight treatment combinations in the $2\times2\times2$ factorial was simulated using RMLANDS. Simulations were run for 1,000 years, at 10 year time steps.

Each treatment combination was simulated once for 1,000 years, with 10-year time steps. The simulation landscape was initialized with the current vegetation type and seral stage obtained from the Landfire Program (http://www.landfire.gov) and temporal dynamics stabilized within 200 years. Output grids of the vegetation cover type and condition plus a group of grids associated with wildfires and pine beetles were saved at each time step for each simulation. We described the composition and configuration of the cover-condition grids at each timestep using FRAGSTATS (McGarigal et al. 2002). Metrics that have been shown to be important predictors of habitat capability for two target species were computed (see below).

We described habitat capability using RMLands output and recently developed HC models for American marten (Wasserman et al. 2008) and flammulated owl (Cushman et al. unpublished data). The habitat suitability model for American marten in northern Idaho suggests that marten select landscapes with high average canopy closure and low fragmentation (Wasserman et al. 2008). Within these unfragmented landscapes, marten select foraging habitat at a fine scale within middle-elevation, late-successional, mesic forests. In northern Idaho, optimum American marten habitat therefore consists of landscapes with low road density, low density of patches and high contrast edges, with high canopy closure and large areas of middle-elevation, late-successional, mesic forest.

As implemented in HABIT@, the model estimates LRA as the product of three Local Resource Indices (LRIs): one for cover, one for adverse edge effects, and one for road intensity. LRI_{cover} is based on the cover-condition grid at the focal cell (Wasserman et al. 2008). LRI_{edge} takes into account the distance to adverse edges, as defined in Wasserman et al. (2008). LRI_{edge} increases with distance to an adverse edge according to a logistic function (Wasserman et al. 2008). LRI_{roads} is based on the distance-weighted road density within a 2,000-m radius circular window. The home range capability for marten is simply the mean LRA across a 630 m radius circular home range (125 ha), inversely scaled by distance. Landscape Capability (LC) is based on the number of 630 m radius home ranges that can be placed on the landscape by tiling non-overlapping home ranges starting with the cells with the highest HRC. Home ranges are then dropped by using the HRC as the probability of retaining each home range (e.g., a home range centered on a cell with a HRC of 0.85 will have a 15% chance of being dropped). Landscape Capability is the total number of home ranges in the landscape for that time step.

The habitat suitability model for flammulated owl includes canopy closure, elevation, edge density, patch richness density, correlation length of warm-dry forest types, landscape percentage area in grass cover types and riparian cover types (Cushman et al. unpublished data). Optimal flammulated owl habitat, as predicted by this model, consists of middle-elevation landscapes, with extensive warm-dry forest, relatively large amounts of grass and intermediate amounts of riparian cover types, intermediate canopy closure, and relatively high landscape heterogeneity, as indicated by the variables patch richness density and edge density.

The flammulated owl model is based on seven environmental variables that predict LRA: canopy cover, elevation, edge density, patch richness density, correlation length of warm-dry forest types, landscape composition by grass cover types, and landscape composition by riparian cover types. LRA is based on implementing the logistic-regression equation for the flammulated owl habitat model in Cushman et al. (submitted). The HRC for flammulated owl is simply the mean LRA across a 630 m radius circular home range (125 ha), inversely scaled by distance. LC is based on the number of 630 m radius home ranges that can be placed on the landscape by tiling non-overlapping home ranges starting with the cells with the highest HRC. Home ranges are then dropped by using the HRC as the probability of retaining each home range (e.g., a home range centered on a cell with a HRC of 0.85 will have a 15% chance of being dropped). Landscape Capability is the total number of home ranges in the landscape for that time step.

We applied the American marten and flammulated owl HABIT@ models to the simulation output for each timestep under each scenario. Figure 9.3 shows a snapshot of a LRI and the HRC for American marten under one of the eight

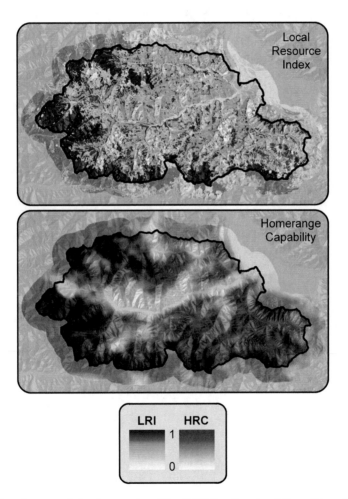

Fig. 9.3 Local resource index for cover-condition (LRI) (*top* figure) and Home Range Capability (HRC) (*bottom* figure) derived from the marten HABIT@ model for the 200 year timestep under the historical climate–no suppression–no vegetation treatment scenario

scenarios after 200 years of the simulation, which is when system dynamics stabilize. For each of the eight scenarios and each species, we calculated the total expected number of home ranges in the landscape (LC) based on HRC for each species at each time step. We summarized the differences among the eight scenarios by computing the mean, median, inter-quartile range, and standard error of LC across 10-year time steps. We tested for significant differences in LC in relation to climate, suppression, treatment, and their interactions using factorial analysis of variance.

9.3 Results

For marten, simulated fire suppression significantly increased Landscape Capability (LC), while simulated future climate significantly decreased LC (Table 9.4), and appears to increase its variability (Fig. 9.4). Similarly, simulated fire suppression significantly increased predicted habitat capability and simulated future climate decreased habitat capability for the flammulated owl (Table 9.4). The dominant effect is increased variability in the home range capability of the study area in the simulated future climate regime (Fig. 9.4). For both species, it appears that LC in the future climate regime will be lower than in the current climate regime regardless of the management scenario implemented. Vegetation fuels treatments did not appear to have a significant effect on LC for either species (Table 9.4)

To clarify the impact of climate change, fire suppression, and vegetation treatment on LC, we present box plots for each of these main effects, across the levels of the others. First, for both marten and flammulated owl, simulated increases in rates of disturbance under future climate nominally reduced the mean and increased the variability of LC in this study area (Fig. 9.5). Similarly, for both marten and flammulated owl, the higher disturbance rates associated with no suppression of wildfire nominally increased mean and variability of LC (Fig. 9.6, Table 9.4). In contrast, simulated vegetation treatments had virtually no effect on the mean or variability of LC for either species (Fig. 9.7, Table 9.4).

Table 9.4 Results of factorial analysis of variance for American marten and flammulated owl Landscape Capability in relation to climate, fire suppression and fuels treatment. For both species, capability is significantly higher under current than future climate and significantly higher under fire suppression than no suppression. There were no significant effects of vegetation treatment or interactions

Term (treatment)	Marten landscape capability		Owl landscape capability	
	Estimate	Prob > |t|	Estimate	Prob > |t|
Intercept	232.985360	<0.0001	36.576241	<0.0001
CLIMATE	−8.581242	0.0002	−1.017071	0.0373
BURN	−9.758519	<0.0001	−1.371524	0.0056
TREAT	1.801136	0.4156	−0.176619	0.7131
Timestep	−0.133488	0.0026	−0.002687	0.0051
CLIMATE*BURN	1.546739	0.7261	−0.002687	0.3836
CLIMATE*TREAT	2.559228	0.5625	−0.838995	0.7497
BURN*TREAT	0.911939	0.8363	−0.306442	0.6611

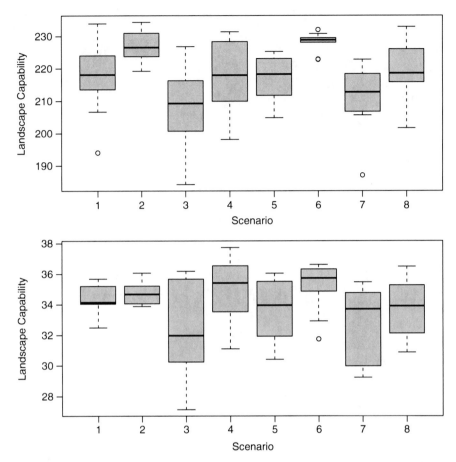

Fig. 9.4 Boxplots for the eight scenarios. (*top*) marten; (*bottom*) flammulated owl. The y-axis is the Habit@ Landscape Capability index based on home range habitat capability in the Prospect Creek study area. The eight scenarios are distributed along the x-axis. Scenarios are numbered as in Table 9.1: *1*: historical climate, no suppression, no vegetation treatment; *2*: historical climate, suppression, no vegetation treatment; *3*: future climate, no suppression, no vegetation treatment; *4*: future climate, suppression, no vegetation treatment; *5*: historical climate, no suppression, vegetation treatment; *6*: historical climate, suppression, vegetation treatment; *7*: future climate, no suppression, vegetation treatment; *8*: future climate, suppression, vegetation treatment

9.4 Discussion

To evaluate the relative expected impacts of climate change, fire suppression, and fuels treatment on wildlife habitat in the northern Rocky Mountains, we used the most recent and robust empirical models of species-habitat relationships for American marten and flammulated owl available. This provides the strongest available understanding of the factors that predict the occurrence of these two species as functions of multi-scale habitat conditions. We coupled multi-scale empirical models

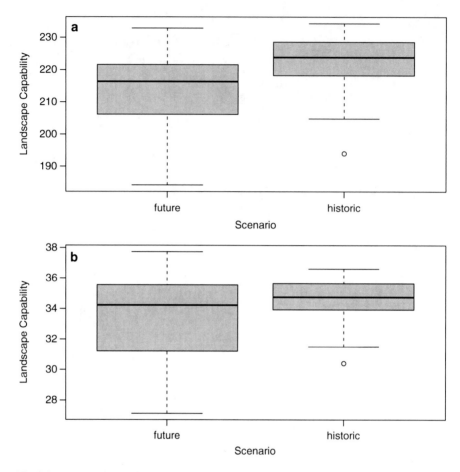

Fig. 9.5 Boxplots for climate change effects. (**a**) marten; (**b**) flammulated owl. The y-axis is the HABIT@ Landscape Capability index based on home range habitat capability in the study area. The x-axis is the climate scenarios (future vs. historic) pooled across the other factors

to spatial simulation of eight scenarios in a factorial framework enabling the evaluation of the relative effects of climate change, fuels treatment, and fire suppression on the habitat capability for each of these species. These scenarios reflect reasonable relative effects of management and potential climate change. In terms of management effects, we simulated both potential effects of fire suppression and of fuels reduction treatments. Our simulation assumed that fire suppression does *not* change the frequency of ignitions or the severity of fires, but does directly affect the probability of fire spread, and thus directly influences the distribution of fire sizes. Our fire suppression scenarios reflect aggressive suppression that may over-represent actual management effectiveness. Thus, we view our fire suppression scenarios as illustrating the largest reasonable effect possible due to suppression. Similarly, the fuels

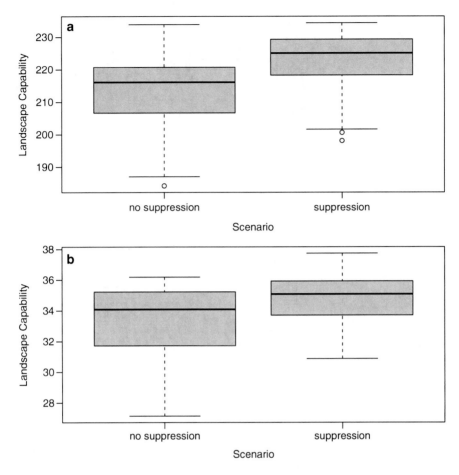

Fig. 9.6 Boxplots for fire suppression effects: (**a**) marten, (**b**) flammulated owl. The y-axis is the HABIT@ Landscape Capability index based on home range habitat capability in the study area. The x-axis is the fire suppression scenarios (nosuppress vs suppress) pooled across the other factors

treatment scenarios reflect aggressive vegetation management. We emulated the current National Forest management focus on fuels reduction and salvage of timber with a goal of treating all eligible lands within the WUI on roughly a 40-year treatment interval. This highly aggressive simulated fuels treatment program is probably more ambitious than would be possible to implement given legal, logistic, and financial limitations facing forests, and thus we view our fuels treatment scenarios as the largest reasonable effect possible due to fuels treatment.

We simulated future climate effects on disturbance regimes as a 10% increase in the frequency and severity of climate conditions that are conducive to burning and bark beetle outbreaks. Climate change is expected to have complex effects on fire

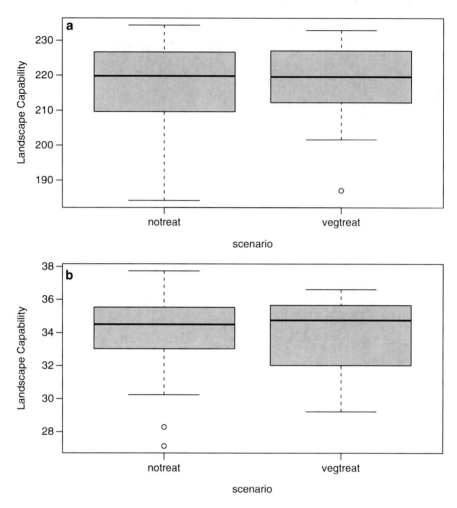

Fig. 9.7 Boxplots for fuels treatment effects: (**a**) marten; (**b**) flammulated owl. The y-axis is the HABIT@ Landscape Capability index based on home range habitat capability in the study area. The x-axis is the vegetation treatment scenarios (notreat vs. vegtreat) pooled across the other factors

and insect disturbance processes at landscape scales (Cushman et al. 2007), making simple parameterization of a single climate-change scenario difficult. Identifying the most likely potential future effect of expected climate change is a major research task requiring the integration of the latest downscaled climate models with sophisticated fire and insect disturbance models (Cushman et al. 2007). This challenging task will likely require the integration of multiple modeling efforts and take a number of years to produce reliable predictions (Cushman et al. 2007). Instead, we chose to implement a relatively simple but nonetheless plausible potential future. Our scenario of a 10% increase in the frequency and severity of conditions favoring wildfire and bark beetles is generally consistent with recent research suggesting that recent climate changes have increased the area burned in wild-fire (Westerling et al. 2006; Littell et al. 2009) and impacted by bark beetles (Logan et al. 2003; Berg et al. 2006). Future climate warming is expected to further increase the area impacted by fire and bark beetles in the Western United States (McKenzie et al. 2004; Running 2006; Westerling et al. 2006). Our choice of a 10% increase in these parameters is admittedly arbitrary but is probably conservative given observed and expected changes in relation to changing climate. Thus, in contrast with the management options, the effects of climate change in our sce-narios probably reflect the minimum expected change due to climate forcing. Therefore, the scenarios simulated are preliminary and hypothetical, in that we did not exhaustively evaluate a range of potential climate change effects or explicitly simulate any particular predictions of future climate from current GCMs. While this is not a definitive exploration of potential climate change effects, we believe it is reasonable and that the results suggest plausible landscape and habitat changes in the future.

The first major conclusion from these simulations is that the expected changes in disturbance regimes due to climate change will likely have a larger effect on landscape habitat capability of these two species than active fire and vegetation management, and that even very aggressive management involving suppression and fuels treatment will not substantially alter climate-driven effects on habitat for these species in the northern Rocky Mountains. This conclusion follows from the simula-tion results, which show that increased fire and insect disturbance under the future climate decreases the median and increases the variability of Landscape Capability for both species, and this holds true even when aggressive fuels treatment and fire suppression are simulated. Our simulation is preliminary, but the results are striking because our simulated climate change scenario was intentionally conservative and our suppression and fuels treatment scenarios were intentionally somewhat more aggressive than realistic.

Simulations illustrate that altered disturbance regimes may have substantial effects on wildlife habitat capability due to changes in the structure and composi-tion of vegetation communities. In this simulation both American marten and flammulated owl appear to be negatively impacted by these changes even though they have very different relationships to vegetation. A priori, one could reason-ably expect that increased fire and insect activity would adversely affect American marten, which is a closed-canopy, late-successional, interior forest species. In contrast, one would expect that increased fire and insect disturbance could benefit flammulated owl, given its preference for intermediate levels of canopy closure and high levels of landscape heterogeneity. However, the LC for both species was reduced slightly under the simulated future climate regime and variability over time increased substantially. The increase in variability may have important implications for population viability, as it is possible that both populations could face severe bottlenecks even if LC only infrequently drops to relatively low levels under these more variable future conditions. These bottlenecks could increase the chance of local extinction due to demographic stochasticity in temporarily reduced populations.

Another implication of these results regards the relative impacts of fire suppression and fuels treatment on habitat capability of these species. In the case of marten, one might expect that fuels treatments and fire suppression would have positive effects by increasing late-successional closed-canopy conditions at middle and high elevations. This was true for suppression, which nominally increased marten habitat capability under both historical and future climate. Fire suppression, however, did not decrease variability in habitat capability over time; indeed, suppression slightly increased variability in habitat capability under the future climate scenario. Fuels treatments did not appear to have any discernible effect on the expected median habitat capability for American marten under either historical or future climate, probably because even extremely aggressive fuels treatments did not substantially alter the area and severity of fire and insect disturbance. In the case of the flammulated owl, one might expect that fire suppression and fuels treatments would have opposite effects on habitat capability, with fire suppression reducing the suitability of low elevation warm-dry forest types for this species by encouraging transition to closed-canopy conditions and homogenization of the landscape mosaic, and increasing habitat suitability by decreasing canopy closure and increasing heterogeneity among patches in the landscape mosaic. These opposing effects were not observed in our simulation results, however. Fire suppression appeared to have a very minor effect on flammulated owl habitat capability, nominally increasing median habitat capability.

Also contrary to expectation, fuels treatment did not have any discernible effect on habitat capability for flammulated owl. In the simulation model, the amount of disturbance is what affects the cover condition most dramatically. The treatment targets had relatively little effect on disturbance regimes, and thus treatments did not markedly affect habitat capability. The area affected by natural disturbance regimes is impressive when compared to what is feasible with management.

9.5   Conclusion

Climate change and natural resources management activities will likely interact in complex ways to affect forest ecosystems and wildlife habitat. Importantly, altered disturbance regimes, fire suppression, and fuels treatments may result in unexpected outcomes. Simulation modeling is a powerful tool available to investigate these interactions and evaluate possible outcomes. In this analysis we explored potential interactions of climate change, fire suppression, and fuels treatment on the habitat capability of two wildlife species with highly contrasting habitat relationships in the northern Rocky Mountains. We expected climate change to negatively affect habitat quality for American marten, and we expected that fire suppression and fuels treatments would benefit this species by increasing the proportion of the landscape in late-successional closed-canopy conditions. Likewise, we expected that climate change would benefit flammulated owl by increasing the amount of low- and mixed-severity disturbance in warm-dry forest types, thereby increasing average canopy openness and landscape heterogeneity. We also expected that fuels treatments would benefit the owl, while fire suppression would lead to reductions in habitat capability. The simulations, however, indicated that many of these expectations were not likely outcomes.

While this exploration is preliminary, we believe it makes three main points. First, climate change and management may interact in complex and nonintuitive ways that defy expectations based on simplistic understanding of each process acting separately. Formal evaluations of the interaction of these multiple processes under realistic scenarios is essential to guide adaptation and mitigation strategies for addressing the effects of climate change. Second, climate-driven changes to disturbance regimes may have larger effects than fire suppression or vegetation treatment. Thus, management tools available to mitigate climate change may have limited power to substantially alter the trajectory of landscape change under altered future disturbance regimes (but see Chap. 10). Third, fuels treatments in particular appear to be of very limited utility in altering habitat capability of these species under current or future climate. This is because they have limited ability to alter overall disturbance regimes, even when they appear to be on aggressive schedules.

We remind the reader that this analysis is preliminary, with simplistic assumptions and representations of how natural disturbance regimes might change with changes in climate and fire management strategies. These are reasonable scenarios that can suggest the kind and approximate magnitude of effects on LC, but they are not comprehensive. Therefore, further research is needed to explore how a range of climate change scenarios and concomitant effects on disturbance regimes interact with a range of management scenarios to affect landscape dynamics and wildlife habitat capability.

The linkage of rigorous, empirical, multi-scale habitat models with spatial simulation of alternative disturbance regimes is a powerful framework to evaluate the potential effects of a range of alternative scenarios on the habitat and populations of wildlife species of interest. For this critical task to be addressed rigorously requires that three major conditions are satisfied. First, reliable spatially explicit models for species of interest must exist. Second, meaningful alternative disturbance scenarios involving climate forcing and realistic alternative management responses must be simulated in a realistic manner using dynamic simulation models such as RMLANDS. Third, the habitat models must be applied to simulation results over time to estimate impacts on habitat under each scenario, requiring that the habitat models and simulation outputs match in terms of attributes and scales of ecological variables. Given sufficient care these conditions can be satisfied for many species and many systems. With this approach, we have the tools in hand to anticipate the potential effects of changing and interacting disturbance regimes on habitat for multiple species.

References

Bachelet, D., J.M. Lenihan, C. Daly, R.P. Neilson, D.S. Ojima, and W.J. Parton. 2001. *MC1: A dynamic vegetation model for estimating the distribution of vegetation and associated ecosystem fluxes of carbon, nutrients, and water.* General Technical Report PNW-GTR-508. Portland: U.S. Forest Service.

Baker, W.L. 1995. Longterm response of disturbance landscapes to human intervention and global change. *Landscape Ecology* 10: 143–159.

Berg, E.E., J.D. Henry, C.L. Fastie, A.D. DeVolder, and S.M. Matsuoka. 2006. Spruce beetle outbreaks on the Kenai Peninsula, Alaska, and Kluane National Park and Reserve, Yukon Territory: Relationship to summer temperatures and regional differences in disturbance regimes. *Forest Ecology and Management* 227: 219–232.

Christensen, N.L. 1988. Succession and natural disturbance: Paradigms, problems, and preservation of natural ecosystems. In *Ecosystem management for parks and wilderness*, eds. J.K. Agee, and D.R. Johnson, 62–81. Seattle: University of Washington Press.

Cushman, S.A., D. McKenzie, D.L. Peterson, J. Littell, and K. McKelvey. 2007. *Research agenda for integrated landscape modelling.* General Technical Report RMRS-GTR-194. Fort Collins: U.S. Forest Service.

Cushman, S.A., B. Hahn, and A. Cilimburg. (unpublished data). Multiple scale habitat selection by flammulated owl in the northern Rocky Mountains. Manuscript in review. *Ecological Applications.*

Dale, V.H., L.A. Joyce, S. McNulty, R.P. Neilson, M.P. Ayres, M.D. Flannigan, P.J. Hanson, L.C. Irland, A.E. Lugo, C.J. Peterson, D. Simberloff, F.J. Swanson, B.J. Stocks, and B.M. Wotton. 2001. Climate change and forest disturbances. *Bioscience* 51: 723–734.

Flannigan, M.D., K.A. Logan, B.D. Amiro, W.R. Skinner, and B.J. Stocks. 2005. Future area burned in Canada. *Climatic Change* 72: 1–16.

Houghton, J.T., Y. Ding, D.J. Griggs, M. Noguer, P.J. van der Linden, and D. Xiaosu, eds. 2001. *Climate Change 2001: The Scientific Basis: Contributions of Working Group 1 to the Third Assessment Report of the Intergovernmental Panel on Climate Change.* Cambridge: Cambridge University Press pp 881.

Iverson, L.R., A.M. Prasad. 2002. Potential redistribution of tree species habitat under five climate change scenarios in the eastern US. *Forest Ecology and Management* 155: 205–222.

Keane, R.E., P. Morgan, and J.D. White. 1999. Temporal patterns of ecosystem processes on simulated landscapes in Glacier National Park, Montana, USA. *Landscape Ecology* 14: 311–329.

Littell, J.S., D. McKenzie, D.L. Peterson, and A.L. Westerling. 2009. Climate and wildfire area burned in western U.S. ecoprovinces, 1916–2003. *Ecological Applications* 19: 1003–1021.

Logan, J.A., J. Regniere, and J.A. Powell. 2003. Assessing the impact of global warming on forest pest dynamics. *Frontiers in Ecology and the Environment* 1: 130–137.

McGarigal, K., S.A. Cushman, M.C. Neel, and E. Ene. 2002. FRAGSTATS: spatial pattern analysis program for categorical maps. http://www.umass.edu/landeco/research/fragstats/fragstats.html. Accessed 5 Mar 2010.

McKenzie, D., D.L. Peterson, and E. Alvarado. 1996. *Predicting the effect of fire on large-scale vegetation patterns in North America.* Research Paper PNW-RP-489. Portland: U.S. Forest Service.

McKenzie, D., Z. Gedalof, D.L. Peterson, and P. Mote. 2004. Climatic change, wildfire, and conservation. *Conservation Biology* 18: 890–902.

Mearns, L.O., R.W. Katz, and S.H. Schneider. 1984. Extreme high-temperature events: Changes in their probabilities with changes in mean temperature. *Journal of Climate and Applied Meteorology* 23: 1601–1613.

Miller, C. 2003. Simulation of effects of climatic change on fire regimes. In *Fire and climatic change in temperate ecosystems of the Western Americas*, eds. T.T. Veblen, W.L. Baker, G. Montenegro, and T.W. Swetnam, 69–94. New York: Springer.

Neilson, R.P. 1995. A model for predicting continental-scale vegetation distribution and water balance. *Ecological Applications* 5: 362–385.

Neilson, R.P., and R.J. Drapek. 1998. Potentially complex biosphere responses to transient global warming. *Global Change Biology* 4: 505–521.

Neilson, R.P., L.F. Pitelka, A.M. Solomon, R. Nathan, G.F. Midgley, J.M. Fragoso, H. Lischke, and K. Thompson. 2005. Forecasting regional to global plant migration in response to climate change. *BioScience* 55: 749–759.

Overpeck, J.T., D. Rind, and R. Goldberg. 1990. Climate-induced changes in forest disturbance and vegetation. *Nature* 343: 51–53.

Peter, S., and K. Vranes. Intergovernmental Panel on Climate Change (IPCC). In Encyclopedia of Earth, Eds. Cutler J. Cleveland, (Washington, D.C.: Environmental Information Coalition, National Council for Science and the Environment). [First published in the Encyclopedia of Earth May 6, 2008; Last revised Date May 6, 2008; Retrieved November 8, 2010 http://www.eoearth.org/article/Intergovernmental_Panel_on_Climate_Change_(IPCC)

Peters, R.L., and T.L. Lovejoy. 1992. *Global warming and biological diversity*. New Haven and London: Yale University Press.

Rogers, P. 1996. *Disturbance ecology and forest management: A review of the literature*. General Technical Report INT-GTR-336. Ogden: U.S. Forest Service.

Running, S.W. 2006. Climate change: Is global warming causing more, larger wildfires? *Science* 313: 927–928.

Solomon, A.M., and R. Leemans. 1997. Boreal forest carbon stocks and wood supply: Past, present, and future responses to changing climate, agriculture, and species availability. *Agricultural and Forest Meteorology* 84: 137–151.

Thompson, R.S., S.W. Hostetler, P.J. Bartlein, and K.H. Anderson. 1998. A strategy for assessing potential future changes in climate, hydrology, and vegetation in the western United States. U.S. Geological Survey Circular 1153.

Westerling, A.L., and T.W. Swetnam. 2003. Interannual to decadal drought and wildfire in the western United States. *Eos Interactions* 84: 545–560.

Westerling, A.L., H.G. Hidalgo, D.R. Cayan, and T.W. Swetnam 2006. Warming and earlier spring increases western U.S. forest wildfire activity. *Science* 313: 940–943. doi:10.1126/science.1128834. Online supplement.

Wasserman, T.N. 2008. *Multi-scale habitat relationships and landscape genetics of Martes Americana in Northern Idaho*. M.S. Thesis. Western Washington University.

Part IV
Landscape Fire Management, Policy, and Research in an Era of Global Change

Chapter 10
Managing and Adapting to Changing Fire Regimes in a Warmer Climate

David L. Peterson, Jessica E. Halofsky, and Morris C. Johnson

10.1 Introduction

Planning and management for the expected effects of climate change on natural resources are just now beginning in the western United States (U.S.), where the majority of public lands are located. Federal and state agencies have been slow to address climate change as a factor in resource production objectives, planning strategies, and on-the-ground applications. The recent assessment by the Intergovernmental Panel on Climate Change (IPCC 2007) and other high-profile reports (e.g., GAO 2007) have increased awareness of the need to incorporate climate change into resource management.

Most of the recent literature on adaptation to climate change has focused on conceptual issues (Hansen et al. 2003), potential actions by local governments and municipalities (Snover et al. 2007), and individual resources and facilities (Slaughter and Wiener 2007). However, efforts to develop strategies that facilitate adaptation to documented and expected responses of natural resources to climate change are now beginning in earnest. For example, the Chief of the U.S. Forest Service recently stated that addressing climate change is one of the top three priorities of the agency (Kimbell 2008). In the most substantive effort to date, the U.S. Climate Change Science Program has developed synthesis and adaptation products for federal land management agencies (Joyce et al. 2008).

The frequency, severity, and extent of wildfire are strongly related to climate (Swetnam and Betancourt 1990; Johnson and Wowchuk 1993; Stocks et al. 1998; Hessl et al. 2004; Gedalof et al. 2005; Heyerdahl et al. 2008; Skinner et al. 2008; Taylor et al. 2008; Littell et al. 2009). Increasing temperatures with climate change

D.L. Peterson (✉)
Pacific Wildland Fire Sciences Laboratory, Pacific Northwest Research Station,
U.S. Forest Service, Seattle, WA 98103-8600, USA
e-mail: peterson@fs.fed.us

D. McKenzie et al. (eds.), *The Landscape Ecology of Fire*, Ecological Studies 213,   249
DOI 10.1007/978-94-007-0301-8_10, © Springer Science+Business Media B.V. 2011

will likely lead to changes in fire regimes in many types of ecosystems (IPCC 2007). Increased spring and summer temperatures with climate change will lead to relatively early snowmelt (Stewart et al. 2005; Hamlet et al. 2007), lower summer soil moisture (Miles et al. 2007) and fuel moisture (Westerling et al. 2006), and longer fire seasons (Wotton and Flannigan 1993; Westerling et al. 2006). These conditions will lead to increased fire frequency and extent (Price and Rind 1994; Gillett et al. 2004; Westerling et al. 2006, Chap. 5). McKenzie et al. (2004) found that for a mean temperature increase of 2°C (expected by mid-twenty-first century), annual area burned by wildfire is expected to increase by a factor of 1.4–5 for most western U.S. states. Dry fuel conditions associated with increased temperatures allow forests to burn whenever an ignition source occurs, with low humidity and high winds contributing to fire spread.

Climate change will alter the effectiveness of fire and fuel management, and therefore necessitates that we adapt how we manage fire and fuels. There are well established scientific principles of fuels management upon which we can rely to inform future strategies. These strategies need to be applied to large landscapes, which are the land units for which managers are responsible and across which fires spread. Adaptation to changing fire regimes and other ecological effects of climate change will help reduce ecosystem vulnerabilities and potentially undesirable effects on ecosystem composition, structure, and function (Millar et al. 2007; Joyce et al. 2008).

Adapting management to changing fire regimes will likely be a major challenge for resource managers in the face of climate change. This chapter outlines general adaptation strategies and specific fire and fuel management options for forest managers under climate change, primarily for dry forests with low-severity and mixed-severity fire regimes (e.g., pinyon pine-juniper [*Pinus* spp., *Juniperus* spp.], ponderosa pine [*Pinus ponderosa*], dry Douglas-fir [*Pseudotsuga menziesii*], mixed conifer, mixed evergreen). We first present strategies and options from the perspective of managers and then expand on some of these from the perspective of research.

10.2   Adapting to the Effects of Climate Change

We initiated science-management collaborations at Olympic National Forest (Washington, USA) and Tahoe National Forest (California, USA) to develop management options that will facilitate adaptation to climate change (Littell et al. N.d.). This was the first attempt to work with national forests to develop specific concepts and applications that could potentially be implemented in management and planning. The focus of this effort was to develop strategies and management options for adapting to climate change across multiple resources, and there was no intention to specifically focus on fire or fuels management. In this chapter, we build on this foundation of general concepts by identifying strategies and management options relevant for managing changing fire regimes across large landscapes.

10.2.1   General Adaptation Strategies

The national forests developed six *general adaptation strategies* (Table 10.1; Littell et al. N.d.) in response to climate change. We have amended these strategies to emphasize their relevance for landscape fire and fuels management in a changing climate:

- Manage for resilience, decrease vulnerability–Fire exclusion has increased understory vegetation and surface fuels in many forests, making them vulnerable to crown fire should wildfire occur. Managing for reduced understory and surface fuels will increase resilience to fire and favor retention of large trees (Dale et al. 2001; Peterson et al. 2005; Joyce et al. 2008). Thinning, surface fuel removal (mechanically or through prescribed burning), and allowing naturally ignited fires to burn (rather than suppressing them) can reduce fuels across sufficient areas to reduce the severity of future wildfires.
- Prioritize climate-smart treatments—Managers have many choices for treating landscapes, but typically have minimal financial and human resources, so prioritizing treatments that are likely to work in a warmer climate may become increasingly necessary (Millar et al. 2007). For example, stand densities may need to be lower in the future to reduce the risk of overstory mortality if fire weather will be

Table 10.1 Summary of general adaptation strategies, and examples of applying those strategies to changing fire regimes

Adaptation strategy	Examples of application to changing fire regimes
Manage for resilience, decrease vulnerability	Reduce stem density and surface fuel in stands where fire exclusion has created vulnerability to crown fire
	Implement fuel treatments across large landscapes in order to modify fire severity and spread
Prioritize climate-smart treatments	Design fuel treatments to be resilient to intense fire behavior that may accompany extreme fire weather in the future
Consider tradeoffs and conflicts	Identify how fuel treatments may affect carbon dynamics, hydrology, and wildlife habitat at various spatial and temporal scales
Manage dynamically and experimentally	Implement various types and intensities of fuel treatments at different spatial and temporal scales and evaluate their effectiveness for reducing crown fire
Manage for process	Plan for the regular occurrence of wildfire at different spatial and temporal scales, rather than only suppressing fire or considering it to be an anomaly
Manage for realistic outcomes	Plan for the regular occurrence of fire, not elimination of fire in wildland-urban interface areas
	Develop collaborative management between public land managers and local residents to modify fuels sufficiently to reduce fire severity if wildfire occurs and to facilitate suppression

more extreme (Dale et al. 2001; Spittlehouse and Stewart 2003). Reduced stand densities would also increase resistance to drought and insect attack.

- Consider tradeoffs and conflicts—Future effects on ecological and socioeconomic sensitivities can result in potential tradeoffs and conflicts for species conservation and other resource values. For example, forest landscapes with periodic thinning and surface fuel treatments may have different carbon dynamics than landscapes without active management in which crown fires would be more likely to occur (Millar et al. 2007; Hurteau et al. 2008).
- Manage dynamically and experimentally—Currently available opportunities (i.e., under current policy) can be used to implement adaptive management over several decades (Dale et al. 2001). For example, different types and intensities of fuel treatments can be used over time and space in order to determine their effectiveness for reducing crown fire.
- Manage for process—Project planning and management can be used to maintain or enhance ecological processes rather than to design specific structures or species composition (Harris et al. 2006). For example, novel mixes of species and spacing can be used following fire in order to reflect likely natural dynamic processes of adaptation.
- Manage for realistic outcomes—Projects that are currently a component of the planning process may have a higher failure rate in a warmer climate, and it will become increasingly important to assess the viability of management goals and desired outcomes (Hobbs et al. 2006). For example, it will never be possible to eliminate fire from wildland-urban interface areas, but land managers can work with local residents to reduce fire hazard to a level that may allow suppression to be effective there, while allowing fire to play a less managed role in other parts of the landscape.

10.2.2   Specific Adaptation Options

The national forests developed nine *specific adaptation options* (Table 10.2). In contrast to the guiding principles provided by general strategies above, adaptation options refer to specific kinds of actions that can be taken at a variety of spatial scales. We have amended the discussion to emphasize the relevance of those options for fire:

- Increase landscape diversity—This option focuses on increasing variety in stand structures and species assemblages over large areas and avoiding "one size fits all" management prescriptions (Millar et al. 2007). This can include applying forest thinning to increase variability in stand structure, increase resilience to stress by increasing tree vigor, and reduce vulnerability to disturbance (Parker et al. 2000; Dale et al. 2001; Spittlehouse and Stewart 2003). Although there is no theory or empirical data at the present time to guide which combinations of stand structures and species will optimize adaptation potential, allowing fires to burn unsuppressed may in some cases help to emulate landscape patterns that

Table 10.2 Summary of specific adaptation options developed by national forests, and examples of applying those options to changing fire regimes

Adaptation option	Examples of application to changing fire regimes
Increase landscape diversity	Thin forest stands to create lower density, diverse stand structures and species assemblages that reduce fire hazard, increase resilience to wildfire (allow overstory survival), and increase tree vigor by reducing competition
Maintain biological diversity	Plant nursery stock from warmer, drier locations than what is prescribed in genetic guidelines based on current seed zones
	Plant mixed species and genotypes, with emphasis on fire resistant species and morphology
Increase resilience at large spatial scales	Implement thinning and surface fuel treatments across large portions of landscapes (e.g., large watersheds) where large wildfires may occur
	Orient the location of treatments in sufficiently large blocks to modify fire severity and fire spread
Treat large-scale disturbance as a management opportunity	Develop plans for management objectives and activities following large fires, including long-term experimentation
Increase management unit size	Focus the spatial scale of management on units (or aggregated units) of hundreds to thousands of hectares in appropriate geographic locations
	Implement fuel treatments across large units and blocks of land to more effectively reduce fire severity and spread
Implement early detection/ rapid response for invasive species	Survey and monitor vegetation following wildfire in order to detect and eradicate undesirable invasive plant species
Match engineering of infrastructure to expected future conditions	Modify drainage systems (e.g., install larger culverts) to accommodate higher water flow resulting from more wildfire
	Design road systems to facilitate efficient fire suppression
Collaborate with a variety of partners	Develop mutual plans for fire and fuels management with adjacent landowners to ensure consistency and effectiveness across large landscapes
Promote education and awareness about climate change	Facilitate discussion among management staff regarding the effects of a warmer climate on fire and interactions with multiple resources
	Educate local residents about how a warmer climate will increase fire frequency, fuel reduction can protect property and collaboration with public land managers will assist broader fuel management objectives

existed during pre-settlement times (Hessburg and Agee 2003). These patterns from natural experiments may hold the greatest adaptation potential.

- Maintain biological diversity—Appropriate species and genotypes can be planted in anticipation of a warmer climate (Smith and Lenhart 1996; Parker et al. 2000; Noss 2001; Spittlehouse and Stewart 2003; Millar et al. 2007), giving more flexibility by diversifying the phenotypic and genotypic template on which climate and competition interact, and to avoid widespread mortality at the regeneration stage. For example, nursery stock from warmer, drier locations than what is prescribed in genetic guidelines based on current seed zones can be planted following a crown fire (Spittlehouse and Stewart 2003).

- Increase resilience at large spatial scales—Proactive management can improve the resilience of natural resources to ecological disturbance and environmental stressors (Dale et al. 2001; Spittlehouse and Stewart 2003; Millar et al. 2007) and reduce the number of situations in which land managers must respond in "crisis mode." For example, if hazardous fuels reduction and allowing some fires to burn unsuppressed reduces fire severity over large areas, then postfire soil erosion can be minimized.

- Treat large-scale disturbance as a management opportunity—Large-scale disturbance causes rapid changes in ecosystems, but also provides opportunities to apply adaptation strategies (Dale et al. 2001; Millar et al. 2007). Carefully designed management experiments for adapting to climate change can be implemented, provided that plans are in place in anticipation of large disturbances. For example, one could experiment with mixed-species tree planting after fire even though the standard prescription might be for a monoculture (Millar et al. 2007). Management experiments need good statistical design, adequate replication, and long-term commitment by managers and scientists to maintain a time series of data that can inform future decisions.

- Increase management unit size—Increasing the size of management units to hundreds or thousands of hectares across logical biogeographic entities such as watersheds will improve the likelihood of accomplishing objectives (Smith and Lenhart 1996). For example, large strategically located blocks of forest land subjected to fuel treatments will reduce fire spread more effectively than smaller dispersed units (Finney 2001). At the present time, there is minimal theory or empirical data to guide the design, size, and spatial patterns of management units, although a closer approximation of patch size created by natural disturbances may be a good place to start.

- Implement early detection/rapid response for invasive species—A focus on treating small problems before they become large unsolvable problems recognizes that proactive management is more effective than delayed implementation (Millar et al. 2007). For example, recently burned areas are often susceptible to the spread of invasive species, which can be detected by monitoring during the first two years after fire (Chap. 8).

- Match engineering of infrastructure to expected future conditions—This refers primarily to road and drainage engineering that can accommodate future changes in hydrology (Spittlehouse and Stewart 2003). However, it might be possible to design road networks to facilitate effective fire suppression in areas that are particularly fire prone.

- Collaborate with a variety of partners—Working with a diversity of landowners, agencies, and stakeholders will develop support for and consistency in adaptation options. For example, national forest managers can work with adjacent state forest managers to agree on fuel treatment plans across large landscapes.

- Promote education and awareness about climate change—It is critical that internal and external education on climate change is scientifically credible and consistent (Spittlehouse and Stewart 2003), with emphasis on the role of active management in adaptation. For example, local residents can be informed that wildfire may be more frequent in a warmer climate, which makes it imperative that they clear brush around homes to reduce fire hazard.

Effective landscape fire and fuel management will require that we consider the potential effects of climate change and adjust activities accordingly. Much of the current dialog among scientists and resource managers about adapting to climate change in general is relevant and applicable to landscape fire and fuel management. Despite considerable uncertainty about the effects of climate change, scientific foundations for adaptation are sufficiently developed to begin the adaptation process. By taking an experimental and learning approach to management it will be possible to be both adaptive and responsive.

10.3   Fuels Management in a Warmer Climate

The expected warming in climate may have implications for the design of fuel treatments in dry forests across the western United States. Climate change will influence fire behavior by increasing temperature, an important factor that controls fire behavior. Temperature regulates several variables that control fuel flammability: relative humidity of the atmosphere, moisture content of dead and live fuel, and wind speed and direction in mountainous terrain (Brown and Davis 1973). Foliar moisture controls fire behavior and thresholds for crown fire initiation (Agee et al. 2002). For example, a closed-canopy stand is typically cooler and has higher humidity than an open stand. These characteristics retain dead and live fuel moisture, which regulates surface fuel temperature and wind speeds (Whelan 1995), although closed-canopy stands often have low canopy base height and high canopy bulk density, both of which increase the probability of crown fire initiation. On the other hand, lowering tree density decreases the probability of crown fire initiation, but may exacerbate fire behavior because solar radiation to the forest floor can desiccate dead and live fuels (Agee and Skinner 2005).

Based on these considerations, fuel treatment guidelines for restoring the resilience of dry forest ecosystems (e.g., Peterson et al. 2005) may need to be adjusted to retain either more or fewer stems per hectare (Harrod et al. 1999; Arno and Allison-Bunnell 2002; Johnson 2008). Forest managers may want to weigh the tradeoffs related to each strategy for their particular project (Peterson and Johnson 2007) and decide which treatment is feasible for addressing the effects of a warmer climate on fuels and fire hazard. Understanding basic concepts of fuels and how to manage them for landscape resilience, and having a way to evaluate effectiveness of fuel treatments, is a good combination for sustainable management at large spatial scales.

10.3.1 Fuel Concepts and Fire Resilience

Fuel is a critical component of both the combustion triangle (fuel, oxygen, heat) and the fire behavior triangle (weather, fuel, topography), which are conceptual aids for understanding the principles of combustion and the elements that influence fire behavior and intensity (Brown and Davis 1973; Pyne et al. 1996). Fuel is classified by its vertical distribution (ground, surface, or aerial) and its general properties within a stand (Ottmar et al. 2007). Ground fuels (e.g., decomposing organic matter, rotting logs) have little influence on wildfire spread. Fire spreads primarily in the surface fuels, which include seedlings and saplings (i.e., trees less than 1.8 m tall), shrubs, herbaceous vegetation, litter, and dead woody material (Brown and Davis 1973). Aerial or crown fuels are composed of live and dead vegetation. Collectively, these fuel layers are referred to as a fuelbed, which represents the average physical characteristics of a relatively homogeneous unit on a landscape with distinct fire environments (Sandberg et al. 2007). Dead woody fuel is classified by fuel moisture time lags (Fosberg and Deeming 1971). In general, small diameter fuels have short time lags and are responsible for fire spread rates. Large diameter fuels have longer time lags and are involved primarily in smoldering. The type of fuel within a fuelbed strongly influences the intensity of wildfire.

   The scientific basis for using fuel treatments to maintain or restore resilience to wildfire in dry forests is well established (Peterson et al. 2005) and has provided support for thinning and surface fuel treatments throughout western North America (Fig. 10.1), including for adaptation to a warmer climate (Joyce et al. 2008). Agee and Skinner (2005) developed four principles of a fire-safe forest: (1) reduce surface fuels, (2) increase height to live crown, (3) decrease crown bulk density, and (4) retain large trees (Table 10.3). Surface fuels can be reduced with treatments such as prescribed fire, pile and burn, and whole-tree harvest. Increasing the height to live crown and decreasing crown bulk density can be achieved by thinning from below (progressively removing trees with the smallest diameter). Fuel reduction treatments designed to leave the large fire resistant trees fulfill the fourth principle of a fire-safe forest. Agee and Skinner (2005) concluded that forests treated according to these principles will be more resilient to wildfires in a warmer climate. In some cases, it may be possible to accomplish fire-safe principles by allowing wildfires to burn unimpeded through areas that have not burned for decades (Miller et al., Chap. 11), although postfire stem density, quantity of fuel removed, and spatial patterns of altered stand and fuel structure cannot be controlled.

   Although resilience to fire can be enhanced with fuel treatments, climate is a major driver of fire regimes (Gedalof et al. 2005; Littell et al. 2009, Chap. 5), and fuel treatment effectiveness is reduced when fires burn under severe conditions (high temperature, high wind speed, low humidity). In some cases, the influence of climate on fire could override fuel treatments, resulting in high-severity fire even in areas where fuels have been reduced. The relative influence of climate versus fuels on fire regimes is specific to the type of ecosystem being considered. For example, boreal forests and subalpine forests typically have fuel loadings that are sufficiently

Fig. 10.1 Removal of smaller trees and surface fuels can potentially reduce the severity of fire behavior and effects in a wildfire. Reduction of stand density and surface fuels is shown in these photos of a ponderosa pine stand on the Lassen National Forest, California, before and after treatment. Lower stand densities and fuels can enhance resilience to fire in a warmer climate by reducing risk of crown fire and protecting overstory trees and forest structure. Photos courtesy of Lassen National Forest

Table 10.3 Principles of fire resistance for dry forests

Principle	Effect	Advantage	Concerns
Reduce surface fuels	Reduces potential flame length	Fire control easier; less torching of individual trees	Surface disturbance less with fire than other mechanical techniques
Increase height to live crown	Requires longer flame length to begin torching	Less torching of individual trees	Opens understory; may allow surface wind to increase
Decrease crown density	Makes tree-to-tree crown fire less likely	Reduces crown fire potential	Surface wind may increase; surface fuels may be drier
Keep big trees of resistant species	Less mortality for same fire intensity	Generally maintains overstory structure	Less economical; may keep trees at risk of insect attack

Adapted from Agee and Skinner (2005)

high to carry fire and potentially propagate crown fires, but high temperature and low humidity are necessary to dry the fuels so they can burn; therefore, climate limits fire regimes in these forests. In contrast, ponderosa pine forests in the American Southwest are hot and dry every summer, but must have sufficient surface fuels to carry fire; therefore, fuels limit fire regimes in these forests. Understanding differences in these relative influences among ecosystems will help to develop and evaluate effective fuel management prescriptions.

10.3.2 Evaluating Effectiveness with Fire Simulation Models

Fire simulation models are valuable for testing the efficacy of fuel treatments, especially given the logistic challenges of conducting large-scale field experiments (Andrews and Queen 2001). For example, Johnson (2008) simulated the effects of thinning and surface fuel treatments on fire hazard using the Fire and Fuels Extension to the Forest Vegetation Simulator (FFE-FVS: Reinhardt and Crookston 2003) on 45,162 stands from dry forests in the western United States. Treatments were patterned after Agee and Skinner's (2005) principles of a fire-safe forest. Stands were evaluated for four thinning densities (125, 250, 500, and 750 trees per hectare [tph]), three surface fuel treatments (leave slash, extract slash, prescribed fire) and no action, resulting in a total of 698,140 projections.

Results indicate that thinning treatments with lower target densities (125 and 250 tph) are more effective at modifying fire behavior than treatments with higher target densities (500 and 750 tph). These results are consistent with those from other studies (Stephens 1998; Harrod et al. 1999; Agee et al. 2000; Pollet and Omi 2002; Martinson and Omi 2003; Finney et al. 2005; Stephens and Moghaddas 2005; Cram et al. 2006; Harrod et al. 2007; Strom and Fulé 2007). Arno and

Allison-Bunnell (2002) suggested that historical surface fire regimes perpetuated ponderosa pine-dominated stands with 75–250 tph, that is, stands of similar density to those simulated in Johnson (2008). In other studies, 125 tph represented historical stands in eastern Washington (Harrod et al. 1999), 100 tph was typical for Southwestern stands (Covington and Moore 1994), and 150 tph was found in old Jeffrey pine-mixed conifer forests in the unmanaged Sierra San Pedro Martir (Mexico) (Stephens and Gill 2005).

Fuel treatment guidelines for dry forests in the western United States have been developed based on output from the simulation model FFE-FVS (Johnson et al. 2007). We use an example from that publication—a forest stand in the Okanogan-Wenatchee National Forest in Washington State—to illustrate how different thinning options can be evaluated (Table 10.4; Fig. 10.2). The stand is composed of 6,154 tph, dominated by Douglas-fir and ponderosa pine. FFE-FVS predicted passive crown fire under severe weather. Before treatment, canopy base height was 0.6 m, and canopy bulk lensity was 0.08 kgm⁻³. The 125 and 250 tph thinning treatments were more effective than the other treatments because they prevented crown fire initiation by reducing ladder fuels within the stand. The 125 and 250 tph thinning treatments generated the highest torching indices, highest canopy base heights, and lowest canopy bulk densities. The 125 tph treatment produced the lowest basal area mortality.

The lower stand densities plus lower surface fuel loads identified above will probably be necessary to confer resilience in dry forests in the face of more severe fire weather. This will tend to reduce the severity of wildfire, and will allow longer periods of time between thinning treatments needed to maintain low fuels. In some forests, caution is needed that stand densities not be reduced to a level that will allow rapid growth of understory vegetation that could increase fire hazard (e.g., Thompson et al. 2007).

Table 10.4 Effects of thinning and surface fuel treatments on fire hazard on a stand in the Okanogan-Wenatchee National Forest, as simulated in the Fire and Fuels Extension to the Forest Vegetation Simulator

Parameters	Thinning treatments (trees ha⁻¹)				
	Initial	125	250	500	750
Torching index (km h⁻¹)	0	130	42	19	27
Basal area mortality (%)	7	20	30	21	70
Canopy bulk density (kg m⁻³)	0.08	0.04	0.05	0.07	0.07
Canopy base height (m)	0.6	12.5	7.0	1.8	1.5
Surface fuels[a] (Mg ha⁻¹)					
0–7.6 cm	6.6	22.0	26.4	30.8	30.8
7.6–15.2 cm	8.8	13.2	17.6	19.8	17.6
15.2–30.4 cm	8.8	8.8	6.6	6.6	4.4
> 30.4 cm	0	0	0	0	0
Litter	4.4	8.8	8.8	11.0	11.0
Duff	26.0	22.0	19.8	17.6	15.4

Adapted from Johnson et al. (2007)

[a] FFE-FVS assigned the initial fuel loading for each fuel component and size class

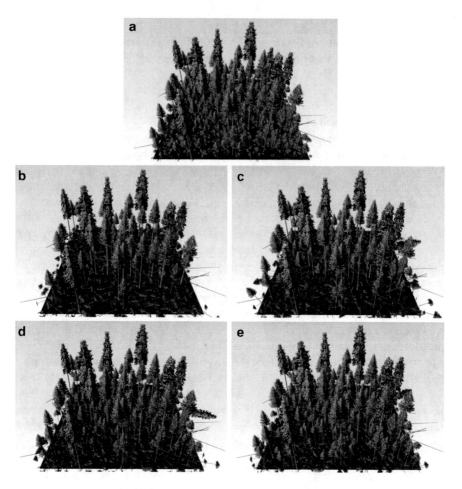

Fig. 10.2 Visualizations of thinning for a stand in the Okanogan-Wenatchee National Forest, as simulated in the Fire and Fuels Extension to the Forest Vegetation Simulator, including Initial conditions and four post-thinning stand densities (trees ha^{-1} = tph). (**a**) Initial conditions, (**b**) thinned to 750 tph, (**c**) thinned to 500 tph, (**d**) thinned to 250 tph, (**e**) thinned to 125 tph. See Table 10.4 for stand and fuel characteristics (Adapted from Johnson et al. (2007))

10.3.3 *Landscape Considerations for Fire and Fuels Management*

Stand-based treatments and evaluations will be more effective when applied in the context of a strategic plan for large landscapes. Therefore, a major challenge in fire management is to determine the optimal placement and size of fuel treatments on the landscape. Fuel treatments are not intended to stop a wildfire, but they can alter fire behavior (Finney and Cohen 2003). Fire managers do not have the capacity or resources to treat all areas that need to be thinned, because of land ownership, conflicting management objectives, and funding limitations (Finney 2007). Given these constraints, decisions about location and size of treatments can be explored with optimization models (e.g., Finney 2007), expert knowledge of local landscapes (Peterson and Johnson 2007), and examination of spatial patterns of forest structure and fuels over large landscapes over time (Fig. 10.3). In general, placement of treatments is designed to create landscape patterns that deter wildfire spread and modify fire behavior, while minimizing area needed for treatment (Finney 2001; Hirsch et al. 2001). Some modeling tools have options for determining the spatial arrangement and placement of fuel treatments. For example, Finney (2007) developed an algorithm to locate the specific treatment areas that reduce fire growth by the greatest amount for target environmental conditions. This type of modeling tool is the first step in developing an application that will help managers to determine the best location to place treatments with the goal of reducing wildfire behavior across a landscape.

Millions of hectares of public and private land would benefit from thinning treatments and surface fuel removal to reduce wildfire behavior (U.S. Forest Service 2000), but they are often not treated because of cost, potential (for prescribed burning) to cause air pollution, lack of safe periods for (prescribed burning) treatment, and esthetic reasons (Rummer 2008). Cost is related to two forms of treatment, *in situ* and extraction. *In situ* operations are designed to change the structure and arrangement of fuel loads and involve activities such as prescribed fire, mastication, or pile-and-burn. Extraction is the removal of fuels and usually costs considerably

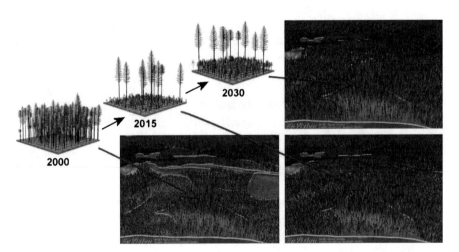

Fig. 10.3 Fuel treatment planning can be improved by quantifying stand structural conditions and fuels across large landscapes over time. Simulation tools can be used to examine the effects, placement, and visual appearance of thinning and fuel treatments throughout stand development. As stand conditions change from pretreatment (2000) to treatment+regeneration (2015) to regrowth (2030), subtle changes in landscape pattern and structure ensue (seen in the three landscape views.)

more than *in situ* methods unless the material removed has economic value. The cost of a project can be calculated from expert opinion, total bid cost, financial records of total enterprise costs, and economic analysis of fixed and variable costs (Keegan et al. 2002; Rummer 2008). Regardless of the methods used for treatment and cost calculation, it may become increasingly difficult for resource managers to treat sufficient area to significantly affect fire spread and behavior in a warmer climate.

Fuel treatments can have unintended consequences on other forest resources. For example, thinning and surface fuel treatments can provide an avenue for propagation of exotic plant species (Crawford et al. 2001; Griffis et al. 2001). Prescribed fire can scorch the crowns of live trees, which may increase stress or tree mortality (Graham et al. 2004). However, the biggest effect of fuel treatments is often on wildlife habitat (Randall-Parker and Miller 2002), with animal species that depend on complex forest structure being negatively affected (Pilliod et al. 2006). For example, a fuelbed structure that prevents crown fire initiation may decrease habitat for species that depend on large patches of dense multi-story forest (e.g., many species of neotropical migrant birds). Alternatively, species that forage in open forest structure (e.g., ungulates) may benefit from fuel treatments. Accounting for this interaction among resources will be a challenging consideration in fuel treatment planning in a warmer climate, because a warmer climate may directly affect those individual resources as well as the interactions.

10.4 Conclusions

The current warming trend in northern latitudes will almost certainly lead to increased area burned by wildfire in most ecosystems, with associated effects on ecosystem structure and function. Fuels will be flammable for longer periods of time. Prolonged droughts and insect attacks may increase fuel loads, leading to increases in fire hazard and fire severity. Exotic plants could further alter fire regimes in some ecosystems (Chap. 8), challenging our ability to manage for resilient and sustainable landscapes. A warmer and drier climate will reduce the effectiveness of fuel treatments in some locations. In these cases, using disturbance events such as wildfire as opportunities to influence species composition for resilience to climate change may be the best adaptation option.

Incorporating potential climate change effects and strategies into management plans will be a key step for agencies and organizations in adapting to climate change. Planning for potential impacts of climate change will increase preparedness, allow for time-efficient response to the effects of climate change, and minimize economic and ecological costs.

Many resource managers consider the current political and regulatory environment to be a severe limitation on adaptation to climate change (Joyce et al. 2008; Littell et al. N.d.). Policies, regulations, and administrative guidelines, though well intended for various conservation objectives, often fail to incorporate climate change and therefore focus on static (e.g., historic range of variation) rather than dynamic resource objectives. Lengthy planning, review, and approval processes can delay timely implementation of management actions (e.g., following a large wildfire) that could facilitate adaptation. Some of these constraints can be overcome by institutionalizing science-management partnerships in order to develop guidelines for addressing fire issues in a warmer climate. Incorporating climate change explicitly into national, regional, and national forest policy would be a major step forward in implementing climate change in established planning processes. "Climate-smart" policies and regulations that provide guidance but allow for local forest-level strategies and management actions that increase resilience and reduce vulnerability to climate change would also promote adaptation. Educational efforts to promote awareness of climate change will help create a more consistent approach within land management agencies and encourage support from stakeholders for fire and fuels management that facilitates adaptation to climate change.

We are optimistic about future opportunities to adapt to climate change with respect to fire. First, a familiar conceptual framework such as adaptive management can be used to facilitate fire and fuels management in a warmer climate. Second, there appears to be a core set of management strategies on which adaptation to climate change can be based (Table 10.1) (Millar et al. 2007; Joyce et al. 2008; Littell et al. N.d.). Third, it appears that resource managers with professional expertise on local landscapes can develop viable options for adapting to climate change if scientists can provide the scientific basis for decision making (Table 10.2). The scientific basis for managing fuels to enhance resilience already exists (Table 10.3) but will need to be continually tested for application to large landscapes. Such testing can initially be done in the simulation environment, but judicious and cautious experimentation by management will likely provide the greatest opportunities for adaptation and learning.

References

Agee, J.K., B. Bahro, M.A. Finney, P.N. Omi, D.B. Sapsis, C.N. Skinner, J.W. van Wagtendonk, and C.P. Weatherspoon. 2000. The use of shaded fuelbreaks in landscape fire management. *Forest Ecology and Management* 127: 55–66.

Agee, J.K., and C.N. Skinner. 2005. Basic principles of forest fuel reduction treatements. *Forest Ecology and Management* 211: 83–96.

Agee, J.K., C.S. Wright, N. Williamson, and M.H. Huff. 2002. Folier unoisture content of Pacific Northwest vegetation and its relation to wildland fire behavior.*Forest Ecology and Management* 167: 57–66.

Andrews, P.L., and L.P. Queen. 2001. Fire modeling and information system technology. *International Journal of Wildland Fire* 10: 343–352.

Arno, S.F., and S. Allison-Bunnell. 2002. *Flames in our forest: disaster or renewal?* Washington: Island Press.

Brown, A.A., and K.P. Davis. 1973. *Forest fire: control and use*, 2nd ed. New York: McGraw-Hill.

Covington, W.W., and M.M. Moore. 1994. Southwestern ponderosa forest structure, changes since Euro-American settlement. *Journal of Forestry* 92: 39–44.

Cram, D.D., T.T. Baker, and J.C. Boren. 2006. *Wildland fire effects in silviculturally treated vs. untreated stands of New Mexico and Arizona.* Research Paper RMRS-RP-055. Ogden: U.S. Forest Service.

Crawford, J.S., C. Wahren, S. Kyle, and W.H. Moir. 2001. Response of exotic plant species to prescribed fire in ponderosa pine forests of northern Arizona. *Journal of Vegetation Science* 12: 261–268.

Dale, V.H., L.A. Joyce, S. McNulty, R.P. Neilson, M.P. Ayres, M.D. Flannigan, P.J. Hanson, L.C. Irland, A.E. Lugo, C.J. Peterson, D. Simberloff, F.J. Swanson, B.J. Stocks, and B.M. Wotton. 2001. Climate change and forest disturbances. *BioScience* 51: 723–734.

Finney, M.A. 2001. Design of regular landscape fuel treatment patterns for modifying fire growth and behavior. *Forest Science* 47: 219–228.

Finney, M.A. 2007. A computational method for optimising fuel treatment locations. *International Journal of Wildland Fire* 16: 702–711.

Finney, M.A., and J.D. Cohen. 2003. Expectation and evaluation of fuel management objectives. In *Fire, fuel treatments, and ecological restoration: Conference proceedings,* eds. P.N. Omi and L.A. Joyce, 353–366. General Technical Report RMRS-P-29. Ogden: U.S. Forest Service.

Finney, M.A., C.W. McHugh, and I.C. Grenfell. 2005. Stand and landscape-level effects of prescribed burning on two Arizona wildfires. *Canadian Journal of Forest Research* 35: 1714–1722.

Fosberg, M.A., and J. Deeming. 1971. *Derivation of the 1- and 10-hour timelag fuel moisture calculations for fire-danger rating.* Research Note-RM-207. Fort Collins: U.S. Forest Service.

Gedalof, Z., D.L. Peterson, and N.J. Mantua. 2005. Atmospheric, climatic and ecological controls on extreme wildfire years in the northwestern United States. *Ecological Applications* 15: 154–174.

Gillett, N.P., A.J. Weaver, F.W. Zwiers, and M.D. Flannigan. 2004. Detecting the effect of climate change on Canadian forest fires. *Geophysical Research Letters* 31: L18211.

Government Accountability Office (GAO). 2007. *Climate change: Agencies should develop guidance for addressing the effects on federal land and water resources.* Report to Congressional Requesters. Washington: Government Accountability Office Report GAO-07-863.

Graham, R.T., S. McCaffrey, and T.B. Jain. 2004. *Science basis for changing forest structure to modify wildfire behavior and severity.* General Technical Report RMRS-GTR-120. Ogden: U.S. Forest Service.

Griffis, K.L., J.A. Crawford, M.R. Wagner, and W.H. Moir. 2001. Understory response to management treatments in northern Arizona ponderosa pine forests. *Forest Ecology and Management* 146: 239–245.

Hamlet, A.F., P.W. Mote, M.P. Clark, and D.P. Lettenmaier. 2007. 20th century trends in runoff, evapotranspiration, and soil moisture in the Western U.S. *Journal of Climate* 20: 1468–1486.

Hansen, L.J., J.L. Biringer, and J.R. Hoffman, eds. 2003. *Buying time: A user's manual for building resistance and resilience to climate change in natural systems.* Berlin: World Wildlife Fund.

Harris, J.A., R.J. Hobbs, E. Higgs, and J. Aronson. 2006. Ecological restoration and global climate change. *Restoration Ecology* 14: 170–176.

Harrod, R.J., B.H. McRae, and W.E. Hartl. 1999. Historical stand reconstruction in ponderosa pine forests to guide silvicultural prescriptions. *Forest Ecology and Management* 114: 433–446.

Harrod, R.J., N.A. Povak, and D.W. Peterson. 2007. Comparing the effectiveness of thinning and prescribed fire for modifying structure in dry coniferous forests. In *The fire environment–innovations, management, and policy: conference proceedings.* Proceedings RMRS-P-46CD. eds. B.W. Butler, and W. Cook, 301–314. Ogden: U.S. Forest Service.

Hessburg, P.F., and J.K. Agee. 2003. An environmental narrative of Inland Northwest U.S. forests, 1800–2000. *Forest Ecology and Management* 178: 23–59.

Hessl, A.E., D. McKenzie, and R. Schellhaas. 2004. Drought and Pacific decadal oscillation affect fire occurrence in the inland Pacific Northwest. *Ecological Applications* 14: 425–442.

Heyerdahl, E.K., D. McKenzie, L. Daniels, A.E. Hessl, J.S. Littell, and N.J. Mantua. 2008. Climate drivers of regionally synchronous fires in the inland Northwest (1651–1900). *International Journal of Wildland Fire* 17: 40–49.

Hirsch, K., V. Kafka, C. Tymstra, R. McAlpine, B. Hawkes, H. Stegehuis, S. Quintilio, S. Gauthier, and K. Peck. 2001. Firesmart forest management: A pragmatic approach to sustainable forest management in fire-dominated ecosystems. *Forestry Chronicle* 77: 357–363.

Hobbs, R., S. Arico, J. Aronson, J.S. Baron, P. Bridgewater, V.A. Cramer, P.R. Epstein, J.J. Ewel, C.A. Klink, and A.E. Lugo. 2006. Novel ecosystems: Theoretical and management aspects of the new ecological world order. *Global Ecology and Biogeography* 15: 1–7.

Hurteau, M.D., G.W. Koch, and B.A. Hungate. 2008. Carbon protection and fire risk reduction: toward a full accounting of forest carbon offsets. *Frontiers in Ecology and the Environment* 6: 493–498.

Intergovernmental Panel on Climate Change (IPCC). 2007. *Climate change 2007: The IPCC fourth assessment report.* Cambridge: Cambridge University Press.

Johnson, M.C. 2008. Analyzing fuel treatments and fire hazard in the Pacific Northwest. Ph.D. dissertation, University of Washington, Seattle.

Johnson, E.A., and D.R. Wowchuk. 1993. Wildfires in the southern Canadian Rocky Mountains and their relationship to mid-tropospheric anomalies. *Canadian Journal of Forest Research* 23: 1213–1222.

Johnson, M.C., D.L. Peterson, and C.L. Raymond. 2007. Guide to fuel treatments in dry forests of the western United States: Assessing forest structure and fire hazard. Service General Technical Report PNW-GTR-686. Portland: U.S. Forest

Joyce, L., G.M. Blate, J.S. Littell, S.G. McNulty, C.I. Millar, S.C. Moser, R.P. Neilson, K. O'Halloran, and D.L. Peterson. 2008. Chapter 3, National forests. In *Preliminary review of adaptation options for climate-sensitive ecosystems and resources – A report by the U.S. Climate Change Science Program and the Subcommittee on Global Change Research*, eds. S.H. Julius and J.M. West, 3.1–3.127. Washington: U.S. Environmental Protection Agency.

Keegan, C., M. Niccolucci, C. Fiedler, J. Jones, and R. Regel. 2002. Harvest cost collection approaches and associated equations for restoration treatments on national forests. *Forest Products Journal* 52: 96–99.

Kimbell, A.R. 2008. Climate change, water, and kids. http://www.fs.fed.us/kidsclimate/index.shtml. 10 July 2008.

Littell, J.S., D. McKenzie, D.L. Peterson, and A.L. Westerling. 2009. Climate and wildfire area burned in western U.S. ecoprovinces, 1916–2003. *Ecological Applications* 19: 1003–1021.

Littell, J.S., D.L. Peterson, C.I. Millar, and K. O'Halloran. [N.d.]. U.S. national forests adapt to climate change through science-management partnerships. Manuscript in review. On file with: Jeremy Littell, CSES Climate Impacts Group, University of Washington, Seattle 98195-5672.

Martinson, E.J., and P.N. Omi. 2003. Performance of fuel treatments subjected to wildfires. In *Fire, fuel treatments, and ecological restoration: conference proceedings*, eds. P.N. Omi and L.A. Joyce, 7–14. Proceedings RMRS-P-29. Ogden: U.S. Forest Service.

McKenzie, D., Z. Gedalof, D.L. Peterson, and P. Mote. 2004. Climatic change, wildfire, and conservation. *Conservation Biology* 18: 890–902.

Miles, E.L., D.P. Lettenmaier, et al. 2007. *HB1303 interim report: A comprehensive assessment of the impacts of climate change on the State of Washington.* Seattle: Climate Impacts Group, University of Washington.

Millar, C.I., N.L. Stephenson, and S.L. Stephens. 2007. Climate change and forests of the future: managing in the face of uncertainty. *Ecological Applications* 17: 2145–2151.

Noss, R.F. 2001. Beyond Kyoto: Forest management in a time of rapid climate change. *Conservation Biology* 15: 578–590.

Ottmar, R.D., D.V. Sandberg, C.L. Riccardi, and S.J. Prichard. 2007. An overview of the Fuel Characteristic Classification System–quantifying, classifying, and creating fuelbeds for resource planning. *Canadian Journal of Forest Research* 37: 2383–2393.

Parker, W.C., S.J. Colombo, M.L. Cherry, M.D. Flannigan, S. Greifenhagen, R.S. McAlpine, C. Papadopol, and T. Scarr. 2000. Third millennium forestry: What climate change might mean to forests and forest management in Ontario. *Forestry Chronicle* 76: 445–463.

Peterson, D.L., and M.C. Johnson. 2007. Science-based strategic planning for hazardous fuel treatment. *Fire Management Today* 67: 13–18.

Peterson, D.L. M.C. Johnson, J.K. Agee, T.B. Jain, D. McKenzie, and E.R. Reinhardt. 2005. *Forest structure and fire hazard in dry forests of the western United States.* General Technical Report PNW-GTR-628. Portland: U.S. Forest Service.

Pilliod, D.S., E.L. Bull, J.L. Hayes, and B.C. Wales. 2006. *Wildlife and invertebrate response to fuel reduction treatments in dry coniferous forests of the western United States: A synthesis.* Ogden: U.S. Forest Service General Technical Report RMRS-GTR-173.

Pollet, J., and P.N. Omi. 2002. Effect of thinning and prescribed burning on crown fire severity in ponderosa pine forests. *International Journal of Wildland Fire* 11: 1–10.

Price, C., and D. Rind. 1994. The impacts of a 2 x CO_2 climate on lightning-caused fires. *Journal of Climate* 7: 1484–1494.

Pyne, S.J., P.L. Andrews, and R.D. Laven. 1996. *Introduction to wildland fire,* 2nd ed. New York: Wiley.

Randall-Parker, T., and R. Miller. 2002. Effects of prescribed fire in ponderosa pine on key wildlife habitat components: preliminary results and a method for monitoring. In *Proceedings of the symposium on ecology and management of dead wood in western forests,* eds. W.F. Laudenslayer, P.J. Shea, B.E. Valentine, C.P. Weatherspoon, and T.E. Lisle, 823–834. General Technical Report PSW-GTR-181. Albany: U.S. Forest Service.

Reinhardt, E., and N.L. Crookston, eds. 2003. *The fire and fuels extension to the forest vegetation simulator.* General Technical Report RMRS-GTR-116. Ogden: U.S. Forest Service.

Rummer, B. 2008. Assessing the cost of fuel reduction treatments: A critical review. *Forest Policy and Economics* 10: 355–362.

Sandberg, D.V., C.L. Riccardi, and M.D. Schaaf. 2007. Reformulation of Rothermel's wildland fire behaviour model for heterogeneous fuelbeds. *Canadian Journal of Forest Research* 37: 2438–2455.

Skinner, C.N., J.H. Burk, M.G. Barbour, E. FrancoVizaino, and S.L. Stephens. 2008. Influences of climate on fire regimes in the montane forests of northwestern Mexico. *Journal of Biogeography* 35: 1436–1451.

Slaughter, R., and J.D. Wiener. 2007. Water, adaptation, and property rights on the Snake and Klamath Rivers. *Journal of the American Water Resources Association* 43: 308–321.

Smith, J.B., and S.S. Lenhart. 1996. Climate change adaptation policy options. *Climate Research* 6: 193–201.

Snover, A.K., L.C. Whitely Binder, J. Lopez, E. Willmott, J.E. Kay, D. Howell, and J. Simmonds. 2007. *Preparing for climate change: A guidebook for local, regional, and state governments.* Oakland: ICLEI—Local Governments for Sustainability.

Spittlehouse, D.L., and R.B. Stewart. 2003. Adaptation to climate change in forest management. *BC Journal of Ecosystems and Management* 4: 1–11.

Stephens, S.L. 1998. Evaluation of the effects of silvicultural and fuels treatments on potential fire behavior in Sierra Nevada mixed-conifer forests. *Forest Ecology and Management* 105: 21–35.

Stephens, S.L., and S.J. Gill. 2005. Forest structure and mortality in an oldgrowth Jeffrey pine mixed conifer forest in northwestern Mexico. *Forest Ecology and Management* 205: 15–28.

Stephens, S.L., and J.J. Moghaddas. 2005. Experimental fuel treatment impacts on forest structure, potential fire behavior, and predicted tree mortality in a California mixed conifer forest. *Forest Ecology and Management* 215: 21–36.

Stewart, I.T., D.R. Cayan, and M.D. Dettinger. 2005. Changes toward earlier streamflow timing across western North America. *Journal of Climatology* 18: 1136–1155.

Stocks, B.J., M.A. Fosberg, T.J. Lynham, L. Mearns, B.M. Wotton, Q. Yang, J.-Z. Jin, K. Lawrence, G.R. Hartley, J.A. Mason, and D.W. McKenney. 1998. Climate change and forest fire potential in Russian and Canadian boreal forests. *Climate Change* 38: 1–15.

Strom, B.A., and P.Z. Fulé. 2007. Pre-wildfire fuel treatments affect long-term ponderosa pine forest dynamics. *International Journal of Wildland Fire* 16: 128–138.

Swetnam, T.W., and J.L. Betancourt. 1990. Fire-Southern Oscillation relations in the southwestern United States. *Science* 249: 1017–1020.

Taylor, A.H., V. Trouet, and C.N. Skinner. 2008. Climatic influences on fire regimes in montane forests of the southern Cascades, California, USA. *International Journal of Wildland Fire* 17: 60–71.

Thompson, J.R., T.A. Spies, and L.M. Ganio. 2007. Reburn severity in managed and unmanaged vegetation in a large wildfire. *Proceedings of the National Academy of Sciences* 104: 10743–10748.

U.S. Forest Service. 2000. *Protecting people and sustaining resources in fire-adapted ecosystems: a cohesive strategy.* Forest Service management response to General Accounting Office Report GAO/RCED-99-65. October 13, 2000. Washington: U.S. Forest Service.

Westerling, A.L., H.G. Hidalgo, D.R. Cayan, and T.W. Swetnam. 2006. Warming and earlier spring increase western U.S. forest wildfire activity. *Science* 313: 940–943.

Whelan, R.J. 1995. *The ecology of fire.* Cambridge: Cambridge University Press.

Wotton, B.M., and M.D. Flannigan. 1993. Length of the fire season in a changing climate. *Forestry Chronicle* 69: 187–192.

Chapter 11
Wilderness Fire Management in a Changing Environment

Carol Miller, John Abatzoglou, Timothy Brown, and Alexandra D. Syphard

11.1 Introduction

Two major factors affecting wilderness fire regimes and their management are climate variability and surrounding land use. Patterns in climate and housing densities are expected to change dramatically in the next several decades (IPCC 2007; Theobald and Romme 2007) with important implications for fire management and policy (Dombeck et al. 2004). Successful protection and stewardship of wilderness means anticipating how these trends will affect fire regimes in the future.

The value and importance of wilderness and other protected areas for global sustainability and preservation of biodiversity have been widely recognized (Mittermeier et al. 2003). In the face of global change, wilderness areas will be critically important for species preservation, watershed protection, and habitat conservation (Barber et al. 2004). Wilderness areas are our best examples of naturally functioning ecosystems where natural fire regimes and landscape dynamics can be observed and studied (Kilgore 1986). As such, they provide useful knowledge about how to manage fire on other lands. The United States has more than 43 Mha of federally designated wilderness. More than 95% of this area is in Alaska (23.2 Mha) and the 11 western states (18.4 Mha) (Landres and Meyer 2000). By definition, these lands are to be managed so that natural ecological processes such as fire and other disturbances can function without human interference. In fact, the Wilderness Act states that these should be untrammeled, self-willed lands where humans practice humility and restraint.

Policy and law support the strategy of allowing lightning-caused fires to burn for their ecological benefits in wilderness (Zimmerman and Bunnell 2000), and it is in large wilderness areas where we have the best chance of restoring natural fire regimes without the need for periodic retreatment or manipulations (Noss et al. 2006). Ideally all lightning-caused fires would be allowed to burn unimpeded, but reality

C. Miller (✉)
Aldo Leopold Wilderness Research Institute, Rocky Mountain Research Station,
U.S. Forest Service, 790 E. Beckwith Ave., Missoula, MT 59801, USA
e-mail: cmiller04@fs.fed.us

D. McKenzie et al. (eds.), *The Landscape Ecology of Fire*, Ecological Studies 213,
DOI 10.1007/978-94-007-0301-8_11, © Springer Science+Business Media B.V. 2011

is much different (Parsons et al. 2003). Successful stewardship of the natural ecological role of fire eludes most wilderness managers and with very few exceptions, the annual area burned by natural fires in wilderness remains far below historical estimates (Parsons and Landres 1998). Despite running counter to the intent of the Wilderness Act, suppression is the management strategy taken on the majority of lightning-caused ignitions in wilderness for myriad biophysical and social reasons (Dale 2006; Black et al. 2007). The practicality of restoring natural fire regimes in wilderness has been questioned, especially in small wilderness areas where the immediate risk of fire escape is high (Husari 1995). Prescribed fire has been proposed to compensate for the lack of natural fire, but it is not an ecological substitute for natural fire and its appropriateness in wilderness has been questioned because it is a deliberate manipulation of the wilderness (Parsons 2000).

In addition to directly impacting natural communities (Backer et al. 2004), suppression interferes with natural dynamics and in some places has contributed to dense forests and large accumulations of dead biomass. A national map of current departures from historical natural fire regimes (Rollins and Frame 2006) suggests that over half of the area within designated wilderness in the 11 western states is moderately or highly departed (Fig. 11.1). In forests that have been greatly affected by fire exclusion, such as the dry forests, management action (e.g., prescribed fire) may be necessary to reduce unnatural fuel accumulations before fires can play their natural role (Agee 2002; Chap. 10). Not all wilderness ecosystems would benefit from more fire, however. Natural and prescribed fire, and even the creation of fuel breaks, can create suitable establishment sites for alien species (Keeley 2006). Following establishment a novel invasive species-fire regime cycle can be launched, leading to increases in fire frequencies in excess of the ability of native plants to recover (Brooks et al. 2004).

The ecological role fire plays in an ecosystem is but one aspect of the context for wilderness fire management. A complex set of biophysical and social factors interact to create this context, which varies widely among wilderness areas. We present climate change and increasing housing densities as two of the most important influences on wilderness fire regimes and their management (Fig. 11.2). This chapter examines potential implications of broad-scale patterns in climate and housing densities on wilderness fire management in the 11 western states in the conterminous United States. We then use two wilderness areas as contrasting examples to illustrate and discuss a diversity of challenges that are likely to arise within the next several decades. We conclude with a discussion of potential management strategies and responses for the future.

11.2 Changing Human Influences

Humans have been influencing fire regimes for millennia, and our modern influence is pervasive (Pyne 1997). Although people do not live inside wilderness areas, the proximity of residential developments to wilderness can influence how wilderness fire regimes are managed (Miller 2003). A major consideration in any fire management

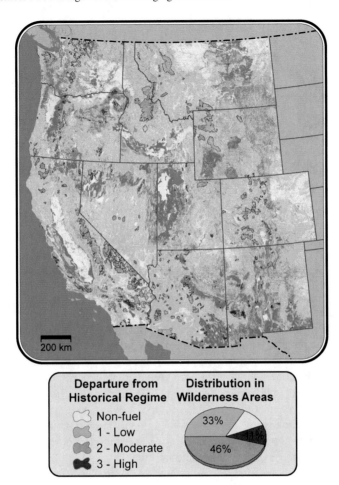

Fig. 11.1 Index of departure from historical fire regimes for the wilderness areas in the western United States. Classes *1*, *2* and *3* indicate *low*, *moderate*, and *high* departures, respectively

decision is the risk that a fire may pose to life and property (USDA and USDI 1998). Indeed, the risk of a wilderness fire escaping and threatening private property is a leading factor in the suppression of wilderness fire (Miller and Landres 2004).

11.2.1   Patterns and Trends

During the past several decades, urban growth and housing development have occurred at unprecedented rates globally and in the United States. In fact, housing development has actually increased faster than population growth, due to factors such as declining average household size and growth in seasonal and retirement

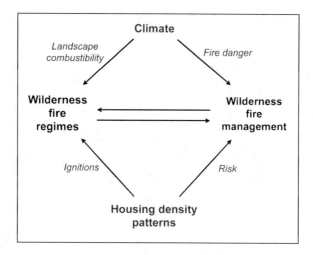

Fig. 11.2 Major influences from climate change and housing growth on wilderness fire management

homes (Liu et al. 2003; Yu and Liu 2007). From another perspective, the amount of urban land area in the United States quadrupled from 1945 to 2002, but during the same time, the population only doubled (Lubowski et al. 2007).

A striking trend in housing development is the disproportionate increase in housing growth in rural areas that have attractive recreational and aesthetic amenities (Crump 2003; Theobald 2004). As more people move away from the city or build seasonal or retirement homes "in the country," wilderness areas are increasingly surrounded by humans and houses. Although the highest housing densities are clustered around metropolitan areas (Fig. 11.3a), the projected rate of growth is high throughout many rural and undeveloped areas (Fig. 11.3b). Southern California has the highest housing densities, but growth projections to 2030 suggest the states of Colorado and Arizona will see the highest proportional increases (Hammer et al. 2004). Trends across the western United States indicate that housing density within 10 km of wilderness areas increased by 17–54% from 1990 to 2000; and by 2030, housing density is projected to increase even more, from approximately 44–103% (Radeloff et al. 2009; Hammer N.d.) (Table 11.1).

A greater number of houses close to wilderness areas elevates the potential fire risk to home owners, and therefore increases the pressure to suppress wilderness fires (Miller and Landres 2004). Suppression alters natural fire regimes in wilderness, which in turn may also affect ecosystem structure and function (Christensen 1988). The low rate of burning in US forests between 1930 and 1980 has been partly attributed to successful suppression efforts, and the increase in area burned since 1980 has been partly attributed to the resulting buildup of fuels and consequent difficulty of fire control (Stephens and Ruth 2005).

Humans can greatly influence when ignitions occur. Across the western United States, the number of human-caused ignitions exceeds the number of lightning ignitions, although the density of human and lightning ignitions varies over space

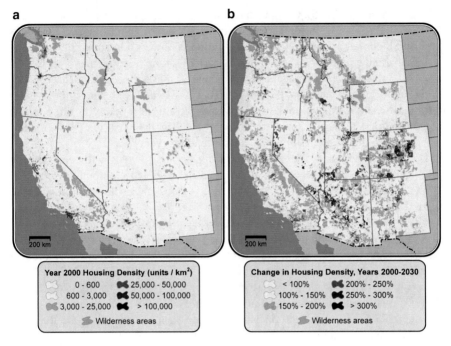

Fig. 11.3 Housing density in (**a**) 2000 and (**b**) projected change in housing density in 2030 for the 11 western U.S. states

Table 11.1 Projected increases in wilderness-proximate housing density by state (Hammer N.d.)

	Percentage increase in average housing density within 10 km of wilderness		
State	1990–2000	2000–2030	2000–2030 (10-year basis)
Arizona	36.5	87.4	29.1
California	20.1	79.4	26.5
Colorado	28.0	82.6	27.5
Idaho	30.8	61.3	20.4
Montana	26.0	57.7	19.2
Nevada	53.7	103.1	34.4
New Mexico	17.9	47.7	15.9
Oregon	17.6	44.4	14.8
Utah	22.9	66.9	22.3
Washington	19.2	50.9	17.0
Wyoming	31.3	67.2	22.4
Western US	26.5	75.3	25.1

and time (Stephens 2005). On one hand, human-caused ignitions probably serve to lengthen the season by introducing ignition sources during times when lightning is absent (Slocum et al. 2007). Conversely, suppression decisions in wilderness tend to squeeze the effective fire season into a narrower time window that is late in the season. Many wilderness managers are uncomfortable with allowing early season ignitions to burn because of the uncertainties associated with managing a long-duration fire (van Wagtendonk 1995). Managers are more likely to allow a late season ignition to burn because they know that it will be extinguished in a few weeks by the first snowfall or other "season-ending event."

Areas where housing development meets or intermingles with undeveloped wildland vegetation are technically defined as the wildland-urban interface (WUI) (Radeloff et al. 2005). Much of the increase in housing development adjacent to wilderness areas can therefore be considered one of two types of WUI as defined by the Federal Register (USDA and USDI 2001). "Intermix WUI" is defined as the intermingling of development with wildland vegetation; the vegetation is continuous and occupies 50% of the area. "Interface WUI" occurs where development abuts wildland vegetation; there is less than 50% vegetation in Interface WUI, but it is within 2.4 km of an area that has 75% vegetation. Housing density in both types of WUI is at least 6.17 housing units/km^2. The WUI has recently been receiving considerable attention in fire policy and management because these are the areas where humans and houses are most susceptible to fire, fighting fires is the most challenging, and human-caused ignitions are most likely to occur (Radeloff et al. 2005). Where WUI is adjacent to designated wilderness areas (or within inholdings), the fire management and decision-making environment is highly complex.

11.2.2   Landscape Scale Implications

Humans can strongly influence the spatial pattern of fire at the landscape scale of a wilderness area. Managers are more likely to suppress lightning-caused fires that start close to the wilderness boundary because of the risk they may pose to values outside the wilderness, or simply the risk of spreading onto land where fires are not permitted to burn for resource benefit (Miller and Landres 2004). Furthermore, natural ignitions *outside* the wilderness boundary that are suppressed may have otherwise burned into wilderness. By eliminating them through suppression, humans can alter the spatial pattern of fire occurrence in wilderness. For example, a modeling analysis for the Selway-Bitterroot wilderness in northern Idaho highlighted places within the wilderness where a natural fire regime would be dependent upon immigration of fires from outside the area (Miller and Parsons 2004) (Fig. 11.4). Such cross-boundary effects of suppression are likely to be even more pronounced in small wilderness areas.

Another way humans influence the landscape patterns of fire is by introducing ignitions. Human-caused ignitions are most likely to occur in areas where there is a high concentration of human activity. Ignitions along transportation corridors have been documented broadly; and the probability of human ignitions is significantly

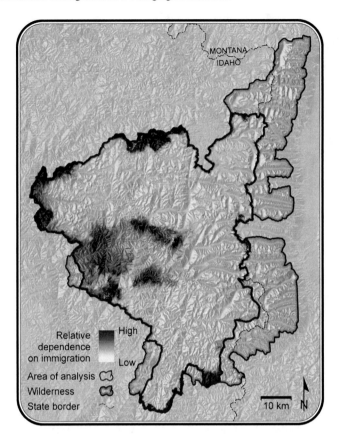

Fig. 11.4   The relative dependence of the natural fire regime on immigration of fires from outside the interior fire management planning zone of the Selway-Bitterroot Wilderness in northern Idaho

higher at shorter distances to human infrastructure, such as housing development or other urban areas (Chou et al.1993; Cardille et al. 2001; Stephens 2005; Yang et al. 2007; Syphard et al. 2008). For example, the vast majority of ignitions that occurred within the last 30 years in the Santa Monica Mountains National Recreation Area occurred directly along roads or developed areas (Syphard et al. 2008). At a coarser scale (58 counties in the state of California), multiple human variables were also very significant in explaining fire frequency, although area burned was more strongly related to biophysical variables (Fig. 11.5) (Syphard et al. 2007). While fire frequency was strongly related to population density, there were also significant spatial relationships between humans and fire at this coarse scale. Fires occurred most frequently when they were close to the WUI, and both fire frequency and area burned were more strongly affected by Intermix WUI and low-density housing than by Interface WUI. This spatial relationship between human activities and fire ignitions suggests that more fires are likely to start at the periphery of wilderness areas if housing development continues to increase in those areas.

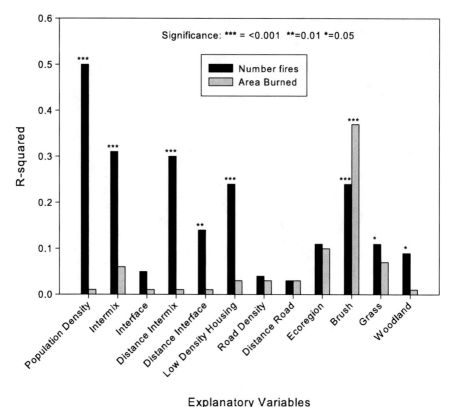

Fig. 11.5 Explanatory variables found for fire occurrence and area burned in California counties. Adjusted R^2 values and significance levels for the explanatory variables are from bivariate regression models for number of fires and area burned in 2000 (Data from Syphard et al. (2007))

The consequences of increasing ignitions associated with population growth and urban development in and around wilderness areas may be complex. On one hand, human ignitions that are not successfully suppressed may offset the effects of suppression and inadvertently serve to help restore fire to some wilderness areas. The obvious problem with this perspective is that these ignitions are also most likely to occur near humans and houses, thereby posing a hazard to lives and property. Another consideration is that increased human ignitions can threaten some ecosystems by increasing fire frequencies. For example, human ignitions and fire frequency in southern California have dramatically increased over the last several decades, more than offsetting the effects of fire suppression (Keeley et al. 1999). When intervals between fire events are too short, many native shrub species are unable to recover, even though they are resilient to the periodic wildfire characteristic of the region's natural fire regime (Keeley and Fotheringham 2001). The problem of increased fire frequency is compounded by spread of exotic grasses in the region. In many areas, these exotic grasses, which favor and facilitate frequent fire, are irreversibly replacing native shrublands (Zedler et al. 1983; Zedler 1995; Halsey 2008).

Another phenomenon that complicates the analysis of human influence on fire is the strong potential for the relationship between people and fire to be non-linear. In California, both area burned and number of fires is highest when population and housing densities are at intermediate levels (Syphard et al. 2007) (Fig. 11.6a). Fire frequency peaks when housing density was more than 6 housing units km^{-2} with greater than 50% vegetation, which corresponds to intermix WUI (Fig. 11.6b). This trend demonstrates that fires initially increase with population and housing density, but then they decline after reaching a certain threshold density. The association between fires and intermediate population or housing density suggests that fire risk, particularly around wilderness areas, is a function of the spatial arrangement of people and fuels. Houses in low density developments may be more difficult and expensive to protect than houses in higher-density clustered developments because fire management resources have to be spread across larger more dispersed

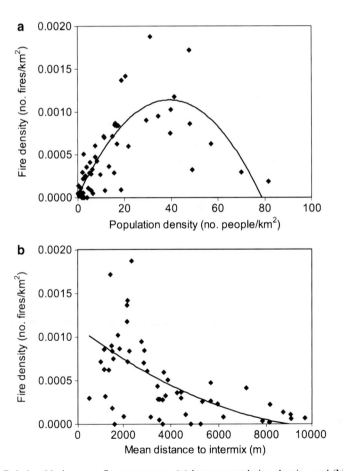

Fig. 11.6 Relationship between fire occurrence, (**a**) human population density, and (**b**) distance to intermix wildland-urban interface (Data from Syphard et al. (2007))

areas, remote areas are more difficult to access, and firefighters have to manage wildland and structural fires simultaneously (Irwin 1987; Cova 2005).

The spatial and temporal patterns of anthropogenic ignition locations are entirely determined by human activities, but the potential for fire to spread and for fires to become large is more a function of biophysical variables, such as vegetation characteristics, climate, and terrain (Pyne et al. 1996). Because fire spread is largely governed by biophysical characteristics, the regional characteristics and conditions of each particular area will largely dictate the potential for a human-caused ignition to influence wilderness, as well as the potential for a wilderness fire to affect lives and property in the WUI.

11.3   Changing Climatic Influences

Climate influences fire regimes across all spatial and temporal scales (Falk et al. 2007), although the strength of these relationships varies among ecosystems (Schoennagel et al. 2004). Climate also influences the fire management environment. When deciding whether to allow lightning-caused fires to burn for their beneficial effects, fire managers regularly utilize climate information to guide decisions (Kolden and Brown 2008), and drought is a specific consideration in the long term planning of such fire incidents (USDA and USDI 2005).

Changes in climate are expected to alter environmental conditions favorable to the spread of large wildfires, thereby affecting wilderness fire regimes and their management. We present new spatial and temporal analyses of climate-induced change on the frequency of extreme fire danger and the length and timing of fire weather season for the western United States. Specifically, we use a fire danger index from the National Fire Danger Rating System (NFDRS) (Cohen and Deeming 1985) which is typically employed by management in strategic decision-making processes. Two contrasting areas—southwestern Colorado and southern California—illustrate regional implications.

11.3.1   Spatial Patterns of Change: Extreme Fire Danger

General circulation models (GCMs) represent the primary tool used to evaluate future climate conditions under the premise that changes in greenhouse gas concentrations and aerosols are the impetus for modulating climate and weather. Future climate and weather generated by GCMs can be used to examine expected changes in fire danger and fire season (Brown et al. 2004). The Intergovernmental Panel on Climate Change (IPCC) produced scenarios that reflect a spectrum of changes in atmospheric concentrations of greenhouse gases that would result from future emissions of greenhouse gases and aerosols (IPCC 2000). For example, the SRES-A1B scenario assumes that by 2100 atmospheric CO_2 concentrations will be double that seen in the late 1990s. This scenario is typically considered to be a mid-range outcome compared to other scenarios.

Evaluating future climate conditions with GCMs is problematic because predictions vary among models. While all models show a pronounced warming signal over the western United States during the twenty-first century, projected changes for other pertinent variables such as relative humidity and precipitation show large model-to-model differences (often of opposing sign) on a regional basis. Because fire danger indexes are strongly related to these other variables, a single GCM should not be used to project changes in fire danger. However, results from individual GCMs provide different realizations of future fire danger and can be used collectively to create a probabilistic range of projections. Furthermore, by using projections from all available GCMs, one can quantify confidence levels as needed in impact assessment studies. We address the problem of discrepancies among models with a multi-model ensemble (MME) approach, using the average of climate variables projected by multiple GCMs to depict a single expectation of changes, as well as an intermodel comparison to show the range of projections among GCMs.

We use monthly output from 15 GCMs from the IPCC's fourth assessment report (IPCC 2007). We considered GCM output from runs forced with greenhouse gas and aerosol concentrations as observed during the late twentieth century (1971–2000) to be our baseline for change, or our *control*. GCM output from runs forced with the SRES-A1B emissions scenario for the mid (2041–2070) and late (2071–2100) twenty-first centuries served as our experimental *treatments*.

To be useful for assessing the impact of climate change on fire danger, the coarse-scale $2 \times 2°$ data from the GCMs needs to be downscaled, or translated to a finer scale. For this downscaling, we used gridded meteorological data at a spatial resolution of 32 km from the North American Regional Reanalysis (NARR) (Mesinger et al. 2006). NARR provides modeled data derived from meteorological stations for the period 1980–2007 for variables such as temperature, precipitation, humidity, and wind. To create downscaled future climate and weather data, we computed difference fields between the experiment runs (middle and late-twenty-first century) and the control run (late twentieth century) from the 15 GCMs on a monthly basis for the relevant meteorological variables. Difference fields are additive for temperature, and multiplicative for wind speed, precipitation, and specific humidity. We then applied these difference fields to the gridded daily weather from NARR for the period 1980–2007 to create downscaled climate/weather grids for the two future time periods (mid- and late-twenty-first century) at 32-km resolution, using inverse distance weighting.

The downscaled set of future climate/weather variables were used as inputs to compute daily gridded (32-km resolution) fire danger indexes (Cohen and Deeming 1985). We also computed fire danger indexes for the historical (1980–2007) period from the NARR data. We present the index known as the Energy Release Component (ERC) computed for the National Fire Danger Rating System (NFDRS) fuel model G (short-needle conifer with heavy dead fuel load). ERC is a relative value reflecting the available energy per unit area within the flaming front at the head of a fire. Computed daily, ERC responds to changes in live and dead fuel moistures, and typically increases during spring and early summer across the western US as dead fuels dry and live fuels cure. ERC values computed for fuel model G have been shown to correlate well with fire activity (Andrews et al. 2003).

We defined extreme fire danger at each grid cell as the 97th percentile ERC value for the historical (1980–2007) period, and examined how its frequency of occurrence might be expected to change in the future across the western US. We used the MME average of climate variables (temperature, precipitation, humidity, wind) to compute daily ERC values, and tallied the number of days per year with extreme fire danger for the mid- and late-twenty-first century. By the mid-twenty-first century (Fig. 11.7a), the number of days per year exceeding historical 97th percentile values is projected to increase by about 45% when averaged across the West. By the late-twenty-first century, projections show that approximately seven additional days per year (areally averaged across the West) will exceed the historic 97th percentile, nearly 65% more than are observed today (Fig. 11.7b).

In addition to computing ERC for the average MME projections, we compared models by computing daily ERC values for projections from each of the 15 GCMs. The average MME provides a projection of future changes that usually performs

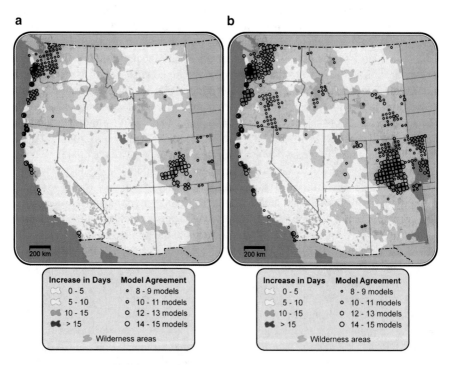

Fig. 11.7 Increase in the number of days per year of extreme fire danger for (**a**) middle and (**b**) late twenty first century compared to the late twentieth century base period for the MME SRES-A1B emission scenario and the number of GCMs that show statistically significant increases in frequency

better than any single GCM (e.g., Riechler and Kim 2008), but it represents only a central tendency of the 15 GCMs, and does not share the functional drivers of the individual models from which it was derived. Therefore, we use the intermodel comparison to help quantify the confidence in the MME projection of ERC. Projections of change from the average MME are largest across regions where there is agreement among the GCMs. For example, in southwestern Colorado, nearly all the GCMs show statistically significant increases in the occurrence of extreme fire danger, thereby suggesting strong agreement among models (Fig. 11.7).

11.3.2 Temporal Patterns of Change: Fire Season

Recent observations indicate some climate-related changes in spring over the West have been taking place (e.g., earlier last freeze date, earlier spring runoff, earlier greenup), but that corresponding changes in autumn (e.g., later first freeze date) have not been observed to a similar extent (IPCC 2007). Other findings include an earlier end to spring precipitation across most of the Southwest, and lower snow: rain ratios in regions of seasonal snowcover (Knowles et al. 2006). Moreover, as increasing temperatures procure an earlier retreat of snowcover, the snow-albedo feedback can accelerate snowcover loss, further advancing the timing of declines in soil and fuel, and by extension, increases in fire danger indexes.

Managers use fire danger ratings to anticipate the level and timing of staffing and resource needs for responding to wildfire events throughout the fire season (NIFC 2009). ERC, in particular, is used to track seasonal trends in fire danger relative to historical weather conditions (Main et al. 1982). The timing of the onset or departure of high or extreme fire danger can be of particular interest because it signifies when fire activity is likely to increase and when fire management resources may begin to become scarce. Fire activity and availability of resources are important influences on whether a wilderness fire is suppressed or not (Miller and Landres 2004).

To highlight potential changes in the length and timing of the fire season, we examined two contrasting areas in more detail: the Weminuche Wilderness in southwestern Colorado and the Cleveland National Forest in southern California. The 197,600 ha Weminuche is the largest wilderness in Colorado. Being a high elevation area (average elevation 3,000 m), its summers are relatively short and cool. Spruce-fir forests and alpine vegetation dominate the area. The Cleveland National Forest has four much smaller wilderness areas totaling 30,500 ha and ranging in elevation from 500–1,700 m. Summers, by contrast, are hot and dry, and the areas are dominated by shrubland and chaparral.

The climatological annual cycle of observed daily mean ERC values for the historical base period (denoted by the blue line) and MME projected daily mean ERC values for the mid-twenty-first century from the MME (red line) are shown for the Weminuche Wilderness and the Cleveland National Forest in Fig. 11.8.

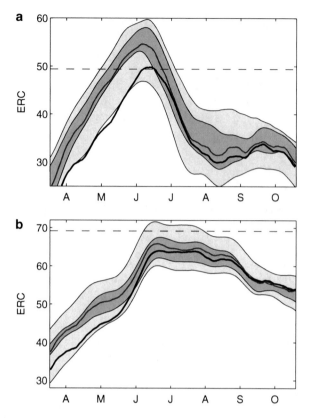

Fig. 11.8 Daily energy release component (ERC) climatology, April (*A*) through October (*O*), of average ERC for the (**a**) Weminuche Wilderness in southwestern Colorado and (**b**) the Cleveland National Forest in southern California. The historical base period (1971–2000) is shown in *black*, the mid-twenty-first century is shown in *red*. The shades of *grey* illustrate the uncertainty among the 15 models: *light* and *dark* shaded bands represent the 10th–90th and 25th–75th percentile ranges among model estimates of daily ERC, respectively. For reference, the horizontal dashed line indicates the historical annual 90th ERC percentile value

Overlain on these average MME results is the envelope of potential outcomes from the 15 GCMs for the mid-twenty-first century.

For the Weminuche (Fig. 11.8a), the steady increase of ERC starting in March occurs as fuels dry out after snowmelt and as precipitation is insufficient to offset the drying process. The ERC values peak in June and July, and then decline due to the onset of the North American monsoon and southerly moisture advection into southwestern Colorado. The second peak in early October is a dry autumn pattern that occurs during the transition between summer monsoonal precipitation and progressive mid-latitude winter cyclones that traverse the West. Three important differences are illustrated by comparing projections of future ERC values to historical values. First, the middle twenty-first century curve is shifted higher compared to the late twentieth century historical curve. Second, the middle twenty-first century curve indicates about a 1 week earlier onset of very high fire danger in the spring, and all but one GCM suggest increases in ERC will occur during late-spring and early summer. Third, the average ERC value for the mid-twenty-first century for the entire month of June exceeds the historical 90th percentile ERC value, a value that is often used to represent very high fire danger, whereas historical ERC values are consistently below this threshold value. All but one GCM indicate increases in ERC during the summer fire season, suggesting confidence in these projections.

Figure 11.8b shows the middle twenty-first century mean ERC time series for the Cleveland National Forest. A pattern of drying throughout the summer is reflected in the increasing ERC which remains elevated through the autumn. Unlike results for the Weminuche Wilderness that essentially show strong intermodel consensus, results for this area in southern California show a wider spread of projections among the GCMs, suggesting a larger degree of uncertainty. As with southwestern Colorado, the curve for the mid-twenty-first century shows an earlier start to the season, but unlike Colorado, it shows a later ending.

11.3.3 Spatio-Temporal Patterns of Change: Landscape Combustibility

Most wilderness areas in the western US are located at higher elevations and may be particularly sensitive to warming trends because of their niche climatological settings (Beniston et al. 1997), and amplified changes in key climatic variables that occur due to positive feedbacks in the climate systems (Giorgi et al. 1997). At landscape scales, the effects of climate change depend on the biophysical conditions of a particular wilderness area and will be reflected in the water balance and its association with fire (Chap. 5).

Downscaled predictions, as described above, are used for the Weminuche Wilderness to compute and map the change in soil water balance for the mid-twenty-first century (2046–2065) assuming the SRES-A1B scenario. Monthly precipitation and monthly minimum and maximum temperatures were averaged for the eight NARR grid cells that overlapped the perimeter of the Weminuche. These were inputs to the weather and soil water submodels in the FACET Model (Urban et al. 2000). The 594,000 ha landscape (the wilderness plus a 10-km buffer) was divided into 3,482 landscape "facets" according to elevation (100-m intervals) and slope-aspect. For each of the elevation-slope-aspect combinations, the FACET Model was used to compute a drought-day index with baseline (1971–2000) and mid-twenty-first century (2046–2065) climate inputs. The drought-day index is a tally of the number of days during the growing season that the soil water is at or below wilting point. FACET adjusts radiation for slope-aspect and precipitation and temperature are adjusted according to locally regressed lapse rates, and so the drought-day index varies with elevation and topographic position. The increases in drought-day values predicted for the mid-twenty-first century relative to the baseline period were then

Fig. 11.9 Predicted increase in the number of drought-days per year for the Weminuche wilderness in southwestern Colorado for the middle twenty-first century (2046–2065) compared to the baseline period (1971–2000)

mapped at a 30-m resolution (Fig. 11.9). Substantial increases are expected, with about 80% of the landscape experiencing a 20% increase or more.

Fire danger and fuel moisture conditions can be expected to follow similar patterns, with portions of the landscape that currently lie near the mean spring freezing elevation showing the greatest change (Diaz et al. 2003). If the upshot of projected changes across higher elevation wilderness areas is an earlier snowmelt (e.g., Lopez-Moreno et al. 2008), earlier onset of declines in fuel moisture could ensue, making larger portions of a wilderness flammable for longer periods of time. In areas that are subject to dry lightning storms (e.g., Rorig and Ferguson 1999), this increase in landscape combustibility could have dramatic impacts on the number, extent, and duration of wilderness fires. This factor has not previously been considered in coarser-scale analyses of potential climate change impacts on wildfire activity (e.g., McKenzie et al. 2004). Some wilderness areas could become understaffed for the workload and level of fire activity that may become the norm in the twenty-first century.

11.4 Future Challenges for Wilderness Fire Management

Even the most pristine and natural wilderness areas are not immune to change or outside influences (Hansen and DeFries 2007). In the next few decades, global climate change and the expansion of the WUI are expected to influence wilderness fire regimes, with important landscape-scale effects and implications for wilderness fire management.

Dramatic changes in housing density for wilderness proximate areas (Fig. 11.3, Table 11.1), will increase the potential for humans to alter natural fire regimes. Simply put, humans start fires, and they put them out, thus affecting when and where fires occur and potentially every aspect of a wilderness fire regime. The degree to which housing growth actually affects wilderness fire regimes in the future will depend on multiple factors, including fire suppression and financial resources, the strong spatial pattern of fire ignition locations, the intermediate relationship between housing or population density and fire risk, and the specific biophysical characteristics of a region. Furthermore, the projections of housing density are dependent on assumptions about population, household, and housing growth rates.

As more people choose to live closer to wilderness areas, the complexity of managing natural fire regimes in wilderness will increase. In some cases, managers will need to carefully consider the costs and benefits of different response strategies. An increase in the number of lives and value of property threatened by wildland fires will increase the complexity of a fire management situation (USDA and USDI 1998), likely decreasing the opportunities for allowing wilderness fires to burn on their own terms. A different effect of WUI expansion is that some wilderness areas could be threatened by the introduction of human-caused ignitions at frequencies and in seasons that naturally would not occur.

Predicted changes in climate will also affect various aspects of wilderness fire regimes, including the seasonality, frequency, extent, spatial pattern, and severity of fires. What has been considered extreme fire danger in the past will become more the norm by the middle of the twenty-first century (Fig. 11.7). Fire seasons will become longer, start earlier in the year (Fig. 11.8) and involve more landscape area (Fig. 11.9). Warmer temperatures and longer fire seasons are likely to increase the area burned by wildland fire and will affect the workload of fire management agencies (McKenzie et al. 2004; Fried et al. 2008).

We focused on fire danger, but climate change will influence wilderness fire regimes in other important ways. For example, the increased observed warming and associated drought (both predicted to continue based on the twenty-first century climate model runs) have increased fire severity, insect outbreaks, and vegetation stress, all leading to increased tree mortality and fuel accumulations (Dale et al. 2001). These synergies, the dependencies of spatial and temporal scale interactions, the complexities of topography, fuels, and weather or climate, along with human actions, make prediction difficult and lead to high uncertainty in modeled outcomes of climate change (Chap. 4).

Although we have discussed implications of changing climate and increasing housing density independent of one another, these trends are likely to interact. Wilderness fire managers will need to account for a changing climate along with a changing human footprint. An expanding WUI and an increase in extreme weather conditions add complexity that could increase fire management costs and threaten firefighter safety (Gebert et al. 2007). The number of ignitions caused by humans in the WUI could increase, and these may be more difficult to suppress during extreme fire danger conditions. Conversely, an increase in area burned, as is expected with climate change, might actually reduce hazardous fuels and concomitant risk to the WUI. Forecasted increases in fire danger indices suggest that we might expect more erratic fire behavior and potentially faster rates of spread. Prescribed fires and natural fires that are allowed to burn in wilderness may be more likely to escape in a future climate, and increased values-at-risk in the WUI may exacerbate the consequences of those escaped fires. As time goes on, the feasibility of allowing wilderness fires to burn unimpeded may decline, particularly in smaller wilderness areas that simply are not big enough for long-duration fires to spread naturally without posing a risk to development. For areas with feedbacks from alien plant invasions (Chap. 8), insect outbreaks, or soil water processes (Chap. 5), decision space could become especially constrained.

In many ways, southern California may be a "perfect storm" of human and climate influences on wilderness fire. Recent wildfires in 2003 and 2007 were particularly destructive in the WUI and put the issue of the WUI in the national spotlight (Keeley et al. 2004). The area ranks highest in the nation for the number of housing units in the WUI, and further expansion and densification can be expected (Hammer et al. 2007). Dominant vegetation types are highly flammable shrublands and chaparral that characteristically experience very high-intensity fires. In many years these fires are driven by weather conditions such as the Santa Ana winds, the frequency of which may increase during the twenty-first century (Miller and Schlegel 2006). Although climate change projections for southern California suggest only modest impacts on fire danger (Figs. 11.7 and 11.8b), the earlier onset of the dry season coupled with increases in summer temperatures may lead to a significant increase in fuel-driven fires (as opposed to wind driven) for southern California, thereby heightening the reality of a 12-month fire season. The region also has exceptional biodiversity and endemism (Stein 2002). Where wilderness areas protect rare habitat and species, the stakes for conservation could be especially high.

Four wilderness areas on the Cleveland National Forest in southern California exemplify the most extreme challenges for wilderness fire management: San Mateo, Agua Tibia, Pine Creek, and Hauser (Fig. 11.10). These wilderness areas are small (the largest is San Mateo, 15,574 ha), and will become increasingly surrounded by WUI of housing densities 6–30 houses km^{-2}. These densities correspond to the modal population density in Fig. 11.6a at which human-caused ignitions may be most numerous (Syphard et al. 2007). Although fire used to be a natural component of these wilderness ecosystems, human-caused wildfires are degrading habitat for sensitive species and unnaturally high fire frequencies are potentially changing certain vegetation types irreversibly (Keeley 2006). None of the four wilderness areas currently allows lightning-caused fires to burn for their resource benefit (U.S. Forest Service 2005), probably because of their proximity to urban areas and concerns about air pollution, extreme fire behavior, and diminishing habitat. With predicted changes in climate and housing density, few, if any, fire management options other than aggressive suppression will remain in these wilderness areas.

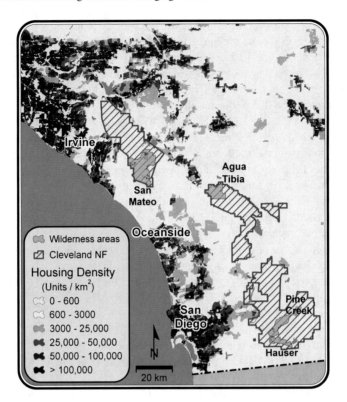

Fig. 11.10 Projected housing density for 2030 near four wilderness areas on the Cleveland National Forest in southern California

11.5 Responding to Change

The spirit and purpose of wilderness can be compromised when a full range of options is not available to wilderness fire managers. Wilderness fire managers need to respond to the new challenges presented by housing growth and climate change, and to what could be increasing constraints on their decisions. Even where constraints have left managers with few options, however, there may still be time to respond with positive results. Coarse-scale analyses such as presented here could help identify where rapid change threatens to reduce management options. An urgent call could then be made for site-specific study and action in these places to develop and implement viable response actions. These analyses can also shed light on the different ways the decision-making environment might change so that we might respond effectively and appropriately.

Our contrasting examples of southwestern Colorado and southern California clearly demonstrate that no single management response strategy is appropriate or desirable. Management objectives and strategies should consider the current and future landscape context of the wilderness, its ecological condition, the role that fire is playing in the ecosystem, and the type and degree of change that can be expected. In southwestern Colorado, fire suppression can be regarded as a threat to the Weminuche Wilderness and is at odds with an untrammeled wilderness where natural processes dominate. In contrast, a threat to wilderness areas in southern California is an unnaturally high fire frequency that compromises natural ecosystem function. Future changes in climate and housing density will affect each wilderness area in different ways and to different degrees. In the Weminuche, the most important changes to the fire regime in the foreseeable future could come from climate factors that alter the fire season and landscape combustibility, whereas in southern California, increases in housing density may be the more immediate concern for wilderness fire managers. Clearly, management approaches need to be tuned to site-specific situations (Della Sala et al. 2004) and should be flexible to account for the uncertainties in forecasts (Millar et al. 2007).

As a framework for developing appropriate management approaches, we contrast two generalized wilderness fire management objectives, summarizing possible management strategies and specific response to change for each (Table 11.2). The first generalized objective is to restore or maintain the process of fire in fire-dependent ecosystems and applies to wilderness areas like the Weminuche where fire has the potential to play a beneficial role in ecosystem function. In this situation, a viable strategy is to allow fires to burn for their natural resource benefits. In the past, this strategy was restricted to lightning-caused fires; in future policy revisions, the ignition source may no longer be a consideration (USDI and USDA 2001). In some places, vegetation and fuel treatments may be warranted to help ensure that when fire is reintroduced, it does not have undesirable impacts on habitat for sensitive species or other values at risk (e.g., human life and private property) that could be threatened by a wilderness fire. The objective of allowing fires to burn for their beneficial effects has traditionally been pursued in large wilderness areas, but could be achieved in smaller wilderness areas. There also may be opportunities to manage for the objective of resource benefits on adjacent lands, thus removing the concern about fires crossing administrative boundaries, and benefiting wilderness areas whose fire regimes are dependent on the immigration of fires from outside wilderness. The second generalized objective is to protect ecosystems from fire. This objective applies to wilderness areas where human ignitions and other ecological changes (e.g., the invasion of non-native grasses) have contributed to fire frequencies that are too high for the persistence of native plant communities. In these situations, the threat fire poses to ecological values may leave only one viable fire management: aggressively suppress all fires.

Effective response to the influences of a changing climate will involve learning to live with the new role that fire may play. In places like the Weminuche, managers and stakeholders may need to learn to accept more stand-replacing fire and early seral vegetation on the landscape. Therefore, specific responses include revising fire management plans, fuel treatment specifications, fire use prescriptions, and preparedness plans to reflect the predicted changes in fire regimes (Table 11.2). In places like southern California, preparedness plans may need to be revised to reflect the growing reality of a year-round fire season.

Table 11.2 Potential fire management strategies and specific responses to meet two generalized wilderness fire management objectives

| | Generalized wilderness fire management objectives | |
	Restore or maintain	Protect
	Restore fire to ecosystems that have been altered by fire suppression or other land use change. Maintain process of fire in ecosystems that have not been altered	*Protect ecosystems that are threatened by fires that are too frequent*
General wilderness fire management strategies	Exploit opportunities to allow lightning-caused ignitions to burn in wilderness. If policy permits, extend to human-caused ignitions Restore vegetation structure and reduce fuels with prescribed fire or mechanical methods if necessary Extend wildland fire use strategy to adjacent nonwilderness lands	Aggressively suppress all fires
Specific responses to climate change	Revise fire and land management plans to reflect climate-mediated changes to fire regimes Modify fuel treatment specifications to ensure they will moderate fire behavior and effects under more extreme fire weather conditions Revise fire use prescriptions to reflect higher fire danger and longer fire seasons	Emphasize preparedness and revise preparedness plans to reflect longer fire seasons, higher fire danger Modify fuel treatment specifications to ensure they will moderate fire behavior and effects under more extreme fire weather conditions
Specific responses to housing growth	Construct fuel breaks to protect WUI communities from wilderness fire Impose building codes and encourage homeowners to create defensible space to reduce risk to homes Use zoning and land use policy to limit housing growth in the proximity of wilderness Educate WUI residents on the benefits of fire and fire management	Construct fuel breaks to protect wilderness from human ignitions in the WUI Emphasize fire prevention education for residents in the WUI and enforce fire restrictions to reduce human ignitions

Potential responses to increasing housing densities include actions that can be taken by managers, communities, and individual homeowners. Improvements in wildfire prevention education and stronger enforcement of fire restrictions may be effective in reducing the number of human ignitions (Fire Prevention Working Group 2004). Creation of defensible space and adoption of building codes that require fire-resistant materials in home construction could reduce the risk that an escaped fire might pose to communities. Fuel breaks constructed outside wilderness might have distinct purposes depending on the general wilderness fire management objective. Where the objective is to restore or maintain fire, fuel breaks may help protect life and property in the interface from naturally caused fires that start in the wilderness. Where fires have become too frequent in wilderness, they may be designed to protect wilderness from human-caused ignitions that start in the WUI. Finally, though unpopular, land-use policies and zoning could limit problematic housing growth where the WUI has not yet encroached on wilderness and where the configuration of prevailing winds, terrain, and vegetation would make houses particularly vulnerable (Finney 2005).

In wilderness, natural processes like fire are supposed to operate freely and shape the landscape without interference from humans. Some wilderness areas are much closer to this goal than others. Effective wilderness stewardship requires that we anticipate future challenges caused by changing climate and housing development patterns, and that we recognize the diversity in management context. In situations where changing climate and human influences preclude fire from playing its natural role, other options for managing vegetation in wilderness may need to be considered, including mechanical thinning, herbicides, and prescribed fires. These manipulative management actions run counter to the original intent of the Wilderness Act and would undoubtedly invite ample controversy. There will be increasing need for science to anticipate consequences of action or inaction by management. The burden of proof for those who would favor manipulating wilderness and degrading its wild character is necessarily high. In cases where that burden can be met, part of society's response strategy will need to include fundamental changes in how we define and manage wilderness areas (Cole et al. 2008).

References

Agee, J.K. 2002. The fallacy of passive management: managing for firesafe forest reserves. *Conservation Biology in Practice* 3: 18–25.

Andrews, P.L., D.O. Loftsgaarden, and L.S. Bradshaw. 2003. Evaluation of fire danger rating indexes using logistic regression and percentile analysis. *International Journal of Wildland Fire* 12: 213–226.

Backer, D.M., S.E. Jensen, and G.R. McPherson. 2004. Impacts of fire-suppression activities on natural communities. *Conservation Biology* 18: 937–946.

Barber, C.V., K.R. Miller, and M. Boness, eds. 2004. *Securing protected areas in the face of global change: Issues and strategies*. Gland and Cambridge: IUCN.

Beniston, M., H.F. Diaz, and R.S. Bradley. 1997. Climatic change at high elevation sites: An overview. *Climatic Change* 36: 233–251.

Black, A.E., M. Williamson, and D. Doane. 2007. Wildland fire use barriers and facilitators. *Fire Management Today* 68: 10–14.

Brooks, M.L., C.M. D'Antonio, D.M. Richardson, J.B. Grace, J.E. Keeley, J.M. DiTomaso, R.J. Hobbs, M. Pellant, and D. Pyke. 2004. Effects of invasive alien plants on fire regimes. *Bioscience* 54: 677–688.

Brown, T.J., B.L. Hall, and A.L. Westerling. 2004. The impact of twenty-first century climate change on wildland fire danger in the western United States: An applications perspective. *Climatic Change* 62: 365–388.

Cardille, J.A., S.J. Ventura, and M.G. Turner. 2001. Environmental and social factors influencing wildfires in the upper Midwest, United States. *Ecological Applications* 11: 111–127.

Chou, Y.H., R.A. Minnich, and R.A. Chase. 1993. Mapping probability of fire occurrence in San Jacinto Mountains, California, USA. *Environmental Management* 17: 129–140.

Christensen, N.L. 1988. Succession and natural disturbance: paradigms, problems, and preservation of natural ecosystems. In *Ecosystem management for parks and wilderness*, eds. J.K. Agee and D.R. Johnson, 62–86. Seattle: University of Washington Press.

Cohen, J.D., and J.E. Deeming. 1985. *The national fire-danger rating system: Basic equations.* General Technical Report GTR-PSW-82. Berkeley: U.S. Forest Service.

Cole, D.N., L. Yung, E.S. Zavaleta, G.H. Aplet, F.S. Chapin III, D.M. Graber, E.S. Higgs, R.J. Hobbs, P.B. Landres, C.I. Millar, D.J. Parsons, J.M. Randall, N.L. Stephenson, K.A. Tonnessen, P.S. White, and S. Woodley. 2008. Naturalness and beyond: Protected area stewardship in an era of global environmental change. *The George Wright Forum* 25: 36–56.

Cova, T.J. 2005. Public safety in the urban-wildland interface: Should fire-prone communities have a maximum occupancy? *Natural Hazards Review* 6: 99–108.

Crump, J.R. 2003. Finding a place in the country. *Environment and Behavior* 35: 187–202.

Dale, L. 2006. Wildfire policy and fire use on public lands in the United States. *Society and Natural Resources* 19: 275–284.

Dale, V.H., L.A. Joyce, S. McNulty, R.P. Neilson, M.P. Ayres, M.D. Flannigan, P.J. Hanson, L.C. Irland, A.E. Lugo, C.J. Peterson, D. Simberloff, F.J. Swanson, B.J. Stocks, and B.M. Wotton. 2001. Climate change and forest disturbances. *Bioscience* 51: 723–734.

Della Sala, D.A., J.E. Williams, C.D. Williams, and J.F. Franklin. 2004. Beyond smoke and mirrors: A synthesis of fire policy and science. *Conservation Biology* 18: 976–986.

Diaz, H.F., J.K. Eischeid, C. Duncan, and R.S. Bradley. 2003. Variability of freezing levels, melting season indicators, and snow cover for selected high-elevation and continental regions in the last 50 years. *Climatic Change* 59: 33–52.

Dombeck, M.P., J.E. Williams, and C.A. Wood. 2004. Wildfire policy and public lands: Integrating scientific understanding with social concerns across landscapes. *Conservation Biology* 18: 883–889.

Falk, D.A., C. Miller, D. McKenzie, and A.E. Black. 2007. Cross-scale analysis of fire regimes. *Ecosystems* 10: 809–823.

Finney, M.A. 2005. The challenge of quantitative risk analysis for wildland fire. *Forest Ecology and Management* 211: 97–108.

Fire Prevention Working Group. 2004. White paper with recommendations for reducing human caused fires and making everyone aware of their responsibility for wildfire prevention. Fire Program Review, Oregon Department of Forestry. URL: http://inr.oregonstate.edu/download/fire_prevention_group_final_paper.pdf. Accessed 25 Jan 2010.

Fried, J.S., J.K. Gilless, W.J. Riley, T.J. Moody, C.S. de Blas, K. Hayhoe, M. Moritz, S. Stephens, and M. Torn. 2008. Predicting the effect of climate change on wildfire behavior and initial attack success. *Climatic Change* 87(Suppl 1): S251–S264.

Gebert, K.M., D.E. Calkin, and J. Yoder. 2007. Estimating suppression expenditures for individual large wildland fires. *Western Journal of Applied Forestry* 22: 188–196.

Giorgi, F., J.W. Hurrell, and M.R. Marinucci. 1997. Elevation dependency of the surface climate change signal: A model study. *Journal of Climate* 10: 288–296.

Halsey, R.W. 2008. *Fire, chaparral, and survival in Southern California.* San Diego: Sunbelt Publications.

Hammer, R.D. [N.d.]. Unpublished data. On file with: Roger Hammer, Department of Sociology, Oregon State University, Corvallis.

Hammer, R.B., S.I. Stewart, R.I. Winkler, V.C. Radeloff, and P.R. Voss. 2004. Characterizing dynamic spatial and temporal residential density patterns from 1940–1990 across the north central United States. *Landscape and Urban Planning* 69: 183–199.

Hammer, R.B., V.C. Radeloff, and J.S. Fried. 2007. Wildland-urban interface housing growth during the 1990s in California, Oregon, and Washington. *International Journal of Wildland Fire* 16: 255–265.

Hansen, A.J., and R. DeFries. 2007. Ecological mechanisms linking protected areas to surrounding lands. *Ecological Applications* 17: 974–988.

Husari, S.J. 1995. Fire management in small wilderness areas and parks. In *Symposium on fire in wilderness and park management*, tech. cords. J.K. Brown, R.W. Mutch, C.W. Spoon, and R.H. Wakimoto, 117–120. General Technical Report INT-GTR-320. Ogden: U.S. Forest Service.

IPCC. 2000. *Special report on emissions scenarios, working group III, Intergovernmental Panel on Climate Change, IPCC.* Cambridge: Cambridge University Press.

IPCC. 2007. *Climate Change 2007: The Physical Science Basis. Contribution of Working Group I to the Fourth Assessment Report of the Intergovernmental Panel on Climate Change.* Cambridge and New York: Cambridge University Press.

Irwin, R.L. 1987. Local planning considerations for the wildland-structural intermix in the year 2000. General Technical Report GTR-PSW-101. Berkeley: U.S. Forest Service.

Keeley, J.E. 2006. Fire management impacts on invasive plants in the western United States. *Conservation Biology* 20: 375–384.

Keeley, J.E., and C.J. Fotheringham. 2001. The historical role of fire in California shrublands. *Conservation Biology* 15: 1536–1548.

Keeley, J.E., C.J. Fotheringham, and M. Morais. 1999. Reexamining fire suppression impacts on brushland fire regimes. *Science* 284: 1829–1832.

Keeley, J.E., C.J. Fotheringham, and M.A. Moritz. 2004. Lessons from the October 2003 wildfires in southern California. *Journal of Forestry* 102: 26–31.

Kilgore, B.M. 1986. The role of fire in wilderness: A state-of-knowledge review. In *National wilderness research conference: Issues, state-of-knowledge, future directions*, compiler R.C. Lucas, 70–103. General Technical Report INT-GTR-220. Fort Collins: U.S. Forest Service.

Knowles, N., M.D. Dettinger, and D.R. Cayan. 2006. Trends in snowfall versus rainfall for the western United States. *Journal of Climate* 19: 4545–4554.

Kolden, C.A., and T.J. Brown. 2008. Using climate for fuels management. Climate, Ecosystem and Fire Applications (CEFA) report 08-01. http://www.cefa.dri.edu/Publications/Climate%20 use%20managed%20fire%20survey%20report.pdf. Accessed 4 Jan 2010.

Landres, P., and S. Meyer. 2000. *National wilderness preservation system database: Key attributes and trends, 1964 through 1999.* General Technical Report RMRS-GTR-18-revised edition. Ogden: U.S. Forest Service.

Liu, J., G.C. Daily, P.R. Ehrlich, and G.W. Luck. 2003. Effects of household dynamics on resource consumption and biodiversity. *Nature* 421: 530–533.

Lopez-Moreno, J.I., S. Guyette, M. Beniston, and B. Alvera. 2008. Sensitivity of the snow energy balance to climatic changes: prediction of snowpack in the Pyrenees in the 21st century. *Climate Research* 36: 203–217.

Lubowski, R., M. Vesterby, and S. Bucholtz. 2007. Land use. In *Agricultural resources and environmental indicators*, eds. K. Wiebe and N. Gollehan, 3–10. Hauppauge: Nova.

Main, W.A., R.J. Straub, and D.M. Paananenn. 1982. FIREFAMILY: Fire planning with historic weather data. General Technical Report GTR-NC-73. St. Paul: U.S. Forest Service.

McKenzie, D., Z. Gedalof, D.L. Peterson, and P. Mote. 2004. Climatic change, wildfire, and conservation. *Conservation Biology* 18: 890–902.

Mesinger, F., G. DiMego, E. Kalnay, P. Shafran, W. Ebisuzaki, D. Jovic, J. Woollen, K. Mitchell, E. Rogers, M. Ek, Y. Fan, R. Grumbine, W. Higgins, H. Li, Y. Lin, G. Manikin, D. Parrish, and W. Shi. 2006. North American regional reanalysis. *Bulletin of the American Meteorological Society* 87: 343–360.

Millar, C.I., N.L. Stephenson, and S.L. Stephens. 2007. Climate change and forests of the future: managing in the face of uncertainty. *Ecological Applications* 17: 2145–2151.

Miller, C. 2003. Wildland fire use: a wilderness perspective on fuel management. Fire, Fuel Treatments, and Ecological Restoration: Conference Proceedings, technical editors P.N. Omi and L.A. Joyce, pp 379–385. Fort Collins, CO: USDA Forest Service, Rocky Mountain Research Station.

Miller, C., and P. Landres. 2004. *Exploring information needs for wildland fire and fuels management.* General Technical Report, RMRS-GTR-127. Fort Collins: U.S. Forest Service.

Miller, C., and D. Parsons. 2004. Can wildland fire use restore natural fire regimes in wilderness and other unroaded lands? Final report to the Joint Fire Science Program, principal investigators Project #01-1-1-05. Missoula: Aldo Leopold Wilderness Research Institute. http://jfsp. nifc.gov/projects/01-1-1-05/01-1-1-05_final_report.pdf. Accessed date for National Interagency Fire Center (2009) is 8 November 2010.

Miller, N.L., and N.J. Schlegel. 2006. Climate change projected fire weather sensitivity: California Santa Ana wind occurrence. *Geophysical Research Letters* 33: L15711.

Mittermeier, R.A., C.G. Mittermeier, T.M. Brooks, J.D. Pilgrim, W.R. Konstant, G.A.B. da Fonseca, and C. Kormos. 2003. Wilderness and biodiversity conservation. *Proceedings of the National Academy of Sciences* 100: 10309–10313.

National Interagency Fire Center (NIFC). 2009. Interagency standards for fire and fire aviation. http://www.nifc.gov/policies/red_book.htm.

Noss, R.F., J.F. Franklin, W.L. Baker, T. Schoennagel, and P.B. Moyle. 2006. Managing fire-prone forests in the western United States. *Frontiers in Ecology and the Environment* 4: 481–487.

Parsons, D.J. 2000. The challenge of restoring natural fire to wilderness. In *Wilderness science in a time of change conference—vol. 5: Wilderness ecosystems, threats, and management*, comps. D.N. Cole, S.F. McCool, W.T. Borrie, and J. O'Loughlin, 276–282. Proceedings RMRS-P-15-VOL-5. Ogden: U.S. Forest Service.

Parsons, D.J., and P.B. Landres. 1998. Restoring natural fire to wilderness: How are we doing? In *Proceedings: 20th tall timbers fire ecology conference: Fire in ecosystem management: Shifting the paradigm from suppression to prescription*, eds. T.L. Pruden and L.A. Brennan, 366–373. Lawrence: Allen Press.

Parsons, D.J., P.B. Landres, and C. Miller. 2003. Wildland fire use: The dilemma of managing and restoring natural fire and fuels in United States wilderness. In *Proceedings of fire conference 2000: The first national congress on fire ecology, prevention, and management*, eds. K.E.M. Galley, R.C. Klinger, and N.G. Sugihara, 19–26. Tallahassee: Tall Timbers Research Station.

Pyne, S.J. 1997. *World fire: The culture of fire on earth*. Seattle: University of Washington Press.

Pyne, S.J., P.L. Andrews, and R.D. Laven. 1996. *Introduction to wildland fire*. New York: Wiley.

Radeloff, V.C., R.B. Hammer, S.I. Stewart, J.S. Freid, S.S. Holcomb, and J.F. McKeefry. 2005. The wildland-urban interface in the United States. *Ecological Applications* 15: 799–805.

Radeloff, V.C., S.I. Stewart, T.J. Hawbaker, U. Gimmi, A.M. Pidgeon, C.H. Flather, R.B. Hammer, and D.P. Helmers. 2009. Housing growth in and near United States protected areas limits their conservation value. *Proceedings of the National Academy of Sciences*. published online before print December 22, 2009, doi:10.1073/pnas.9011131107.

Riechler, T., and J. Kim. 2008. How well do coupled models simulate present-day climate? A comparison of three generations of coupled models. *Bulletin of the American Meteorological Society* 89: 303–331.

Rollins, M.G., and C.K. Frame. 2006. *The LANDFIRE prototype project: Nationally consistent and locally relevant geospatial data for wildland fire management*. General Technical Report RMRS-GTR-175. Fort Collins: U.S. Forest Service.

Rorig, M.L., and S.A. Ferguson. 1999. Characteristics of lightning and wildland fire ignition in the Pacific Northwest. *Journal of Applied Meteorology* 38: 1565–1575.

Schoennagel, T., T.T. Veblen, and W.H. Romme. 2004. The interaction of fire, fuels, and climate across Rocky Mountain forests. *Bioscience* 54: 661–676.

Slocum, M.G., W.J. Platt, B. Beckage, B. Panko, and J.B. Lushine. 2007. Decoupling natural and anthropogenic fire regimes: A case study in Everglades National Park, Florida. *Natural Areas Journal* 27: 41–55.

Stein, B.A. 2002. *States of the union: Ranking America's biodiversity*. Arlington: NatureServe.

Stephens, S.L. 2005. Forest fire causes and extent on United States Forest Service lands. *International Journal of Wildland Fire* 14: 213–222.

Stephens, S.L., and L.W. Ruth. 2005. Federal forest-fire policy in the United States. *Ecological Applications* 15: 532–542.

Syphard, A.D., V.C. Radeloff, J.E. Keeley, T.J. Hawbaker, M.K. Clayton, S.I. Stewart, and R.B. Hammer. 2007. Human influences on California fire regimes. *Ecological Applications* 17: 1388–1402.

Syphard, A.D., V.C. Radeloff, N.S. Keuler, R.S. Taylor, T.J. Hawbaker, S.I. Stewart, and M.K. Clayton. 2008. Predicting spatial patterns of fire in a southern California landscape. *International Journal of Wildland Fire* 17: 602–613.

Theobald, D.M. 2004. Placing exurban land-use change in a human modification framework. *Frontiers in Ecology and the Environment* 2: 139–144.

Theobald, D.M., and W.H. Romme. 2007. Expansion of the US wildland-urban interface. *Landscape and Urban Planning* 83: 340–354.

Urban, D.L., C. Miller, P.N. Halpin, and N.L. Stephenson. 2000. Forest gradient response in Sierran landscapes: The physical template. *Landscape Ecology* 15: 603–620.

U.S. Forest Service. 2005. Land management plan. Part 2 Cleveland National Forest strategy, R5-MB-077. Pacific Southwest Region. http://www.fs.fed.us/r5/scfpr/projects/lmp/docs/cleveland-part2.pdf. Accessed 25 Jan 2010.

USDA and USDI. 1998. *Wildland fire and prescribed fire management policy: Implementation procedures reference guide*. Boise: National Interagency Fire Center.

USDA and USDI. 2001. Urban wildland interface communities within the vicinity of federal lands that are at high risk from wildfire. *Federal Register* 66: 751.

USDA and USDI. 2005. *Wildland fire use implementation procedures reference guide*. Boise: National Interagency Fire Center.

USDI and USDA. 2001. *Review and update of the 1995 federal wildland fire management policy*. Boise: National Interagency Fire Center.

van Wagtendonk, J.W. 1995. Large fires in wilderness areas. In *Proceedings: Symposium on fire in wilderness and park management*, tech coords, eds. J.K. Brown, R.W. Mutch, C.W. Spoon, and R.H. Wakimoto, 113–116. General Technical Report INT-GTR-320. Ogden: U.S. Forest Service.

Yang, J., H.S. He, S.R. Shifley, and E.J. Gustafson. 2007. Spatial patterns of modern period human-caused fire occurrence in the Missouri Ozark Highlands. *Forest Science* 53: 1–15.

Yu, E., and J. Liu. 2007. Environmental impacts of divorce. *Proceedings of the National Academy of Sciences of the United States of America* 104: 20629–20634.

Zedler, P.H. 1995. Are some plants born to burn? *Trends in Ecology & Evolution* 10: 393–395.

Zedler, P.H., C.R. Gautier, and G.S. McMaster. 1983. Vegetation change in response to extreme events: The effect of a short interval between fires in California chaparral and coastal scrub. *Ecology* 64: 809–818.

Zimmerman, G.T., and D.L. Bunnell. 2000. The federal wildland fire policy: opportunities for wilderness fire management. In *Wilderness science in a time of change conference–Volume 5: wilderness ecosystems, threats, and management*, comps. D.N. Cole, S.F. McCool, W.T. Borrie, and J. O'Loughlin, 288–297. Proceedings RMRS-P-15-VOL-5. Ogden: U.S. Forest Service.

Chapter 12
Synthesis: Landscape Ecology and Changing Fire Regimes

Donald McKenzie, Carol Miller, and Donald A. Falk

12.1 Introduction

Fire is a ubiquitous ecosystem process, and one that is expected to respond rapidly and unpredictably in a changing climate. The effects of altered fire regimes will be felt in many if not most of Earth's ecosystems (Gillett et al. 2004; Millar et al. 2007; Bowman et al. 2009; Parisien and Moritz 2009). Consequently, fire managers will be challenged to envision and enable landscapes of the future in which ecological function is maintained, and adaptation strategies will have to be creative and dynamic (Chap. 10). In this chapter we recap key lessons from the preceding contributions and comment on the state of the art in landscape fire theory, application, and management, with an eye toward specifying a research agenda for the landscape ecology of fire. Specifically, we explore the possibility that the energy-regulation-scale (ERS) framework (Chap. 1) could be applied to a wide variety of issues in fire ecology. We then present our (short list of) candidates for "key concepts" in the landscape ecology of fire.

12.2 What Have We Learned about the Landscape Ecology of Fire?

In this book, we have mostly eschewed coverage of two mainstays of landscape ecology that have been covered extensively elsewhere: empirical analysis of spatial pattern based on remote sensing (Lillesand et al. 2003) and landscape fire simulation models (Mladenoff and Baker 1999; Turner et al. 2001; Keane et al. 2004). We have instead devoted a substantial section to theoretical considerations and questions of scale. Theoretical frameworks are an important complement to empirical

D. McKenzie(✉)
Pacific Wildland Fire Sciences Laboratory, U.S. Forest Service,
400 N 34th St., #201, Seattle, WA 98103-8600, USA
e-mail: dmck@u.washington.edu

D. McKenzie et al. (eds.), *The Landscape Ecology of Fire*, Ecological Studies 213,
DOI 10.1007/978-94-007-0301-8_12, © Springer Science+Business Media B.V. 2011

data analysis and simulation modeling (O'Neill et al. 1986; Brown et al. 2002; Falk et al. 2007; West et al. 2009). Advances in theory must keep pace with progress in empirical analysis, driven by the creative application and extension of statistical methods in ecology and fire science across landscape scales (Díaz-Avalos et al. 2001; Moritz et al. 2005; Wu et al. 2006; Kellogg et al. 2008) and in landscape simulations (Keane and Finney 2003). This momentum is driven not only by increased computing power, but also, and perhaps more importantly, by more judicious use of model evaluation (Scheller and Mladenoff 2007; Kennedy et al. 2008).

Section I (Concepts and Theory) suggests how new conceptual and theoretical models may enable us to think across scales and anticipate "no-analog" conditions for future fire regimes. For example, by defining landscape fire in terms of energy and regulation, McKenzie et al. (Chap. 1) offer a simple universal language for extrapolation into no-analog conditions. Indeed the energy in the Earth system (Pielou 2001), expressed in climate dynamics and evolving as global warming continues, propagates directly into landscape fire dynamics. McKenzie and Kennedy (Chap. 2) and Moritz et al. (Chap. 3) show that we can borrow tools liberally from other disciplines—physics, engineering, complex systems, and physiology—while increasing the robustness of core analyses within landscape ecology by quantifying relationships across scales. Landscape ecologists are known to claim that our field is "theory-challenged," or at the least limited to phenomenological observations and constrained by the complexity of "middle-number" systems (O'Neill et al. 1986; Chap. 1). Attention to scaling relations might overcome the seeming intractability of the middle-number domain and answer pressing questions about resilience and perhaps even sustainability of landscapes in a changing climate (Chap. 3).

Section II (Climate Context) brings global and regional climatology into the landscape domain via the cross-scale applicability of energy-water relations (Milne et al. 2002; Peters et al. 2004). Top-down controls on fire regimes (climate variability at multiple spatial and temporal scales—Chap. 4) are a direct manifestation of Earth's energy system. Both land-surface and ocean couplings with the atmosphere provide a coarse-scale manifestation of the energy-regulation polarity introduced by McKenzie et al. (Chap. 1). Littell and Gwozdz (Chap. 5) downscale this global polarity into the regional (ecosection) domain (Fig. 5.1). Water-balance deficit (DEF)(Stephenson 1990; Lutz et al. 2010), the outcome of the interaction between kinetic energy in radiation and potential energy in fuels and the damping (regulation) of moisture, is the best predictor of fire extent at regional scales (see also Littell et al. 2009). Although the role of DEF in the ERS framework of McKenzie et al. (Chap. 1) has yet to be quantified, it shows promise for linking broad-scale fire climatology to the more complex and variable dynamics of landscape fire severity and spatial pattern and the ultimate consequences for landscape structure and composition (i.e., landscape memory—Peterson 2002).

Section III (Landscape Dynamics and Interactions) maps the multiple influences of spatial processes and climate onto real ecosystems, while highlighting the importance of fire's interactions with other physical and ecological processes. These studies illustrate multiple processes interacting with changing landscape fire regimes, including ways that these interactions may be transformed, possibly abruptly and in unexpected ways, in a rapidly changing climate. Smithwick (Chap. 6) presents a model of biogeochemical resilience based on the interactions of fire with the often unseen elements on a landscape, e.g., nutrient pools *vs.* spatial patterns of trees. These biogeochemical processes and pools may be just as important in determining future ecosystem trajectories (and with them resilience to abrupt change) as more obvious features such as vegetation structure and composition. Swetnam et al. (Chap. 7) show how the deep temporal record in fire-scarred trees can be mapped into the spatial domain of landscape ecology. By combining traditional dendroecological methods of identifying local fire years with the tools of geospatial modeling, fires and fire regimes can be reconstructed spatially, and their landscape properties analyzed. For example, interpolation of point records into landscape surfaces offers a window into the configuration of past landscapes. Such reconstructions open the door to understanding landscape dynamics, such as the influence of post-fire mosaics that create "landscape memory" and affect subsequent disturbances (Fig. 12.1). Keeley et al. (Chap. 8) focus on the effects of invasive species on fire

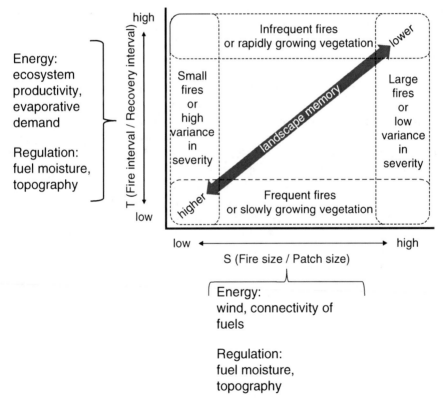

Fig. 12.1 The spatio-temporal domain of landscape memory in ecosystems affected by fire. Spatial and temporal gradients affecting landscape memory are subject to energy and regulation across scales (Chap. 1). Calibration of the "S" and "T" axes will change depending on the spatial extent in question, and so is subject to scaling laws. Fire size is related to the slope of the interval-area relation, and point fire interval to its Y-intercept (Falk et al. 2007)

regimes in a uniquely complex and variable landscape (California, USA). They highlight the dependencies and interactions that make future projections at landscape scales uncertain at best. Invasive species are "game changers" in many ecosystems, particularly in the many cases of pyrophilic species that can alter fuel complexes and thus the extent, severity, and seasonality of landscape fire. Cushman et al. (Chap. 9) address the effects of landscape fire regimes on wildlife habitat in the central Rocky Mountains (USA), where fire regimes may change significantly in the future (Keane et al. 2004). Two lessons emerge from their simulations. First, the influences of warming temperatures are likely to dominate habitat changes in this forested landscape, overriding the effects of even the most aggressive treatments to resist change. Second, ensemble projections provide much better estimates of future ranges of variation than single scenarios, whether the modeling tools be at global (GCMs) or landscape (fire and vegetation) scales.

Section IV (Landscape Fire Management, Policy, and Research in an Era of Global Change) applies these ideas about fire as a landscape process to pressing issues in ecosystem management. This is the human dimension, where the central challenge is to develop new options for guiding landscape fire regimes in an era of rapidly changing land use and climate, such that both ecosystems and human populations can adapt. The urgency of understanding the dynamics of fire-human interactions and indentifying paths for adaptation is in large measure a result of the rapid changes in regional and global climate projected for the twenty-first century. Outcomes of the workshops reported by Peterson et al. (Chap. 10) on adapting to climate change suggest that although there is a bewildering array of problems facing land and natural resource managers, there is also a wealth of experience and creativity in this human resource, which has observed and documented fire on diverse landscapes over many years. These authors also highlight the need for thinking "out of the box" about options for active management of fire regimes, not to replicate some historical or desired condition *per se* but to anticipate landscape structures that will be resilient to fires in a warming world. Miller et al. (Chap. 11) call our attention (gently) to the elephant in the room of futuring: human population growth. The need to reintroduce wildfire in many wilderness landscapes is well documented, but as the interface between human dwellings and wilderness becomes more extensive and complex, encouraging wilderness fire becomes more delicate, contentious, and constrained. Out-of-the-box thinking may be necessary to find solutions that work across boundaries. In combination with global climate change, with the promise of larger and more intense fires, human population and land use exacerbate all problems of fire management.

12.3   Research Needs

Where do we go from here? What follows is conceptual synthesis of major research directions, rather than a list of specific projects. For a recent example of a detailed research agenda for landscape fire, see Cushman et al. (2007).

- What would a fully developed theory of landscape fire look like? We proposed an initial framework in Chap. 1 that could take advantage of scaling relations in key variables such as heat flux from combustion, potential energy in fuels, and topographic variance. Such a theory might be excused for failing to predict the behavior of individual fires (joining most fire behavior models under some conditions), yet give estimates of aggregate properties of fire-affected landscapes such as patch-size distributions or spatial variability in fuel loadings.
- Bring our understanding of the energy-water (or energy-regulation) dynamic in fire climatology down to landscape scales. Climate-model downscaling *per se* reaches a limit at about 4–12 km resolution (Salathé et al. 2007); downscaling from weather station records has been successful to about 0.5 km (Daly et al. 2008). If water relations are the key to fire-climate modeling, however, as Littell and Gwozdz (Chap. 5) suggest, then it is theoretically possible to estimate landscape variability in fuel moisture at scales relevant to landscape fire (30–100 m). We also should be careful about inferring that water-balance deficit will have similar effects on fire in ecosystems of widely different aridity (McKenzie and Littell (in press).
- Improve our ability to predict and quantify high-energy events (Romme et al. 1998; Peters et al. 2004; Gedalof et al. 2005; Scheffer et al. 2009). Transient high-energy events, which are most difficult to predict, have the most long-lasting effects on landscape pattern and process (Romme et al. 1998). There is a difference between predicting individual events, which is especially challenging because of their largely stochastic nature as landscape events, and predicting their propensity or frequency, which should be more a function of mean-field conditions than of a fortuitous alignment of necessary and sufficient conditions.
- Improve our understanding of how the spatio-temporal structures in fire history reflect landscape fire dynamics (Moritz 2003; McKenzie et al. 2006; Falk et al. 2007; Scholl and Taylor 2010). McKenzie and Kennedy (Chap. 2) and Swetnam et al. (Chap. 7) show how fire-scar data can reveal both landscape pattern of fires and the nature of controls (e.g., top-down *vs.* bottom-up) on historical fire regimes, in the absence of a record of either the weather or the fuel abundance and condition associated with any particular fire. If we can infer controls on fire regimes, we can better predict how fire-prone landscapes may change as the controls change, and where management intervention is more likely to succeed. Fire scar evidence can be better integrated with other sources of information about unmanaged fire regimes, such as age structure reconstructions and charcoal analysis.
- Fill major geographic gaps in our understanding of fire regimes. This book was limited in scope to western North America, where fire regimes are particularly well documented. Would the same questions as asked in this book be appropriate for very different systems such as the Eurasian boreal forest (Conard and Ivanova 1997; Stocks et al. 1998; Gustafson et al. 2010) or the Australian Jarrah forest (Bell and Koch 2006) (e.g., see Fig. 1.4), which are clearly shaped by fire?
- Find fire-induced tipping points for the reorganization of ecosystems, particular those that might be reached as a result of climate change. Keeley et al.

(Chap. 8) and others before them (Zedler et al. 1983) point to the sensitivity of California chaparral to increasing fire frequency; there is a tipping point beyond which the dominant vegetation changes irreversibly. Are there similar tipping points elsewhere, perhaps associated with fire severity, fire extent, fire-insect interactions, biogeochemistry (Chap. 6), or multiple stresses associated with fire (McKenzie et al. 2009). For example, high elevation mountainous landscapes could see much more area become flammable earlier in the fire season (Chap. 11), thereby altering fire regimes with unknown consequences for vegetation.

• Are there truly "landscape" scales in fire ecology? This is common parlance of course, but ill-defined. There may be ways to quantify the inherent scales of fire regimes, however. For example, Moritz et al. (Chap. 3) identify a "meso-domain" within which scaling laws in fire-size distributions follow power laws (see also Reed and McKelvey 2002). Analogously, McKenzie et al. (Chap. 1) posit domains of maximum ecological complexity associated, albeit loosely, with spatial scales (Fig. 1.5).

12.4   Concluding Thoughts

Fire is an integral part of landscape process, memory, and resilience, as opposed to an external perturbation (despite our liberal use of the term "disturbance" throughout). As a contagious process, it "bleeds" across scales and requires a specification of ecosystem dynamics that is robust across scales. We hope that our readers come away, at a minimum, with new perspectives (and research ideas) in three key areas.

First, landscape memory is the cumulative outcome of landscape fire dynamics (Peterson 2002; Chap. 3). Fire's legacy on the landscape is clear in some locations while subtle in others, long-lasting in some while transient in others (Fig. 12.1). Deconstructing landscape memory illuminates fire history and its interactions with ecosystem processes over time and space. This deconstruction should be engineered in a way that enables projections of alternate futures by tuning parameters estimated therein.

Second, the interactions of top-down and bottom-up regulation of fire regimes provide a coherent framework for problems across scales, and thus a potential foundation for a theory of landscape fire. The ERS framework, or some analogue, provides a deep mechanism in ecosystem energetics, and thus physics, for the tangible expression of top-down and bottom-up regulation in real ecosystems. There is much more room for empirical, modeling, and theoretical work to create a mature model of what regulates fire regimes.

Third, scaling laws in fire regimes provide an expression of complex dynamics. Scale is featured in every landscape ecology book, and indeed landscape fire research is closing in on Levin's (1992) oft-quoted goal of understanding how ecological processes change across scales. Quantitative scaling laws can at least complement, and in some cases replace, hierarchical models (Chap. 1). This is likely to be especially true when contagious disturbance (fire) is a significant element of landscape dynamics.

References

Bell, D.T., and J.M. Koch. 2006. Post-fire succession in the northern jarrah forest of Western Australia. *Australian Journal of Ecology* 5: 9–14.

Bowman, D.M.J.S., J.K. Balch, P. Artaxo, W.J. Bond, J.M. Carlson, M.A. Cochrane, C.M. D'Antonio, R.S. DeFries, J.C. Doyle, S.P. Harrison, F.H. Johnston, J.E. Keeley, M.A. Krawchuk, C.A. Kull, J.B. Marston, M.A. Moritz, I.C. Prentice, C.I. Roos, A.C. Scott, T.W. Swetnam, G.R. van der Werf, and S.J. Pyne. 2009. Fire as an earth system process. *Science* 324: 481–484.

Brown, J.H., V.K. Gupta, B.-L. Li, B.T. Milne, C. Restrepo, and G.B. West. 2002. The fractal nature of nature: power laws, ecological complexity, and biodiversity. *Philosophical Transactions of the Royal Society B* 357: 619–626.

Conard, S.G., and G.A. Ivanova. 1997. Wildfire in Russian boreal forests - potential impacts of fire regime characteristics on emissions and global carbon balance estimates. *Environmental Pollution* 98: 305–313.

Cushman, S.A., D. McKenzie, D.L. Peterson, J.S. Littell, and K.S. McKelvey. 2007. *Research agenda for integrated landscape modeling*. General Technical Report RMRS-GTR-194. Fort Collins: U.S. Forest Service.

Daly, C., M. Halbleib, J.I. Smith, W.P. Gibson, M.K. Doggett, G.H. Taylor, J. Curtis, and P.A. Pasteris. 2008. Physiographically sensitive mapping of temperature and precipitation across the conterminous United States. *International Journal of Climatology* 28: 2031–2064.

Díaz-Avalos, C., D.L. Peterson, E. Alvarado, S.A. Ferguson, and J.E. Besag. 2001. Space–time modelling of lightning-caused ignitions in the Blue Mountains, Oregon. *Canadian Journal of Forest Research* 31: 1579–1593.

Falk, D.A., C. Miller, D. McKenzie, and A.E. Black. 2007. Cross-scale analysis of fire regimes. *Ecosystems* 10: 809–823.

Gedalof, Z., D.L. Peterson, and N.J. Mantua. 2005. Atmospheric, climatic and ecological controls on extreme wildfire years in the northwestern United States. *Ecological Applications* 15: 154–174.

Gillett, N.P., F.W. Zwiers, A.J. Weaver and M.D. Flannigan. 2004. Detecting the effect of climate change on Canadian forest fires. *Geophysical Research Letters* 31: L18211. doi:10.1029/2004GL020876.

Gustafson, E.J., A.Z. Shvidenko, B.R. Sturtevant, and R.M. Scheller. 2010. Predicting global change effects on forest biomass and composition in southcentral Siberia. *Ecological Applications* 20: 700–715.

Keane, R.E., and M.A. Finney. 2003. The simulation of landscape fire, climate, and ecosystem dynamics. In *Fire and climatic change in temperate ecosystems of the Western Americas*, eds. T.T. Veblen, W.L. Baker, G. Montenegro, and T.W. Swetnam, 32–68. New York: Springer.

Keane, R.E., G. Cary, I.D. Davies, M.D. Flannigan, R.H. Gardner, S. Lavorel, J.M. Lenihan, C. Li, and T.S. Rupp. 2004. A classification of landscape fire succession models: Spatially explicit models of fire and vegetation dynamics. *Ecological Modelling* 256: 3–27.

Kellogg, L.-K.B., D. McKenzie, D.L. Peterson, and A.E. Hessl. 2008. Spatial models for inferring topographic controls on low-severity fire in the eastern Cascade Range of Washington, USA. *Landscape Ecology* 23: 227–240.

Kennedy, M.C., E.D. Ford, P. Singleton, M. Finney, and J.K. Agee. 2008. Informed multi-objective decision-making in environmental management using Pareto optimality. *Journal of Applied Ecology* 45: 181–192.

Levin, S.A. 1992. The problem of pattern and scale in ecology. *Ecology* 73: 1943–1967.

Lillesand, T., R.W. Kiefer, and J. Chipman. 2003. *Remote sensing and image interpretation*. New York: Wiley.

Littell, J.S., D. McKenzie, D.L. Peterson, and A.L. Westerling. 2009. Climate and wildfire area burned in western U.S. ecoprovinces, 1916–2003. *Ecological Applications* 19: 1003–1021.

Lutz, J.A., J.W. van Wagtendonk, and J.F. Franklin. 2010. Climatic water deficit, tree species ranges, and climate change in Yosemite National Park. *Journal of Biogeography*. doi:10.1111/j.1365-2699.2009.02268.x:1-15.

McKenzie, D., and J.S. Littell. (in press). Climate change and wilderness fire regimes. *International Journal of Wilderness*.

McKenzie, D., A.E. Hessl, and L.-K.B. Kellogg. 2006. Using neutral models to identify constraints on low-severity fire regimes. *Landscape Ecology* 21: 139–152.

McKenzie, D., D.L. Peterson, and J.S. Littell. 2009. Global warming and stress complexes in forests of western North America. In *Developments in Environmental Science*, Vol. 8, *Wild Land Fires and Air Pollution*, eds. A. Bytnerowicz, M. Arbaugh, A. Riebau, and C. Anderson, 319–337. Amsterdam: Elsevier Science.

Millar, C.I., N.L. Stephenson, and S.L. Stephens. 2007. Climate change and forests of the future: Managing in the face of uncertainty. *Ecological Applications* 17: 2145–2151.

Milne, B.T., V.K. Gupta, and C. Restrepo. 2002. A scale-invariant coupling of plants, water, energy, and terrain. *EcoScience* 9: 191–199.

Mladenoff, D., and W.L. Baker. 1999. *Spatial modeling of forest landscapes: approaches and applications*. New York: Cambridge University Press.

Moritz, M.A. 2003. Spatio-temporal analysis of controls of shrubland fire regimes: Age dependency and fire hazard. *Ecology* 84: 351–361.

Moritz, M.A., M.E. Morais, L.A. Summerell, J.M. Carlson, and J. Doyle. 2005. Wildfires, complexity, and highly optimized tolerance. *Proceedings of the National Academy of Sciences* 102: 17912–17917.

O'Neill, R.V., D.L. deAngelis, J.B. Waide, and T.F.H. Allen. 1986. *A hierarchical concept of ecosystems*. Princeton: Princeton University Press.

Parisien, M.A., and M.A. Moritz. 2009. Environmental controls on the distribution of wildfire at multiple spatial scales. *Ecological Monographs* 79: 127–154.

Peters, D.P., R.A. Pielke Sr., B.T. Bestelmeyer, C.D. Allen, S. Munson-McGee, K.M. Havstad, and H.A. Mooney. 2004. Cross-scale interactions, nonlinearities, and forecasting catastrophic events. *Proceedings of the National Academy of Sciences* 101: 15130–15135.

Peterson, G.D. 2002. Contagious disturbance, ecological memory, and the emergence of landscape pattern. *Ecosystems* 5: 329–338.

Pielou, E.C. 2001. *The energy of nature*. Chicago: University of Chicago Press.

Reed, W.J., and K.S. McKelvey. 2002. Power-law behavior and parametric models for the size-distribution of forest fires. *Ecological Modelling* 150: 239–254.

Romme, W.H., E.H. Everham, L.E. Frelich, and R.E. Sparks. 1998. Are large infrequent disturbances qualitatively different from small frequent disturbances? *Ecosystems* 1: 524–534.

Salathé Jr., E.P., P.W. Mote, and M.W. Wiley. 2007. Review of scenario selection and downscaling methods for the assessment of climate change impacts on hydrology in the United States Pacific Northwest. *International Journal of Climatology* 27: 1611–1621.

Scheffer, M., J. Bascompte, W.A. Brock, V. Brovkin, S.R. Carpenter, V. Dakos, H. Held, E.H. van Nes, M. Rietker, and G. Sugihara. 2009. Early-warning signals for critical transitions. *Nature* 461: 53–59.

Scheller, R.M., and D.J. Mladenoff. 2007. An ecological classification of forest landscape simulation models: tools and strategies for understanding broad-scale forested ecosystems. *Landscape Ecology* 22: 491–505.

Scholl, A.E., and A.H. Taylor. 2010. Fire regimes, forest change, and self-organization in an old-growth mixed-conifer forest, Yosemite National Park, USA. *Ecological Applications* 20: 691–692.

Stephenson, N.L. 1990. Climatic control of vegetation distribution: The role of the water balance. *The American Naturalist* 135: 649–670.

Stocks, B.J., M.A. Fosberg, T.J. Lynham, L. Mearns, B.M. Wotton, Q. Yang, J.Z. Jin, K. Lawrence, G.R. Hartley, J.A. Mason, and D.W. McKenney. 1998. Climate change and forest fire potential in Russian and Canadian boreal forests. *Climatic Change* 38: 1–13.

Turner, M.G., R.H. Gardner, and R.V. O'Neill. 2001. *Landscape ecology in theory and practice: Pattern and process*. New York: Springer.

West, G.B., B.J. Enquist, and J.H. Brown. 2009. A general quantitative theory of forest structure and dynamics. *Proceedings of the National Academy of Sciences* 106: 7040–7045.

Wu, J., K.B. Jones, H. Li, and O.L. Loucks. 2006. *Scaling and uncertainty analysis in ecology*. Dordrecht: Springer.

Zedler, P.H., C.R. Gautier, and G.S. McMaster. 1983. Vegetation change in response to extreme events: the effect of a short interval between fires in California chaparral and coastal scrub. *Ecology* 64: 809–818.

Index